Highlights in Theoretical Chemistry

Vol. 3
Series Editors: Ch.J. Cramer • D.G. Truhlar

For further volumes:
http://www.springer.com/series/11166

Jean-Philip Piquemal • Kenneth D. Jordan
Volume Editors

From Quantum Mechanics to Force Fields

A Topical Collection from Theoretical
Chemistry Accounts

With contributions from

Jiali Gao • Mark S. Gordon • Kenneth D. Jordan • Thomas Keyes
José N. Canongia Lopes • Marco Masia • Takeshi Nagata
Sandeep Patel • Jean-Philip Piquemal • Paul L.A. Popelier
Pengyu Ren • Mathieu Salanne • Pär Söderhjelm • Feng Wang
Darrin M. York

 Springer

Volume Editors
Jean-Philip Piquemal
Sorbonne Université, Lab. de Chimie Théorique and CNRS
UPMC
Paris Cedex 05
France

Kenneth D. Jordan
Department of Chemistry
University of Pittsburgh
Pittsburgh
USA

Originally Published in Theor Chem Acc, Volume 131 (2012)
© Springer-Verlag Berlin Heidelberg 2012

ISSN 2194-8666   ISSN 2194-8674 (electronic)
ISBN 978-3-662-51175-6   ISBN 978-3-642-34450-3 (eBook)
DOI 10.1007/978-3-642-34450-3
Springer Heidelberg New York Dordrecht London

Contents

Theor Chem Acc (2012) 131:1207
DOI 10.1007/s00214-012-1207-x

EDITORIAL

From quantum mechanics to force fields: new methodologies for the classical simulation of complex systems

Jean-Philip Piquemal · Kenneth D. Jordan

Published online: 31 March 2012
© Springer-Verlag 2012

Molecular dynamics simulations have become one of the key methods for simulating the properties of biological systems and complex materials. Underlying such simulations are the force fields that describe how the energy varies as a function of geometry. As a result, the reliability of the simulations depends on the quality of the force field. In many cases, this requires the explicit incorporation of polarization. For these reasons, accurate electronic structure methods are playing an increasingly important role in the parameterization of force fields.

This special issue of Theoretical Chemistry Accounts illustrates recent trends in the design and applications of accurate force fields. Of course, the field is vast and is growing rapidly; thus, we cannot claim completeness within such a collection. However, the following 15 papers reflect the present questions including the strategies for (i) the inclusion of the polarization energy and (ii) an optimal parametrization of model and highlight the directions to follow as new exciting fields of application emerge.

Published as part of the special collection of articles: From quantum mechanics to force fields: new methodologies for the classical simulation of complex systems.

J.-P. Piquemal (✉)
UPMC, Paris 6, Laboratoire de Chimie Théorique and CNRS,
Sorbonne Université, Campus Jussieu,
Tour 12-13, 4ème étage, CC 137, 4 Place Jussieu,
UMR 7616, 75252 Paris Cedex 05, France
e-mail: jpp@lct.jussieu.fr

K. D. Jordan (✉)
Department of Chemistry, University of Pittsburgh,
219 Parkman Avenue, Pittsburgh, PA 15260, USA
e-mail: jordan@pitt.edu

Three papers discuss the optimization and parametrization of new models.

Markutsya et al. present a coarse-grained model for β-D-glucose based on force matching.
Wu et al. propose an automation process for the AMOEBA polarizable force field parameterization of small molecules.
Pinnick et al. detailed an approach enabling fast convergence of ab initio free energy perturbation calculations with the adaptive force matching method.

Three papers put in perspectives the actual importance of the polarization energy.

Kumar et al. discussed the polarizing forces of water.
Guàrdia et al. detailed the importance of the polarization damping and its consequences on ion solvation dynamics.
Söderjelm focuses on polarization effects in protein–ligand calculations and shows that it could extend farther than the actual induction energy.

Five papers review or propose new models explicitly incorporation polarization.

Han et al. present a new optimization of the explicit polarization X-Pol potential using a hybrid density functional.
Giese et al. illustrate the grand challenges in density functional expansion methods.
Nagata et al. propose new strategies to obtain analytic gradients in order to perform molecular dynamics simulations using the fragment molecular orbital (FMO) method with effective potentials.
Bauer et al. review recent applications and developments of the charge equilibration force fields.

  Springer

Mills et al. illustrate their new polarizable multipolar electrostatics approach based on the Kriging machine learning method and apply it to alanine.

Finally, four papers present models that are applied to difficult systems and fields of application.

Canongia Lopes et al. propose a generic and systematic force field for ionic liquids modeling.
Wang et al. described a new polarizable model capturing the energetics and dynamics of carbon monoxide.
Marjolin et al. extend polarizable models to the accurate treatment of lanthanide and actinides in solution allowing the computation of hydration free energies.

Salanne et al. discussed the inclusion of many-body effects in models for ionic liquids.

While the development of new models will continue, it is clear that electronic structure calculations and advanced parametrization strategies will play a key role in force field design for years to come. We hope that the 15 papers of this Special Issue will be of lasting values.

Theor Chem Acc (2012) 131:1137
DOI 10.1007/s00214-012-1137-7

REGULAR ARTICLE

Polarisable multipolar electrostatics from the machine learning method Kriging: an application to alanine

Matthew J. L. Mills · Paul L. A. Popelier

Received: 9 May 2011 / Accepted: 29 August 2011 / Published online: 18 February 2012
© Springer-Verlag 2012

Abstract We present a polarisable multipolar interatomic electrostatic potential energy function for force fields and describe its application to the pilot molecule MeNH-Ala-COMe (AlaD). The total electrostatic energy associated with 1, 4 and higher interactions is partitioned into atomic contributions by application of quantum chemical topology (QCT). The exact atom–atom interaction is expressed in terms of atomic multipole moments. The machine learning method Kriging is used to model the dependence of these multipole moments on the conformation of the entire molecule. The resulting models are able to predict the QCT-partitioned multipole moments for arbitrary chemically relevant molecular geometries. The interaction energies between atoms are predicted for these geometries and compared to their true values. The computational expense of the procedure is compared to that of the point charge formalism.

Keywords Quantum chemical topology · Force field · Multipole moment · Polarisation · Atoms in molecules · Machine learning

Published as part of the special collection of articles: From quantum mechanics to force fields: new methodologies for the classical simulation of complex systems.

M. J. L. Mills
Manchester Interdisciplinary Biocentre (MIB),
131 Princess Street, Manchester M1 7DN, UK

P. L. A. Popelier (✉)
School of Chemistry, University of Manchester,
Oxford Road, Manchester M13 9PL, UK
e-mail: pla@manchester.ac.uk

1 Introduction

Molecular dynamics and mechanics methods are used to investigate a wide variety of chemical problems, including but not limited to peptide structure [1], semiconductors [2] and the origins of life [3]. There is a competition between the accuracy of the results of such studies and the computational cost of carrying them out. Quantum mechanical (QM) methods may be employed in cases where electronic detail is required [4]. Explicit treatment of the electrons makes the cost of QM simulations very high and as such they are limited to use with small systems over short time periods [5]. Hybrid methods may be employed to reduce this cost, wherein small parts of the system considered chemically important are treated explicitly and the rest of the system is treated approximately [6, 7]. For cases where QM detail is not required, the whole system may be treated using an approximate potential energy function. The most commonly employed class of such functions are referred to as force fields (FF). FF energy expressions make the study of longer simulations or larger systems feasible by sacrificing some of the accuracy of the description of the potential energy surface (PES) of the results for reduced computational loads. The continuing increases in computer power along with the introduction of molecular dynamics codes that can take advantage of GPUs [8] have allowed the recent extension of the domain of application of FFs to much longer simulation times and system sizes than have previously been available [9, 10].

There are many FFs commonly used in biochemical research. Popular potentials include CHARMM [11, 12], AMBER [13, 14], GROMOS [15] and OPLS [16, 17]. Certain aspects differ between them all, but they share a common framework. The central shared assumption is that the potential energy of a chemical system can be written as a function of the relative positions of its nuclei. This potential energy function is not derived from first principles but is guided by chemical intuition. Typically, it takes the form of a combination of bond-stretch, angle-bend and torsional rotation functions (grouped together as the 'bonded' part) along with an expression for the electrostatic, short-range repulsion and long-range attraction interactions (the 'non-bonded' part). The bond-stretch, angle-bend and torsion terms are usually expressed in the form given in Eq. 1

$$E_{\text{bonded}} = \sum_{n=2}^{n_{\max}} \left(\sum_B C_R^n (B - B_0)^n + \sum_\theta C_\theta^n (\theta - \theta_0)^n \right) + \sum_\varphi \sum_{m=1}^{m_{\max}} C_\varphi^m \cos m\varphi, \tag{1}$$

where B, θ and φ refer to bond lengths, angles and dihedrals, respectively, a subscript 0 denotes a reference value and the Cs are a set of constants to be determined. The values of $m_{\max}$ and $n_{\max}$ refer to the order of the expansion for each term and vary between FF, with higher values expected to represent the energetics of each physical phenomenon more accurately. Work has been carried out on the utility of including coupling terms for a more accurate description of the bonded energy [18]. Various forms for the bond-stretch energy have been proposed and compared [19]. Improper torsional terms may also be added to describe deviation from equilibrium geometries in planar systems [20]. The non-bonded expression typically consists of a Lennard-Jones function paired with a charge–charge representation of the electrostatic interaction, or

$$E_{\text{non-bonded}} = \sum_{i<j} \left(\frac{A_{ij}}{r_{ij}^{12}} - \frac{B_{ij}}{r_{ij}^{6}} + \frac{q_i q_j}{\varepsilon r_{ij}} \right). \tag{2}$$

Here, q_i represents the partial charge assigned to the atom i, r_{ij} is the distance between atoms i and j, ε is the relative permittivity that allows for screening of the electrostatic interaction and A_{ij}, B_{ij} are constants found by application of combining rules to predefined atomic values [21]. These interactions are computed between atoms separated by three or more bonds. Special terms may be added to cover additional physical phenomena such as hydrogen bonding. The parameters present in the bonded equations are typically found by fitting to a set of data. This set may be collected experimentally either [22] by application of ab initio methods to small molecules [23] or by a combination of both [24]. The parameterisation process requires the predefinition of a set of distinct atom types. These are typically chosen to represent atoms within a broad class of functional groups. The transferability of parameters to models outside the fitting set is strongly dependent on which atom types are included.

Force fields as described have achieved success in allowing the investigation into a wide range of chemical problems. However, they have been shown to be deficient in predicting results for important classes of molecules, such as nucleic acid base pairs [25]. In some cases, the electrostatic polarisation has been found to be entirely responsible for binding in biochemical systems [26]. The fitting of parameters to bulk properties can cause the lack of a correct description at the more detailed level of clusters. The FF geometries of small molecular clusters may be in strong disagreement with their ab initio counterparts [27]. Such methods suffer from poor transferability, and methods that predict correct macroscopic behaviour from correct microscopic behaviour should be less affected by this issue. Correct bulk behaviour should emerge from such methods.

Bond-stretch and angle-bend terms are transferable between chemical environments with a low associated error. The non-bonded terms must therefore be responsible for the majority of the lack of transferability found in FF. The electrostatic interaction term in particular is affected [28]. In typical FF, charges are assigned to atom types by fitting the ab initio electrostatic potential around a series of molecules [29]. This method reduces transferability between conformations as in reality the charge distribution is parametrically dependent on the nuclear positions, but this is not recreated when fixed charges are employed.

Capturing the change in electrostatic properties with molecular conformation is the goal of polarisable FF. There are polarisable versions of several major FFs, including AMBER [30] and CHARMM [31]. The polarisation methods employed typically fall into one of three groups: point polarisable dipoles [32], Drude oscillators [33] or fluctuating charges [34]. The first two methods require iteration to self-consistency, which adds computational cost to a force field calculation. All methods suffer from the polarisation catastrophe. This occurs when polarisable electrostatic moments are at a short separation, leading to a divergent evaluation of their interaction energy. This problem is typically avoided by use of an arbitrary damping function [35].

This communication focuses on the replacement of the electrostatic term with a more detailed form that incorporates polarisation by application of a machine learning method. Applications of machine learning methods to the design of potentials are relatively sparse. Applications of neural networks have recently been reviewed by ourselves [36]. Of particular relevance are applications to the simulation of liquids [37, 38] and to the computation of interatomic potentials [39–41]. Kriging (Gaussian Process Regression) has also been used to model potentials for solids [42] and has been applied to QSAR problems [43]. Of the investigated methods, Kriging provides the best cost-to-accuracy ratio. An application of the current procedure to ethanol was recently described [44] and is here generalised and applied to an amino acid pilot system. *The method proposed here avoids the iterative calculation of the polarisation and does not require a (short-range) damping function.*

Point charges, as employed in FF, are isotropic, whereas the electron distribution of a chemical system is not. Consideration of the ab initio expression for the total molecular energy clarifies this issue. The total molecular energy can be written as a function of reduced density matrices, as specified by Eq. 3, which is valid for closed-shell Hartree–Fock wave functions,

$$E_{\text{tot}} = -\frac{1}{2} \int d\mathbf{r}_1 \nabla_1^2 \rho(\mathbf{r}_1, \mathbf{r}_2)_{\mathbf{r}_1 = \mathbf{r}_2}$$
$$+ \frac{1}{2} \iint d\mathbf{r}_1 d\mathbf{r}_2 \frac{\rho_{\text{tot}}(\mathbf{r}_1)\rho_{\text{tot}}(\mathbf{r}_2)}{r_{12}}$$
$$- \frac{1}{4} \iint d\mathbf{r}_1 d\mathbf{r}_2 \frac{\rho(\mathbf{r}_1, \mathbf{r}_2)\rho(\mathbf{r}_2, \mathbf{r}_1)}{r_{12}} \tag{3}$$

The first term is the electronic kinetic energy, the second the Coulomb interaction and the third describes the exchange interaction. Partitioning the total energy into intra-atomic and inter-atomic contributions, as detailed in Eq. 4, enables the exploration of a mapping between ab initio energy contributions and the classical FF expressions of Eqs. 2 and 3,

$$E_{\text{tot}} = \sum_A E_{\text{kin}}^A + \frac{1}{2} \sum_A E_{\text{coul}}^{AA} - \frac{1}{4} \sum_A E_X^{AA} + \frac{1}{2} \sum_{\substack{A,B \\ B \neq A}} E_{\text{coul}}^{AB}$$
$$- \frac{1}{4} \sum_{\substack{A,B \\ B \neq A}} E_X^{AB} \tag{4}$$

If we restrict the inter-atomic Coulomb interactions in Eq. 4 (the fourth group of terms) to be between atoms separated by more than three bonds, then we obtain the analogue of the electrostatic energy as given by the last term in Eq. 2. An expansion in terms of spherical harmonics of $1/r_{12}$ appearing in the electrostatic energy expression of Eq. 3 (the second term) allows the interaction energy between two atoms A and B to be expressed in terms of multipole moments and a geometric interaction tensor $T_{l_A m_A}^{l_B m_B}$, with the introduction of a convergence criterion, or

$$E_{\text{Coulomb}} = \sum_A \sum_{B \neq A} \sum_{l_A=0}^{l_{\max}} \sum_{l_B=0}^{l_{\max}} \sum_{m_A=-l_A}^{l_A} \sum_{m_B=-l_B}^{l_B} Q_{l_A m_A} T_{l_A m_A}^{l_B m_B} Q_{l_B m_B} \tag{5}$$

Q_{lm} is the mth component of a multipole of rank l. A moment of rank l has $2l + 1$ individual components numbered from $-l$ to $+l$. For example, the dipole moment is of rank 1 and has three components, $m =$

$(-1, 0, 1)$. $T_{l_A m_A}^{l_B m_B}$ is dependent on the internuclear separation and mutual orientation of local coordinate systems centred on the nuclei. We note here that analytical first and second derivatives of this potential are required for geometry optimisation and molecular dynamics calculations. These are known analytically within the rigid body formalism [45], and enable a robust investigation into PESs [46]. A future communication will detail the analytical form of the derivatives for the method described herein. The general formulae for geometry-dependent multipoles have been recently described [47]. It is useful to introduce a rank for an atom–atom interaction, denoted L. This rank measures the number of multipole moments used to compute the energy and is equal to $l_A + l_B + 1$. In previous studies, we have shown that $L = 5$ is necessary to reproduce both the broad structural features of a liquid-like system (i.e. water [48]) and the structural details of hydrated biomolecules [27]. Computing the complete $L = 5$, energy would require the building of 25 models, one for each multipole moment of each atom. The building of Kriging models is the most time-consuming part of the process; therefore, we restrict the models to monopole, dipole and quadrupole moments only (i.e. nine components in total). No additional technology is required to treat the higher moments, and they may be included where CPU time is available. The model building time does not increase with the rank of a multipole moment. The decision to restrict the included moments to up to quadrupole means that $l_{\max}$ in Eq. 5 is set to 2. Hence, octupole-dipole, octupole-charge and hexadecapole-charge terms (and their inverse counterparts) are not included in the energy evaluation of Eq. 5 used to test the models. Using the true octupole and hexadecapole moments, we can investigate the contribution of these terms to a total 1–4 and higher interaction energy. The mean of the unsigned values of the sum of these three energy contributions amounts to 8 kJ mol^{-1} for 100 test conformations. This means that, on average, 8 kJ mol^{-1} is un-assessed in terms of Kriging polarisation.

The convergence criterion states that the expansion will converge when the magnitude of a particular vector is smaller than the inter-nuclear distance. This vector is defined as the difference between two position vectors, each marking the location of an electron with respect to a nucleus. In general, this convergence criterion allows the application of Eq. 5 to 1–4 and higher terms; although in some cases, 1–4 interactions may not meet this criterion. In such cases, it is possible to move the predicted multipole moments to different centres by application of a "shift" procedure [49]. The resulting shifted moments allow the calculation of initially non-convergent 1–4 interactions on the same footing as the 1–5 and higher interactions [50]. An algorithm for determining the optimal shift has recently been described [51]; however, it is not applied here as it requires higher moments to give accurate results. The separation of nuclear and electronic coordinates allows the separate computation of the multipole moments and their subsequent combination to compute energies. Multipole moments provide a more accurate picture of the ab initio electrostatics than atom-centred charges. The values of the multipole moments for a particular system may be determined by several methods. Most common amongst these is the Distributed Multipole Analysis [52], whilst several alternative methods have been defined and compared [53, 54]. FF that include multipolar electrostatics and polarisation have been described [55], but are yet to find use amongst computational biochemists. As such, the number of cases that validate the additional expense of including a more detailed picture of electrostatics is still limited [56].

A method that allows strict atomic partitioning is required in order to keep within the FF framework. The idea central to QCT is the partitioning of a system into subspaces by means of a gradient vector field. The methods of QCT thus include the partitioning of a molecular charge density into regions that can be described as atoms by consideration of the gradient vector field of the charge density [57, 58]. QCT is rooted in quantum mechanics [59], using atomic theorems to define atomic properties. The use of the term QCT has been justified previously [60, 61]. Atomic properties are described by integrals over the resulting atomic *basins*, where the integrand is a point property. For example, the integration over the basin of 1 returns the atomic volume, N. When 1 is replaced by the electron density, the result is the atomic monopole moment, or charge. The partitioning is applicable to chemical wave functions in general. Figure 1 shows the QCT atoms of an extended conformation of AlaD. The boundaries between atoms are clear. When an interatomic surface is not present, atoms extend to infinity and in practical applications have to be capped by an isosurface in $\rho(\mathbf{r})$.

The atomic multipole moments (AMM) can be expressed as integrals over the atomic basins.

$$Q_{lm}(\Omega) = \int_\Omega d\mathbf{r}\,\rho_{\text{tot}}(\mathbf{r})R_{lm}(\mathbf{r}) \qquad (6)$$

ρ_{tot} is the total ground-state electron density and R_{lm} is a regular spherical harmonic function, which can be complex. Ω denotes integration over an atomic basin. We work with real spherical harmonics; therefore, Eq. 6 has to be modified according to the procedure described in Stone's monograph [62]. The AMMs have a parametric dependence on the nuclear positions and a direct dependence on the electron density, i.e. AMMs are conformationally dependent [63]. For a given conformation, there is a

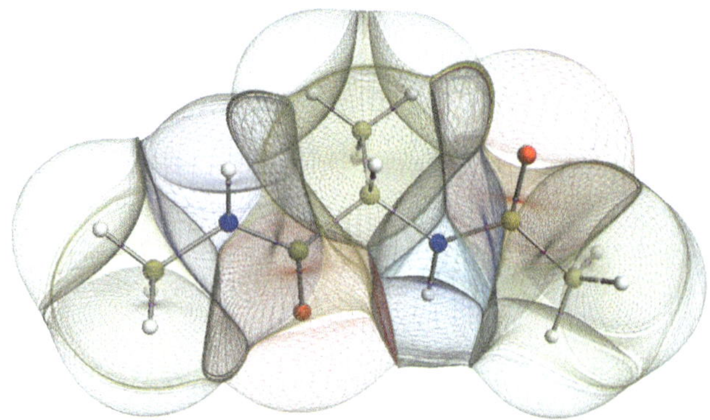

Fig. 1 The QCT atomic partitioning of the AlaD molecule. Atoms are capped by the $\rho = 0.001$ a.u. isosurface

single value for each AMM in a topologically partitioned molecule. We can therefore describe polarisation as the change in the value of a multipole moment of a QCT atom when the molecular nuclear positions (and hence the internal coordinates) are altered. This approach replaces a previously explored avenue, where polarisation was treated in the context of *long-range* Rayleigh–Schrödinger perturbation theory [64]. There the change in an AMM (typically only the dipole moment) is expressed in terms of an electric field through a polarisability (tensor). In the current method, we do not express polarisability explicitly. Instead, *we construct a direct mapping between an AMM of interest to nuclear coordinates of its environment*. This method is valid at short range because it draws its data from (super)molecular wave functions that obey the Pauli principle and whose electron densities lead to well-defined non-overlapping atoms. An additional advantage of the use of QCT is that any system can be decomposed into discrete atoms, and thus the distinction between inter- and intramolecular polarisation can be removed. That is, an atom is polarised by all other atoms in the system, whether this is composed of a single molecule (as in the current case) or a cluster of molecules. Previous work [37] has described the application of the current method to clusters of rigid water molecules, where the AMMs of a central water molecule were modelled as dependent on the position and orientation of a number of neighbouring water molecules (from 1 to 4). Provided a consistent set of coordinates can be defined for every structure in a training set, Kriging models can be built for the AMMs that model inter- and intramolecular polarisation on the same footing.

Figure 2 illustrates the concept of a moment directly varying as a function of varying nuclear positions. This figure displays the value of the monopole moment of C_α of AlaD for a series of values of the protein backbone dihedral angles φ and ψ. C_α is chosen as its AMMs are expected to vary most significantly with these central angles. It should be noted that the monopole moment Q_{00} coincides with the *net* atomic charge because Eq. 6 involves the *total* charge density, which contains the nuclear charge as well. Rotations were carried out around these angles starting from the extended structure in Fig. 1 without optimisation of the resulting structures. An angular step of 10° was employed in both dihedral angles. The charge remains positive in all cases, but varies over the torsion space between almost neutral ($Q_{00} = 0.001e$) and nearly a full electronic charge ($Q_{00} = 0.884e$). In general, the relationship between Q_{00} and φ, ψ is particularly complicated, but also smooth. We cannot confidently map this plot onto a Ramachandran map because internal coordinates other than φ, ψ are not altered (or relaxed). In fact, a small number of conformations were so strained that a wave function could not be obtained for them with GAUSSIAN's default parameters. Rather than commenting on local extrema, we just point out the broad periodicity over 180° in φ and ψ.

Replacement of the isotropic charge–charge electrostatic interaction energy by Eq. 5 is not immediately feasible due to the expense involved in evaluating a wave function for each conformation of interest and integrating the electron density over atomic basins. It is first necessary to find an alternative analytical form for the AMMs that avoids such an evaluation at every step. In this work, this is achieved by modelling the dependence of the AMMs on the internal coordinates of a molecule. A prescription for defining the internal coordinates of a molecule is required. We use an atomic local frame (ALF) to express the Cartesian coordinates of a molecule from an atomic point of view. Each atom of the molecule has a different ALF. The frame is centred on the atom in question. The x-axis is chosen to point along the line from the origin to the heaviest atom bonded to it in the molecular graph. The xy-plane is defined so as to contain both the heaviest and second heaviest atoms bonded to the origin atom. In the case of terminal atoms, the x-axis is defined by the only atom bonded to the origin atom, and the xy-plane is defined to contain the x-axis atom and its own heaviest bonding partner. Where two or more duplicate elements are bonded to a central atom, the Cahn-Ingold-Prelog rules are employed to decide priority. If all atoms are equal according to these rules, then the lowest numbered amongst them is used. For example, the atom H18 has the frame H18–N17–C6. Being a terminal hydrogen atom, its x-axis is defined by its bonding partner N17. This atom has two other bonding partners, C6 and C19. As C6 is directly bonded to O10 and C3, with C19 bonded to H20, H21 and H22, the former is of higher priority. The AMMs are computed in these frames.

Internal coordinates are used to build the Kriging models as for a system of N atoms, there are 3 N-6 such coordinates. The dimensionality of the feature space is important when building models. Higher-dimensional relationships are more difficult to model, so N should be minimised where possible. However, Kriging is able to cope well with high values of d. For the AlaD system the dimensionality is 60. The internal coordinates used in this work are the magnitudes of the position vectors of the atoms defining the x-axis and xy-plane and the angle between these two vectors. The position of any atom *not* used to define the ALF is described using spherical polar coordinates. Machine learning methods can be used to build models for the relationship between AMMs and nuclear geometry. The Kriging method [65, 66] (also termed Gaussian Process Regression) is employed here for this purpose. Internal coordinate vectors for a set of appropriate examples and the corresponding AMMs constitute the training data of the models. The result of the model building is an analytical form for the AMMs as a function of nuclear position,

$$y(\mathbf{x}) = \mu + \sum_{i=1}^{N_{\text{train}}} w_i \exp\left[-\sum_{j=1}^{d} \theta_j \left| x_j^i - x_j \right|^{p_j} \right] \tag{7}$$

Here, μ is the global trend in the data and w_i is the Kriging weight for a particular training example i, computed from the training data. θ_j and p_j are the jth components of the $3N - 6$ dimensional Kriging vector parameters $\boldsymbol{\theta}$ and $\mathbf{p}$. These are vectors with $3N - 6$ elements that describe the weight each feature is given in making a prediction ($\boldsymbol{\theta}$) and the power to which the difference between *test* and *training* values of features are raised. There are therefore two times the number of coordinates parameters ($2(3N - 6)$) to be determined. In Eq. 7, x_j^i is an internal coordinate for a *training* set geometry, and x_j is the corresponding value of that coordinate for a *test* geometry. For such a *test* geometry, the Kriging model assumes an AMM to be composed of the global trend, added to a weighted sum over all the training examples of the correlation between the structures of the

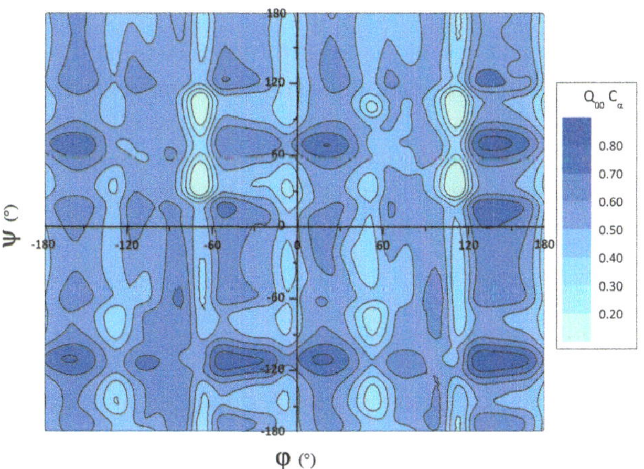

Fig. 2 Contour map of the relationship between the protein dihedral angles φ, ψ and the monopole moment (Q_{00}) of the AlaD α-carbon

test and each *training* conformation. The correlation is here measured by the exponential power correlation function, although alternatives may be used in its place. It should be noted that for conformations far outside the training set, the magnitude of the feature differences is large. As this value increases, the exponential factor becomes smaller and the predicted moments tend towards the global trend, which is an estimate of the average over all conformations. Conformationally averaged AMMs have been previously suggested as a method for the inclusion of a better electrostatic description than the point charge method [67]. The fact that the Kriging models reduce to this representation for unseen conformations is an additional advantage of using Eq. 7 to model the polarisation of AMMs. The building of a Kriging model involves maximisation of the logarithm of the likelihood function L of the training data with respect to these parameters,

$$\log L(\boldsymbol{\theta}, \mathbf{p}, \mu, \sigma | y^i; i = 1, 2, \ldots, N_{\text{train}})$$
$$= -\frac{N_{\text{train}}}{2} \log(\sigma^2) - \frac{1}{2} \log(|\mathbf{K}|)$$
$$- \frac{(\mathbf{y} - \mathbf{1}\mu)^T \mathbf{K}^{-1} (\mathbf{y} - \mathbf{1}\mu)}{2\sigma^2} \quad (8)$$

$\mathbf{K}$ is the correlation matrix whose elements are the values of the exponential power correlation function for pairs of examples in the training set, $\mathbf{y}$ is the vector of training outputs and μ, σ^2 are the mean and variance of the training data. The vector $\mathbf{1}$ is has N_{train} components that are all equal to 1. The elements of $\mathbf{K}^{-1}(\mathbf{y} - \mathbf{1}\mu)$ are the Kriging weights referred to in Eq. 7. The computational expense of the training lies in inverting $\mathbf{K}$ for each step of the optimisation; therefore, search methods that require small numbers of steps are generally preferred. A particle swarm optimisation was used to find the optimal parameters $\boldsymbol{\theta}$ and $\mathbf{p}$ for each AMM. The results of this optimisation were then subjected to a Nelder–Mead downhill search to refine their values. A more detailed account of Kriging as applied to AMMs can be found in [37].

The quality of the final models depends on the number of training examples, N_{train}, and how well the AMM hypersurfaces can be modelled as Gaussian processes. The ultimate test of the method lies in the accuracy of geometry optimisation and molecular dynamics calculations. Carrying out such a test requires expressions for the electrostatic forces and parameterisation of Eq. 1 against the electrostatic method defined here. In addition, a mathematical form for the long-range attractive and short-range repulsive interactions must be chosen. Whether this will prevent a close agreement with the underlying QM method for such tests is an open question as of yet. We work under the assumption that improved electrostatics will allow a closer agreement with the QM method than the point charge formalism. The accuracy of the isolated method can be investigated by creating a series of test cases, generated by the same method as the training cases. Comparison of the predicted AMMs (for each atom) to the true values found by the evaluation of Eq. 6 by numerical integration provides a direct measure of the predictive power of the models. The force field electrostatic energy can be computed using the predicted and true moments, measuring the feasibility of the replacement of the charge–charge interaction in FF with the machine learning–based polarisable multipole method described herein.

In future work, we will investigate the transferability of Kriging models by a feature selection process. The prediction formula in Eq. 7 requires that the full set of internal coordinate features be present. In order to use the model to predict the AMMs of atoms in environments other than those of the training molecules, the list of features needs to be truncated. For example, if we wish to predict AMMs for an alanine α-carbon in a protein using the models built with AlaD, we can only include the features present in both systems, i.e. the coordinates of the side chain and peptide bond atoms. The values of $\boldsymbol{\theta}$ can be interpreted as weights for each feature, giving their relative importance to a predicted AMM. The features can be ranked in the order of importance by consideration of their $\boldsymbol{\theta}$ values, and models can be rebuilt using a set of features that describe the environment in decreasing detail as features are removed in this rank order. As atom types emerge naturally from this analysis, they do not have to be imposed onto systems by arbitrary considerations. Measuring the effect of this feature selection on the accuracy of the moment predictions allows the quantification of the transferability of the QCT AMM models. Such a prescription allows the models built here for the AlaD molecule to be used to predict the electrostatics of larger molecules, for example longer polyalanine chains. These models will provide accurate local polarisation due to the features that are included in the transferred Kriging models. Any polarisation effect from the presence of atoms not in the originally modelled molecule will not be captured by the transferred models. For this reason, investigations are underway into the length of amino acid chain that can be feasibly modelled, as well as the extent of the protein environment that must be included in order for the AMMs to become constant with respect to including more environment.

2 Computational details

All ab initio calculations were performed using GAUSSIAN03 [68]. Structures were created and inspected with GaussView3.0 [69]. Atomic integrations were carried out with the programs AIMall [70] and MORPHY01 [71, 72], with the GUI [73] of the latter used for visualisation of atomic basins. Kriging models were built and tested using the in-house programs EREBUS and NYX, respectively. Marvin [74] was used in the generation of force field minimum energy structures. Molecular images were produced with VMD [75, 76].

3 Generation of training data

Testing the method requires a pilot system of appropriate relevance to the intended final application. Alanine acts as such a system, providing the structural elements common to all amino acids. We cap the single amino acid structure with –NHMe and –COMe to produce AlaD. The connectivity and numbering system for this molecule is shown in Fig. 3, along with the φ and ψ angles that are used in Figs. 2 and 4. It should be noted that the proposed methodology is not dependent on any protein-specific factors. It may therefore be considered generic and can be applied, for example, to biomolecules (e.g. proteins, DNA/RNA), liquids and solvated ions, and can be used to compute the interactions between any combination of these species.

The quality of the final Kriging models is dependent on the sampling method used to select the training data. It is necessary to include nuclear arrangements similar to those that will later occur in the context of prediction. Here, we test the ability of Kriging to predict nuclear arrangements generated by the same process as the training data, so this similarity is guaranteed. As future applications lie in FFs, the method of structure generation must be relevant to molecular dynamics or geometry optimisation. Low-frequency normal modes can often be used to describe large-scale movements in proteins [77] and structural changes in smaller molecules. We therefore first perform a search to find as many of the minima on the PES as possible. The AlaD molecule was drawn and then optimised at the HF/6-31G** level of theory. The resulting geometry was used to seed a conformational search on the Dreiding force field PES. 500 structures were produced via this method. These structures were then re-optimised at the HF/6-31G** level. This low level was employed for three reasons, at the

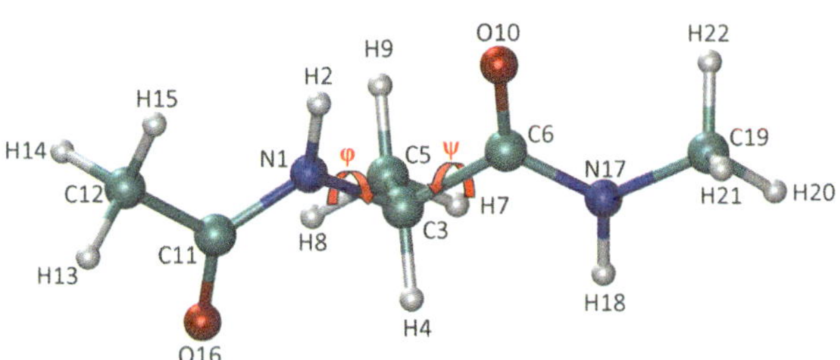

Fig. 3 Structure of the AlaD molecule and the numbering system employed

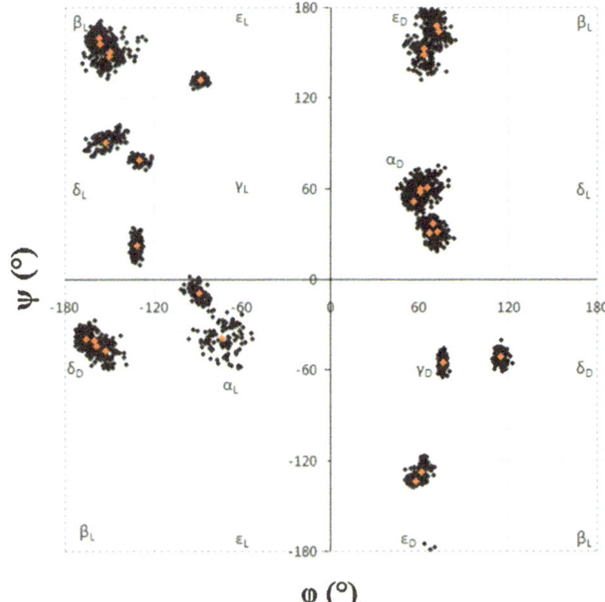

Fig. 4 φ, ψ-Projection plot of the minimum energy structures of AlaD (shown in *orange*) and the structures distorted along the normal modes of vibration (shown in *black*)

current stage of FF development. First, Hartree–Fock lacks dispersion energy, which allows us to separate out the Coulomb part of the ab initio energy [78]. This is important for the parameters of the empirical part of the potential, subject to a future publication in progress, where the Lennard-Jones function is employed to compute the 1–4 and more distant 1–n dispersion interactions. Secondly, HF/6-31G** also allows for fast computation of the required training and test data, where the small basis set compensates for the systematic error known to be present in HF. The accuracy of the method described herein is not strongly dependent on the underlying ab initio method. Therefore, we can extend conclusions drawn at this level to higher levels. Thirdly, the interatomic Coulomb energies are larger in magnitude at HF level compared to a method that incorporates electron correlation, using the same basis set. Indeed, the HF method is known to produce much more polar wave functions than post-HF or DFT methods. The energy prediction errors that will be discussed below are therefore worst upper limits in terms of absolute value. For the worst predicted test conformation, the monopole moment for N1, for example, changes from $-1.57e$ to $-1.05e$ (or $\sim 30\%$) on moving from HF/6-31G** to B3LYP/6-311+G(2d,p) (the latter evaluated at the HF/6-31G** geometry). From Eq. 9a, one can deduce that smaller absolute values of an AMM lead to smaller errors in interaction energy. Indeed, the first two terms in Eq. 9a show that an AMM prediction error is amplified (i.e. multiplied) by the magnitude of an AMM on a probe atom. Fourthly and finally, given that we can make general comments about the method based on HF wave functions, we can also make comparisons to HF results for long polyalanine chains in the very near future, connecting to the results in this paper. At higher levels of theory, comparison to chains of significant length is prohibited by the scaling of ab initio geometry optimisation. We note here that, in total, 29 minima were found for the molecule on the HF/6-31G** PES.

Population of the basins around these minimum energy structures is achieved by employing the normal modes of vibration. Normal modes were computed for each of the minima. The dataset was created by randomly distorting each minimum along a normal mode by a random factor, 100 times for each minimum. This produced 2,900 distorted structures. The distorted structures are shown projected onto the φ, ψ plane as black points in Fig. 4, along with the minima shown as orange points. This projection shows the extent of the distortions in φ, ψ space, but in reality, each conformation differs from another in all internal coordinates. The combination of minima and normal mode distortions means that the training set contains information related to conformational change. For example, rotation about an amide bond would be well predicted in terms of AMMs as minima corresponding to the starting and end structures are present in the training set, as well as a set of intermediate structures generated by the normal mode distortion procedure. For FF applications, it is possible to restrict the number of minima used in the model building process to minima belonging to a set of chemically prevalent structures. Certain regions of the Ramachandran map may not appear with a significant probability during a molecular dynamics simulation. In addition to this, minima may be structurally similar in terms of their φ and ψ angles, as shown by the clusters of orange points in Fig. 4. Such structures may differ in other significant coordinates (particularly torsion angles of the type H18–N17–C6–O10), but some are related by rotations of the capping methyl groups which may bias the models towards particular backbone structures. Reduction in the number of minima used to populate the training set is possible by consideration of the RMSD between pairs of structures in a chosen coordinate space and should result in more accurate Kriging models. This is because for a given training set size, there will be more distorted structures per minimum and the interpolation effort can be focused on interpolating between the distortions about a particular minimum rather than between the different minima themselves. A careful study of which of the full set of minima should be included will be carried out prior to the application of the currently proposed method to proteins.

The 2,900 distorted structures were randomly numbered, and wave functions for each of the first 1,000 examples were computed. It will turn out, by experimentation, that 1,000 examples were sufficient to train and test the models. The final 1,900 structures were hence discarded. The AMMs were then determined with respect to the principle axes of inertia. Prior to model building, the AMMs must be rotated to the ALF of the atom they are centred on [79]. Determining the AMMs in a common reference frame and then rotating them to their ALFs saves $N_{\text{train}} * (N_{\text{atom}} - 1)$ wave function evaluations compared to the alternative of rotating the molecule to each ALF prior to carrying out the ab initio calculations and subsequent atomic integration. A total of 22,000 atomic integrations were carried out. The average time for a single point computation was 127 s, and integrations averaged 16 s per atom for H, 47 s for C, 99 s for N and 58 s for O. The approximate total CPU time to produce the data necessary to build the Kriging models was 254 h. A local Linux computer cluster composed of approximately 200 compute nodes was used to generate the results. Therefore, approximately 2 h 30 min was required to prepare the model builds in real time. The majority of the total CPU time for the entire process was expended on building the Kriging models. The values of the Kriging vector parameter **p** (Eq. 7) may be optimised in the maximisation of the likelihood to take any real number value in the range $0 < p \leq 2$, but may also be fixed at either 1 or 2. The reduction in flexibility of the correlation function caused by fixing **p** results in a poorer fit, but significant savings in the computation time needed to build the models can be achieved as the number of parameters is halved. For the sequentially built models, those with optimised values of **p** took an average of 15 h per AMM. The average times with **p** fixed at 1 and 2 were 1 h 20 min and 55 m, respectively. Overall, if **p** is optimised, then the Kriging CPU time is 3,564 CPU h, or 93% of the total time to build the models. Using the Linux cluster, this represents 18 h in reality. For $p = 1$, the percentage spent building models becomes 50%, and for $p = 2$, it is 44%. Work towards an efficient FORTRAN implementation of the Kriging model building process is ongoing. A significant speedup of the building of models for individual AMMs will make feasible increased values of l_{max} in Eq. 5. The current program (EREBUS) is written in the JAVA language. Of the data production CPU time, 13% represents ab initio calculation and the remaining 87% is spent on atomic integration. These relative percentages will fluctuate as higher levels of theory are employed and algorithmic efficiency improvements are made.

4 Results and discussion

Testing the final models requires us to determine their ability to reproduce the values of Eq. 5 for arbitrary chemically relevant conformations of AlaD. We can write the difference between the true and predicted energy, for a given multipole–multipole interaction, in terms of the true AMMs and their prediction errors as follows,

$$\Delta E_{l_A m_A}^{l_B m_B} = T_{l_A m_A}^{l_B m_B} (Q_{l_A m_A} \Delta Q_{l_B m_B} + Q_{l_B m_B} \Delta Q_{l_A m_A} \\ - \Delta Q_{l_A m_A} \Delta Q_{l_B m_B}) \tag{9a}$$

where

$$\Delta Q_{lm} = Q_{lm}^{\text{true}} - Q_{lm}^{\text{predicted}} \tag{9b}$$

noting that

$$T_{l_A m_A}^{l_B m_B} \propto 1/r^{l_A + l_B + 1} \tag{10}$$

where r is the internuclear distance.

There are three issues to consider when examining the interaction energy error for a pair of atoms. First is the errors in the predicted moments, second the magnitude of the true moments and third the magnitude of the interaction tensor. The AMM prediction errors are foremost as they can be controlled by the improvement of the models. It is clear that large errors in the prediction of the AMMs will lead to large errors in the prediction of E_{AB}. Larger-valued AMMs will cause the inflation of the first two terms inside the brackets. This may happen, for example, in the case of oxygen atoms, which carry a high net charge or for very polar carbon atoms. The value of the interaction tensor also influences the error. For atoms that are well separated in space, the magnitude of $T_{l_B m_B}^{l_A m_A}$ is diminished compared to atoms that are in close proximity. In the latter case, $T_{l_B m_B}^{l_A m_A}$ may be very large and interaction energies are subsequently large. In addition, the inverse power dependence of $T_{l_B m_B}^{l_A m_A}$ on the values of l_A and l_B implies that the contribution to the total error decreases with increasing $L = l_A + l_B + 1$. *Attempts to improve the accuracy of the method should therefore be focused on the modelling of the lower rank AMMs.*

The optimal number of training examples is unknown a priori. Previous tests have shown that 500 examples are sufficient for systems with numbers of atoms similar to AlaD. To investigate this, Kriging models were built with sequentially increasing training set sizes, beginning at 100 and increasing in steps of 50, up to a maximum of 500. This method allows the inspection of the developing accuracy of the models as more examples are added to the training set. Training examples were filtered based on the magnitude of their integration errors, measured by the values of $L(\Omega)$. Ideally, this quantity should be zero for all atoms; deviation from this value implies inaccurate integration of AMMs. The cut-off was set at 0.001 au. The remaining data of the original 1,000 examples were used as an *external* test set, amounting to 100 test geometries.

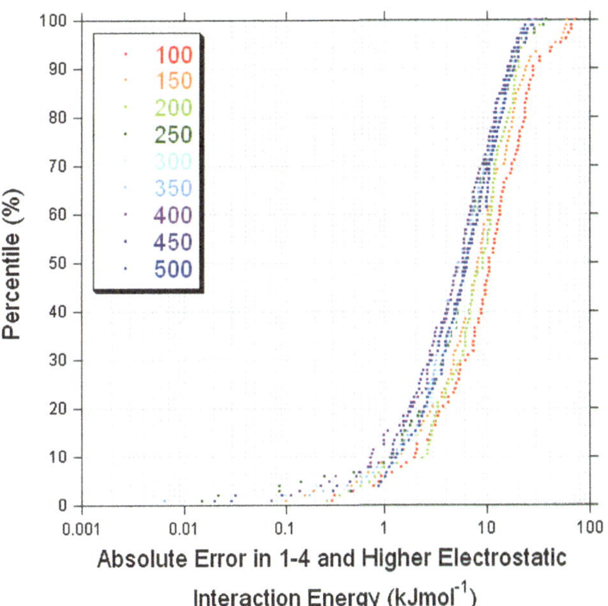

Fig. 5 Development of absolute total 1–4 and higher electrostatic interaction energy errors as training set size is sequentially increased. The average absolute interaction energy is 83 kJ mol^{-1} and has range from 0.04 to 248.5 kJ mol^{-1}

Figure 5 shows the S-curves for each training set size when used to predict the total 1–4 and higher electrostatic interaction energies of these test conformations. The curves show the $\log_{10}$ of the *absolute* interaction energy error against their *percentile values*. For example, the 50th absolute energy error percentile is the error that 50% of the predicted geometries are within. The interaction energy error can be defined as the difference between the energy evaluated in Eq. 5 using the true AMMs and using the predicted AMMs, where A and B are separated by three or more bonds. Alternatively, the same interaction energy error can be obtained by summing the energy difference evaluated in Eq. 9a, over all l_A, l_B, m_A and m_B values, and over all atoms A and B separated by three or more bonds. It should be emphasised again that the absolute value of this interaction energy error is plotted on the x-axis of Figs. 5, 6, 7 and 8. *Improvement of the models is manifested in leftward movement of the S-curve.* The randomly selected example geometries forming the initial training set of 100 distorted geometries (red curve) result in significant errors compared to the optimal models, with a maximum of 73 kJ mol^{-1} and an average of 15 kJ mol^{-1}. Increasing the training set size by adding 50 examples each time produces a visually distinguishable improvement of the models up to a training set size of 250 examples. Thereafter, the progression is small and does not represent an improvement in every case. This suggests that 250 examples are required to produce the optimal models for the atoms of AlaD. Smaller training set sizes reduce the time needed to build all of the

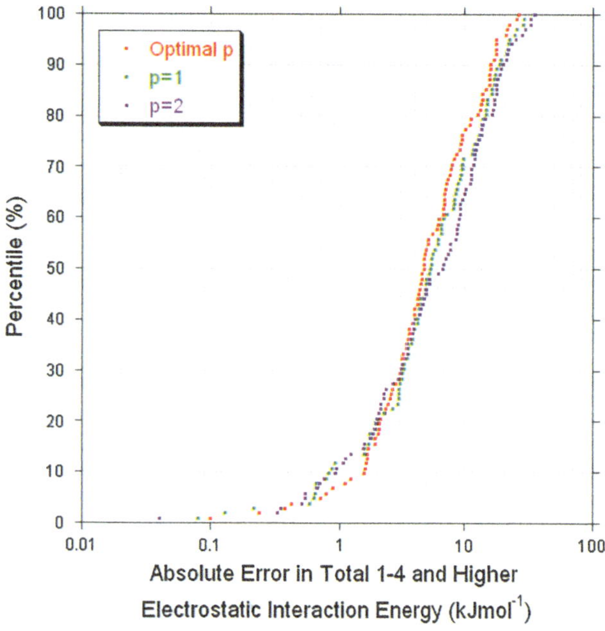

Fig. 6 Absolute total 1–4 and higher electrostatic interaction energy errors for optimal training set size models where **p** is optimised (*red*), fixed at 1 (*green*) and fixed at 2 (*purple*). The average absolute interaction energy is 83 kJ mol^{-1} and has range from 0.04 to 248.5 kJ mol^{-1}

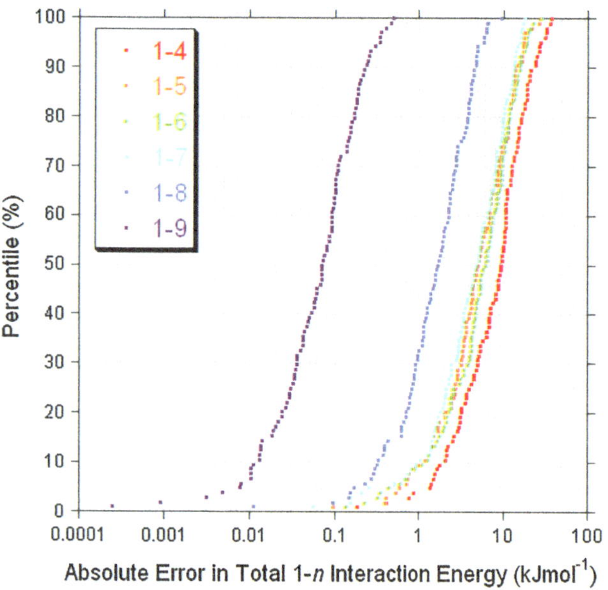

Fig. 7 Absolute total 1–n electrostatic interaction energy errors for optimal training set size models for $n = 4$–9

AMM models for a system, reducing the number of wave functions required, integrations over the electron density along with both the number of steps in the sequential building of Kriging models and the dimensionality of the correlation matrix, **K**. In future applications, the number of calculations need not be as many as those carried out here.

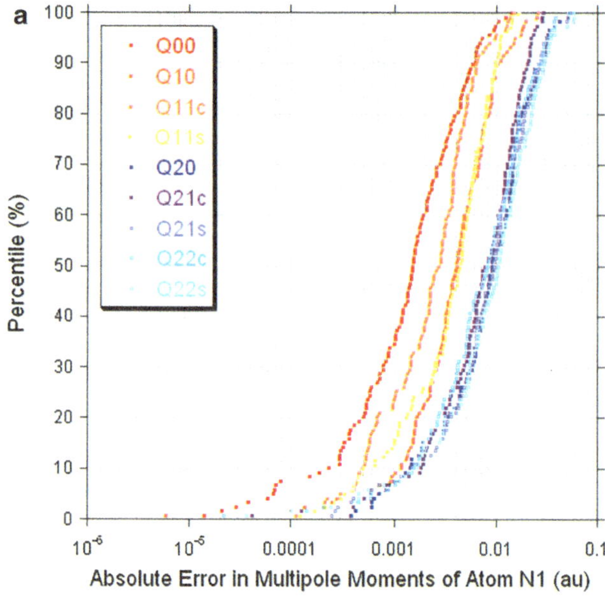

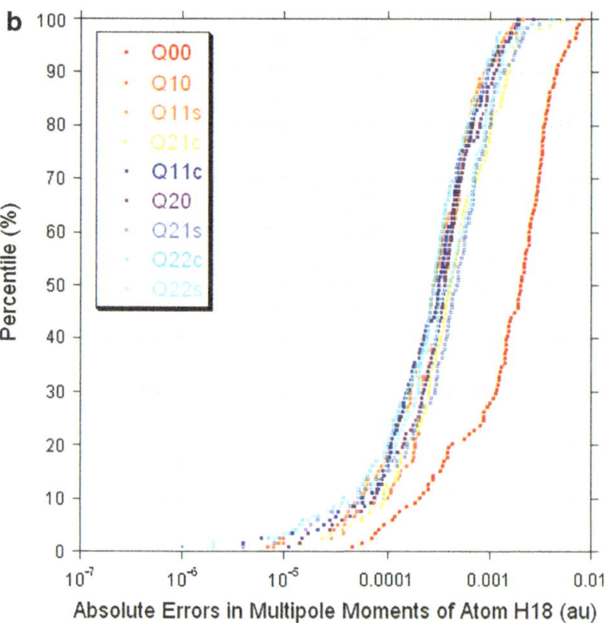

Fig. 8 S-curves of the absolute error in prediction of the nine multipole moments of atoms **a** N1 and **b** H18 for each structure in the test set. MM errors are given in appropriate atomic units (see caption for Table 2)

We are not limited to using the same number of training examples for each model, thus the final AMM models were chosen by consideration of the r^2 correlation coefficient of each when used to predict the training set. This procedure was carried out by EREBUS. The optimal models for each AMM can then be combined to predict optimally accurate values for AMMs, and from these interaction energies can be obtained.

Figure 6 shows the absolute total interaction energy error S-curves for the optimal training set size models where $\mathbf{p}$ is optimised (red) and fixed at 1 (green) and 2 (purple). Fixing the values of $\mathbf{p}$ at 1 improves the lowest 20% of the prediction errors compared to the optimal $\mathbf{p}$. Specifically, Fig. 6 shows that the green curve ($p = 1$) intersects the red curve (optimal $\mathbf{p}$) at the point with coordinates (2 kJ mol^{-1}, 20%); moving beyond this point, from the left to right, one finds that the red curve is superior to the green one. Similarly, fixing the values of $\mathbf{p}$ at 2 (purple curve) improves the lowest 30% of the prediction errors compared to the optimal $\mathbf{p}$. However, the upper part of the curve deteriorates for both $\mathbf{p} = (1, 1, \ldots, 1)$ and $\mathbf{p} = (2, 2, \ldots, 0.2)$ with the maximum error (not easily perceptible from the graph) increasing by 8.7 kJ mol^{-1} for $\mathbf{p} = (1, 1, \ldots, 0.1)$ and 8.3 kJ mol^{-1} for $\mathbf{p} = (2, 2, \ldots, 0.2)$. As the low percentile errors remain within the generally quoted chemical accuracy of 4 kJ mol^{-1} in all cases, we choose to proceed with the models wherein $\mathbf{p}$ is an optimised parameter, gaining accuracy in the higher percentiles but increasing total CPU time, as discussed above. In future applications that require significantly greater numbers of models to be built, fixing $\mathbf{p}$ may become necessary in order to make the building of those models computationally feasible.

The average absolute error in the total 1–4 and higher ($1-n$, $n \geq 4$) electrostatic interaction energy when predicting using the $\mathbf{p}$-optimal models is 7 kJ mol^{-1}, with a maximum of 26 kJ mol^{-1} and a minimum of 0.1 kJ mol^{-1}. The standard deviation over the test cases is 6 kJ mol^{-1}. The range of absolute interaction energies corresponding to these errors is from 0.04 to 248 kJ mol^{-1}, with an average value of 83.3 kJ mol^{-1}. These total energy errors are "low-detail" tests of the method, involving summation over all atom pairs A, B and AMMs. An understanding of how the method can be improved can be gained by inspecting the results at a greater level of detail.

To further break down the total energy errors into their components, we can remove the summation over all atom pairs and consider groups of individual atom–atom interactions. Table 1 shows the average results for the $1-n$ interactions between pairs of elements for values of n found in the AlaD system, along with the average values of the interaction energies for each combination. Figure 7 shows the S-curves for the total $1 - n$ interaction energies summed over atom pairs.

There are 174 1–4 and higher interactions in the full set, divided by element-element criteria into 10 subsets: 57 H–H interactions, 50 C–H, 24 O–H, 18 N–H, 9 C–C, 8 O–C, 4 N–C, 2 N–O, 1 O–O and 1 N–N. Dividing the interactions according to n produces 35 subsets in total. In general, the average interaction errors decrease with the value of n, as can be seen from Fig. 7. The only significant exception to this rule is for the N–C interactions. The general pattern occurs because the dependence on n is a consequence of

Eqs. 9a, 9b and 10. In general, higher values of n imply greater average values of R, reducing the magnitude of $T^{l_A m_A}_{l_B m_B}$ in the calculation of the prediction errors. In some cases, the molecule may fold in such a way as to nullify this by making certain 1–n distances (where $n > 4$) shorter than 1–4 distances. Deviation can also be caused by the fact that the AMMs of two atoms of the same element may not be predicted equally as accurately as each other over the test set. The 1–5N–C distances in the test set are on average 15% longer than the 1–4 distances. Therefore, the second explanation must be true, and this is borne out by an examination of the AMM contributions to the energy error.

Equation 9a suggests that the most fundamental elements in the total energy errors are the errors in the individual ΛMMs as these can be decreased in magnitude by employment of better models. Figure 8 shows the S-curves for the absolute errors in prediction of the AMMs for the example atoms N1 and H18, whilst Table 2 collects the prediction errors on each AMM of each atom averaged over the test set.

For N1 (Fig. 8a), the monopole is the AMM with the smallest magnitude prediction errors, followed by the three dipole components and then the five quadrupole components. This pattern is repeated for all non-hydrogen atoms whose AMMs vary little with distortion of the nuclear skeleton. For instance, in some cases, the charge may have errors close to the dipole components, e.g. Q_{10} is better predicted on average than Q_{00} for atom C5. In general for these atoms, an inverse relationship exists between prediction accuracy and moment rank, l.

This relationship is not observed for H18 (Fig. 8b), wherein the absolute errors in the prediction of the monopole moment are significantly worse than the rest of the moments. This pattern is repeated across all hydrogen atoms in AlaD. The charge is predicted with the same order of magnitude absolute average error for all atoms, with the maximum at 0.0039e (C18) and minimum at 0.0018e (H8). The higher AMMs have an order of magnitude greater average error in prediction for non-hydrogen atoms compared to hydrogen. To place these errors in context, the average true values of each moment are included in Table 2.

Whilst the ultimate genesis of the total energy errors is in the moment predictions, it is interesting to separate their influence into atomic contributions. This is achieved by setting all moments to their true values except for those of one atom, which is predicted using its optimal models. The test set total energies were computed 22 times, each atom being treated as described. That is, the true atoms of 21 atoms interacting with the predicted moments on one atom compared to the completely true energies. Table 3 collects the resulting errors in the total energy for each atom. The

Table 1 Average absolute error in total 1–n ($n \geq 4$) electrostatic interaction energy $\overline{|\Delta E|}$ for element–element interactions grouped by n, along with average magnitudes of each group of interaction energies $\overline{|E|}$

Elements	1–n	$\overline{|E|}$	$\overline{|\Delta E|}$	$|\Delta E|_{min}$	$|\Delta E|_{max}$	σ
H–H	1–4	14.66	1.31	0.0060	7.43	1.18
	1–5	33.34	0.97	0.0043	3.81	0.81
	1–6	66.82	0.81	0.00085	4.14	0.68
	1–7	6.55	0.62	0.018	4.24	0.59
	1–8	6.88	0.53	0.010	2.27	0.40
	1–9	1.10	0.10	0.00024	0.48	0.098
H–C	1–4	484.17	7.07	0.20	33.42	5.74
	1–5	71.49	3.13	0.023	15.98	2.95
	1–6	345.07	3.07	0.063	15.88	3.06
	1–7	27.88	2.71	0.058	19.03	2.64
	1–8	12.14	0.71	0.0092	3.84	0.65
H–N	1–4	106.67	3.75	0.064	15.63	3.31
	1–5	538.37	3.14	0.0038	12.54	2.51
	1–6	37.77	2.62	0.018	19.45	2.65
	1–7	23.98	1.86	0.064	8.57	1.44
H–O	1–4	702.46	4.25	0.076	13.60	3.37
	1–5	180.07	4.61	0.0089	22.87	4.03
	1–6	650.10	3.87	0.064	20.76	3.65
	1–7	195.78	2.23	0.045	10.77	2.07
	1–8	29.89	1.93	0.020	12.46	1.86
C–C	1–4	1,709.98	4.00	0.12	14.10	3.41
	1–5	162.07	2.04	0.011	8.81	1.64
	1–6	385.89	2.06	0.042	7.78	1.67
	1–7	28.96	0.58	0.011	2.56	0.46
N–C	1–4	147.69	1.80	0.012	7.26	1.55
	1–5	1,413.28	3.52	0.046	11.59	2.73
	1–6	75.92	1.42	0.012	6.02	1.11
O–C	1–4	1,090.56	4.12	0.0038	16.20	3.42
	1–5	1,055.32	2.30	0.016	7.68	1.63
	1–6	70.09	1.33	0.011	7.07	1.23
	1–7	273.35	1.72	0.033	8.36	1.52
N–O	1–4	978.73	2.12	0.041	8.11	1.89
	1–6	712.95	2.09	0.034	16.38	2.79
N–N	1–4	1,071.55	2.51	0.033	11.69	2.36
O–O	1–6	634.34	1.65	0.014	14.10	2.47

All energies are given in kJ mol^{-1}. Each line corresponds to a set

quantities these errors are measured against are the same as the energies plotted in Fig. 6, where the average absolute total interaction energy is 80 kJ mol^{-1} and the range is approximately 250 kJ mol^{-1}.

Then, N–C ordering for the 1–n interactions as discussed above is explained by the contributions of atoms N1 and N17. The 1–4 N–C interaction is between atoms N17 and C5, whilst the two 1–5 interactions occur between atoms N1–C19 and N17–C11. N1 makes a significantly higher contribution to the total energy error than does N17, the former 2.31 kJ mol^{-1} to the latter 0.61 kJ mol^{-1} on average. In addition, C19 makes the highest contribution of all C atoms with 2.20 kJ mol^{-1}. The combination of the errors on the atoms N1 and C19 therefore outweigh the increase in the interaction distance in influencing the energy errors.

The errors in Table 3 can be somewhat correlated to the chemical environment of the molecule. For example, the methyl group formed by atoms H7, H8, H9 and C5 shows a trend where H9 is worse predicted than H7 and H8. The molecular distances between these H atoms and O16 are smaller for H9, suggesting a stabilising interaction is formed between this pair of atoms. This results in a more polarised H9 with larger higher moments than H7 and H8. The greater magnitude of the moments means that their prediction errors will be larger for models that are equally as accurate in percentage terms. This example is not

Table 2 Average absolute atomic multipole moment values $\overline{|Q|}$ and corresponding average absolute prediction errors $\overline{|\Delta Q|}$ over the test set examples

	Q_{00}		Q_{10}		Q_{11c}		Q_{11s}	
	$\overline{|Q|}$	$\overline{|\Delta Q|}$	$\overline{|Q|}$	$\overline{|\Delta Q|}$	$\overline{|Q|}$	$\overline{|\Delta Q|}$	$\overline{|Q|}$	$\overline{|\Delta Q|}$
Nitrogen								
N1	1.54	0.0025	0.19	0.0055	0.14	0.0034	0.11	0.0050
N17	1.55	0.0031	0.12	0.0042	0.14	0.0042	0.13	0.0044
Oxygen								
O10	1.40	0.0020	0.31	0.0020	0.44	0.0025	0.24	0.0022
O16	1.41	0.0027	0.36	0.0017	0.42	0.0056	0.23	0.0021
Carbon								
C3	0.58	0.0029	0.02	0.0037	0.56	0.0037	0.09	0.0039
C5	0.22	0.0031	0.02	0.0028	0.02	0.0032	0.03	0.0042
C6	1.84	0.0029	0.63	0.0034	0.34	0.0027	0.43	0.0032
C11	1.84	0.0024	0.33	0.0023	0.68	0.0025	0.27	0.0025
C12	0.16	0.0032	0.03	0.0036	0.05	0.0038	0.02	0.0027
C19	0.67	0.0039	0.40	0.0050	0.25	0.0053	0.29	0.0052
C–Hydrogen								
H4	0.04	0.0030	0.08	0.0009	0.03	0.0012	0.05	0.0011
H7	0.05	0.0022	0.01	0.0007	0.11	0.0005	0.01	0.0007
H8	0.05	0.0018	0.08	0.0006	0.03	0.0010	0.06	0.0007
H9	0.05	0.0024	0.08	0.0012	0.02	0.0013	0.06	0.0010
H13	0.03	0.0028	0.04	0.0007	0.03	0.0010	0.08	0.0009
H14	0.02	0.0022	0.07	0.0008	0.05	0.0008	0.04	0.0009
H15	0.02	0.0024	0.06	0.0007	0.06	0.0007	0.05	0.0007
H20	0.04	0.0025	0.05	0.0008	0.06	0.0011	0.05	0.0011
H21	0.03	0.0028	0.05	0.0011	0.05	0.0012	0.05	0.0013
H22	0.03	0.0026	0.04	0.0011	0.05	0.0012	0.06	0.0010
N–Hydrogen								
H2	0.45	0.0025	0.11	0.0005	0.07	0.0005	0.09	0.0004
H18	0.46	0.0023	0.09	0.0005	0.08	0.0004	0.08	0.0004

	Q_{20}		Q_{21c}		Q_{21s}		Q_{22c}		Q_{22s}	
	$\overline{|Q|}$	$\overline{|\Delta Q|}$	$\overline{|Q|}$	$\overline{|\Delta Q|}$	$\overline{|Q|}$	$\overline{|\Delta Q|}$	$\overline{|Q|}$	$\overline{|\Delta Q|}$	$\overline{|Q|}$	$\overline{|\Delta Q|}$
Nitrogen										
N1	0.34	0.0120	0.15	0.0097	0.58	0.0120	0.59	0.0110	0.27	0.0140
N17	0.36	0.0086	0.36	0.0081	0.41	0.0093	0.37	0.0092	0.51	0.0087
Oxygen										
O10	0.06	0.0040	0.08	0.0039	0.08	0.0041	0.08	0.0049	0.10	0.0044
O16	0.07	0.0082	0.03	0.0038	0.10	0.0028	0.12	0.0097	0.05	0.0035
Carbon										
C3	0.21	0.0065	0.04	0.0052	0.07	0.0065	0.48	0.0075	0.08	0.0064
C5	0.06	0.0061	0.04	0.0043	0.05	0.0062	0.04	0.0062	0.04	0.0059
C6	0.16	0.0041	0.37	0.0037	0.19	0.0029	0.32	0.0031	0.33	0.0030
C11	0.34	0.0020	0.22	0.0032	0.34	0.0024	0.26	0.0027	0.11	0.0034
C12	0.06	0.0070	0.07	0.0062	0.07	0.0068	0.10	0.0062	0.06	0.0065
C19	0.17	0.0078	0.26	0.0080	0.31	0.0085	0.10	0.0088	0.17	0.0094
C–Hydrogen										
H4	0.10	0.0012	0.06	0.0015	0.13	0.0014	0.02	0.0016	0.05	0.0019
H7	0.12	0.0012	0.03	0.0008	0.01	0.0010	0.18	0.0010	0.03	0.0011

Table 2 continued

	Q_{20}		Q_{21c}		Q_{21s}		Q_{22c}		Q_{22s}	
	$\overline{\lvert Q\rvert}$	$\overline{\lvert\Delta Q\rvert}$	$\overline{\lvert Q\rvert}$	$\overline{\lvert\Delta Q\rvert}$	$\overline{\lvert Q\rvert}$	$\overline{\lvert\Delta Q\rvert}$	$\overline{\lvert Q\rvert}$	$\overline{\lvert\Delta Q\rvert}$	$\overline{\lvert Q\rvert}$	$\overline{\lvert\Delta Q\rvert}$
H8	0.10	0.0009	0.12	0.0011	0.15	0.0006	0.02	0.0008	0.05	0.0007
H9	0.12	0.0019	0.10	0.0018	0.14	0.0014	0.03	0.0016	0.04	0.0013
H13	0.09	0.0010	0.05	0.0018	0.10	0.0009	0.12	0.0017	0.07	0.0011
H14	0.11	0.0008	0.10	0.0010	0.08	0.0008	0.07	0.0012	0.09	0.0012
H15	0.08	0.0008	0.11	0.0010	0.10	0.0007	0.06	0.0010	0.09	0.0011
H20	0.08	0.0014	0.09	0.0013	0.08	0.0011	0.09	0.0012	0.10	0.0013
H21	0.08	0.0015	0.08	0.0016	0.09	0.0015	0.87	0.0015	0.85	0.0014
H22	0.08	0.0013	0.08	0.0010	0.07	0.0012	0.09	0.0013	0.10	0.0012
N–Hydrogen										
H2	0.02	0.0007	0.01	0.0007	0.02	0.0006	0.01	0.0008	0.01	0.0006
H18	0.01	0.0005	0.01	0.0007	0.01	0.0007	0.01	0.0004	0.01	0.0006

All values are in the appropriate atomic units (i.e. e for Q_{00}, eBohr for Q_{1m} and eBohr^2 for Q_{2m})

Table 3 Contributions of all multipole moment models of each atom individually to the absolute error in the total 1–4 and higher electrostatic energy prediction $\overline{\lvert\Delta E\rvert}$

	$\overline{\lvert\Delta E\rvert}$	$\lvert\Delta E\rvert_{\min}$	$\lvert\Delta E\rvert_{\max}$	σ
N1	2.31	0.054	11.27	2.25
H2	1.55	0.0030	6.78	1.35
C3	0.69	0.010	2.25	0.54
H4	1.51	0.10	4.99	1.16
C5	1.13	0.024	4.25	1.06
C6	1.38	0.0046	7.01	1.28
H7	0.54	0.0024	1.97	0.42
H8	0.59	0.019	2.00	0.49
H9	0.82	0.022	8.42	1.22
O10	0.64	0.0077	2.96	0.60
C11	0.11	0.0024	0.50	0.10
C12	1.66	0.00070	7.08	1.31
H13	2.51	0.025	12.18	2.11
H14	2.07	0.0072	9.38	1.86
H15	2.27	0.042	9.07	1.92
O16	1.70	0.0042	19.38	3.37
N17	0.61	0.013	3.41	0.58
H18	1.57	0.014	5.02	1.21
C19	2.20	0.0090	8.32	1.83
H20	1.37	0.029	6.01	1.21
H21	1.52	0.0061	6.11	1.26
H22	1.43	0.024	12.17	1.76

All energies are reported in kJ mol^{-1}. The average absolute interaction energy is 83 kJ mol^{-1} and has range from 0.04 to 248.5 kJ mol^{-1}

affected by the second influential observation, which is that the number of interactions is not constant for each element–element interaction pair at a specific value of n. For example, there are 7 H–H 1–5 interactions, compared to 8 O–H 1–5 interactions in this table. The capping methyl groups are involved in the greatest number of interactions in the molecule, so their absolute error summed over these interactions is much higher. This explains the greater errors observed for H13, H14, H15, H20, H21 and H22 in the AlaD system.

We close this section by some comments on the use of CPU time in the evaluation of the energy with the current approach of high-rank polarisable AMMs. The size of systems treatable by this method depends on the time taken to evaluate the energy. Comparing this to the time for the corresponding charge–charge energy calculation provides a quantitative measure of the computational expense of including Eq. 5 in a force field. Evaluation of Eq. 5 requires prediction of the AMMs and subsequent evaluation of the interaction tensor for each term in the summation. The latter can be carried out using either explicit formulae [80] or a recurrence relation [81]. Predicting the AMMs requires a set of Cartesian coordinates that are then transformed into internal coordinates in an atomic local axis system. These coordinates are normalised and used in Eq. 7 to produce normalised values for the AMMs, which are subsequently unnormalised. The time-dependence lies in the system size (number of internal coordinates) and the number of training examples used to build a particular model. For the calculations described herein, the average time to produce $(22 \times 9 = 198)$ AMMs was 140 ms. Of this, the time to compute internal coordinates was negligible ($<0.01\%$). Eighty percent of the AMM prediction time corresponds to the reading of Kriging model files prior to making predictions. If these files were read and stored in the RAM once, instead of reading them from disk when needed, the cost would be significantly reduced.

Summation of the energy expression with the interaction tensor expressed explicitly, required on average 0.5 ms, whilst the recurrence relation gave an average of 0.2 ms. This is in comparison with calculations carried out using only the monopole moments and the Coulomb law, which average 0.02 ms. The ratio of these values should be machine-independent, and therefore we state inclusion of this method in a force field framework will incur approximately an order of magnitude increase in time required for the evaluation of the electrostatics. However, one should not forget that the energy evaluation is dominated by the monopole–monopole interactions for the systems we eventually apply the proposed method to (i.e. those with thousands of atoms). High-rank multipolar interactions only need to be evaluated at short-range. There is further scope for optimisation of the method, both in moment prediction and in energy evaluation.

5 Conclusions

The machine learning method Kriging has been employed in the modelling of the dependence of the QCT atomic multipole moments of the system usually referred to in the literature as alanine dipeptide or AlaD. The Kriging models can be used to compute atom–atom electrostatic interaction energies instead of the point charge models of traditional FF. The inclusion of multipole moments gives a more detailed description of the electrostatics than the point charge model. The use of Kriging allows a natural and direct description of polarisation as the response of an electron distribution to conformational change, avoiding

the short-range polarisation catastrophe. The multipole moments are completely determined by the gradient vector field of the electron density, and the use of QCT provides smooth multipole moment surfaces without discontinuities. The method as described is general. It may be applied to any system of interest given a set of internal coordinates that describe the geometry. The underlying ab initio level of theory may be replaced with any other level of theory that allows computation of an accurate wavefunction, with the added cost being purely in CPU time required to generate the data. More multipole moments can be added to each atom to improve the electrostatic description without alteration in the method. For AlaD, the maximum error in the total electrostatic energy is 26 kJ mol^{-1} and the average error is 7 kJ mol^{-1}. These values may be improved by consideration of the sampling method used to create the training set and should improve with the use of post-HF and DFT methods to generate the underlying data set.

The function may be employed inside the force field framework with subsequent optimisation of the remaining empirical parameters (e.g. force constants for bond-stretch and angle-bend terms). The computational cost of the method in application is approximately one order of magnitude greater than the point charge method. First derivatives of the function will soon be available, allowing application to molecular dynamics and mechanics experiments.

Acknowledgments The authors would like to thank the EPSRC for financial support. Part of the computational element of this research was achieved using the High Throughput Computing (CONDOR) facility of the Faculty of Engineering and Physical Sciences, the University of Manchester.

References

1. Hegefeld WA, Chen SE, DeLeon KY, Kuczera K, Jas GS (2010) J Phys Chem A 114(47):12391–12402
2. Perez-Angel CE, Seminario JM (2011) J Phys Chem C 115(14):6467–6477
3. Swadling JB, Coveney PV, Greenwell CH (2010) J Am Chem Soc 132(39):13750–13764
4. Car R, Parrinello M (1985) PhysRevLett 55(22):2471–2474
5. Remler DK, Madden PA (1990) Mol Phys 70(6):921–966
6. Warshel A, Levitt M (1976) J Mol Biol 103:227–249
7. Hu H, Yang W (2009) J Mol Struct THEOCHEM 898(1–3):17–30
8. Stone JE, Phillips JC, Freddolino PI, Hardy DJ, Trabuco LG, Schulten K (2007) J Comput Chem 28(16):2618–2640
9. Voelz VA, Bowman GR, Beauchamp K, Pande VS (2010) J Am Chem Soc 132(5):1526–1528
10. Khalili-Araghi F, Jogini V, Yarov-Yarovoy V, Tajkhorshid E, Roux B, Schulten K (2010) Biophys J 98(10):2189–2198
11. Brooks BR, Bruccoleri RE, Olafson BD, States DJ, Swaminathan S, Karplus M (1983) J Comp Chem 4:187–217
12. Foloppe N, MacKerell AD Jr (2000) J Comp Chem 21(2):86–104
13. Weiner SJ, Kollman PA, Case DA, Singh UC, Ghio C, Profetajr S, Wiener P (1984) J Am Chem Soc 106:765–784

14. Cornell WD, Cieplak P, Bayly CI, Gould IR, Merz KM Jr, Ferguson DM, Spellmeyer DC, Fox T, Caldwell JW, Kollman PA (1995) J Am Chem Soc 117:5179–5197

15. Oostenbrink C, Villa A, Mark AE, van Gunsteren WF (2004) J Comput Chem 25(13):1656–1676

16. Jorgensen WL, Swenson CJ (1985) J Am Chem Soc 107:569–578

17. Kaminsky GA, Friesner RA, Tirado-Rives J, Jorgensen WL (2001) J Phys Chem B 105(28):6474–6487

18. Halgren TA (1996) J Comp Chem 17:490–519

19. Dinur U, Hagler AT (1994) J Comput Chem 15(9):919–924

20. Tuzun RE, Noid DW, Sumpter BG (1997) J Comput Chem 18(14):1804–1811

21. Al-Matar AK, Rockstraw DA (2004) J Comput Chem 25(5):660–668

22. Lifson S, Warshel A (1968) J Chem Phys 49(11):5116–5129

23. Ewig CS, Berry R, Dinur R, Hill JR, Hwang MJ, Li H, Liang C, Maple J, Peng Z, Stockfisch TP, Thacher TS, Yan L, Xiangshan N, Hagler AT (2001) J Comput Chem 22(15):1782–1800

24. Halgren TA (1995) J Comput Chem 17(5–6):490–519

25. Banas P, Hollas D, Zgarbova M, Jurecka P, Orozco M, Cheatham TE III, Sponer J, Otyepka M (2010) J Chem Theory Comput 6(12):3836–3849

26. Tong Y, Mei Y, Ji CG, Li YL, Chang GJ, Zhang JZH (2010) J Am Chem Soc 122(14):5137–5142

27. Shaik MS, Devereux M, Popelier PLA (2008) Mol Phys 106:1495–1510

28. Ponder JW, Case DA (2003) Adv Protein Chem 66:27–85

29. Bayly CI, Cieplak P, Cornell WD, Kollman PA (1993) J Phys Chem 97(40):10269–10280

30. Cieplak P, Caldwell J, Kollman P (2001) J Comput Chem 22(10):1048–1057

31. Patel S, Brooks CL III (2004) J Comput Chem 25(1):1–15

32. Rick SW, Stuart SJ (2002) Potential and algorithms for incorporating polarizability in computer simulations. In: Lipkowitz KB, Boyd DB (eds) Reviews in computational chemistry, vol 18. Wiley-VCH, New York, pp 89–146

33. Harder E, Anisimov VN, Vorobyov IV, Lopes PEM, Noskov SY, MacKerell Jr AD, Roux B (2006) J Chem Theory Comput 2:1587–1597

34. Hemmingsen L, Amara P, Ansoborlo E, Field MJ (2000) J Phys Chem A 104:4095–4101

35. Thole BT (1981) Chem Phys 59:341–350

36. Handley CM, Popelier PLA (2010) J Phys Chem A 114:3371–3383. doi:10.1039/b905748j

37. Handley CM, Hawe GI, Kell DB, Popelier PLA (2009) Phys Chem Chem Phys 11:6365–6376. doi:10.1039/b905748j

38. Houlding S, Liem SY, Popelier PLA (2007) Int J Quantum Chem 107(14):2817–2827

39. Behler J, Parrinello M (2007) Phys Rev Letts 98:146401–146404

40. Hobday S, Smith R, Belbruno J (1999) Model Simul Mater Sci Eng 7:397–412

41. Sanville E, Bholoa A, Smith R, Kenny SD (2008) J Phys Condens Matter 20:285219

42. Bartok AP, Payne MC, Kondor R, Csanyi G (2010). Phys Rev Lett 104(13):136403–136406

43. Hawe GI, Alkorta I, Popelier PLA (2010) J Chem Inf Model 50:87–96

44. Mills MJL, Popelier PLA (2011) Comput Theor Chem "Special issue: ESPA 2010": in pages

45. Popelier PLA, Stone AJ (1994) Mol Phys 82:411–425

46. Popelier PLA, Stone AJ, Wales DJ (1994) Farad Discuss 97:243–264

47. Elking DM, Perera L, Duke R, Darden T, Pedersen LG (2010) J Comput Chem 31(15):2702–2713

48. Liem SY, Popelier PLA, Leslie M (2004) Int J Quantum Chem 99:685–694

49. Joubert L, Popelier PLA (2002) Mol Phys 100:3357–3365

50. Rafat M, Popelier PLA (2006) J Chem Phys 124:144102–144108

51. Solano CJF, Pendás AM, Francisco E, Blanco MA, Popelier PLA (2010) J Chem Phys 132:194110

52. Stone AJ (1981) Chem Phys Lett 83(2):233–239

53. Pilme J, Piquemal J-P (2008) J Comput Chem 29:1440–1449

54. Volkov A, Coppens P (2004) J Comput Chem 25:921–934

55. Devereux M, Plattner N, Meuwly M (2009) J Phys Chem A 113(47):13199–13209

56. Ponder JW, Wu C, Pande VS, Chodera JD, Schnieders MJ, Haque I, Mobley DL, Lambrecht DS, DiStasio RAJ, Head-Gordon M, Clark GNI, Johnson ME, Head-Gordon T (2010) J Phys Chem B 114:2549–2564

57. Bader RFW (1990) Atoms in molecules. A quantum theory. Oxford University Press, Oxford

58. Popelier PLA (2000) Atoms in molecules. An introduction. Pearson Education, London

59. Bader RFW, Popelier PLA (1993) Int J Quantum Chem 45(2):189–207

60. Popelier PLA, Bremond EAG (2009) Int J Quantum Chem 109:2542–2553

61. Popelier PLA, Aicken FM (2003) Chem Phys Chem 4:824–829

62. Stone AJ (1996) The theory of intermolecular forces. Clarendon, Oxford

63. Koch U, Popelier PLA, Stone AJ (1995) Chem Phys Lett 228:253–260

64. in het Panhuis M, Popelier PLA, Munn RW, Angyan JG (2001) J Chem Phys 114:7951–7961

65. Krige DG (1951) J Chem Metal Min Soc S Afr 52:119–139

66. Cressie N (1993) Statistics for spatial data. Wiley, New York

67. Plattner N, Meuwly M (2009) J Mol Model 15(6):687–694

68. GAUSSIAN03, Frisch MJ, Trucks GW, Schlegel HB, Scuseria GE, Robb MA, Cheeseman JR, Montgomery JAJ, Vreven JT, Kudin KN, Burant JC, Millam JM, Iyengar SS, Tomasi J, Barone V, Mennucci B, Cossi M, Scalmani G, Rega N, Petersson GA, Nakatsuji H, Hada M, Ehara M, Toyota K, Fukuda R, Hasegawa J, Ishida M, Nakajima T, Honda Y, Kitao O, Nakai H, Klene M, Li X, Knox JE, Hratchian HP, Cross JB, Adamo C, Jaramillo J, Gomperts R, Stratmann RE, Yazyev O, Austin AJ, Cammi R, Pomelli C, Ochterski JW, Ayala PY, Morokuma K, Voth GA, Salvador P, Dannenberg JJ, Zakrzewski VG, Dapprich S, Daniels AD, Strain MC, Farkas O, Malick DK, Rabuck AD, Raghavachari K, Foresman JB, Ortiz JV, Cui Q, Baboul AG, Clifford S, Cioslowski J, Stefanov BB, Liu G, Liashenko A, Piskorz P, Komaromi I, Martin RL, Fox DJ, Keith T, Al-Laham MA, Peng CY, Nanayakkara A, Challacombe M, Gill PMW, Johnson B, Chen W, Wong MW, Gonzalez C, Pople JA (2003) Gaussian Inc, Pittsburgh

69. GaussView3.0. (2003) Semichem Inc, Gaussian Inc, Pittsburgh

70. Keith TA (2011) AIMAll. 11.04.03 edn. http://aim.tkgristmill.com

71. Popelier PLA (1996) Comput Phys Commun 93(2–3):212–240

72. Popelier PLA (1994) Chem Phys Lett 228(1–3):160–164

73. Rafat M, Devereux M, Popelier PLA (2005) J Mol Graph Model 24:111–120

74. Marvin (2010) 5.3.1 edn. ChemAxon (http://www.chemaxon.com)

75. Humphrey W, Dalke A, Schulten K (1996) J Mol Graph 14:33–38

76. Stone J (1998) An efficient library for parallel ray tracing and animation. University of Missouri, Rolla

77. Jensen F, Palmer DS (2011) J Chem Theory Comput 7(1):223–230

78. Jensen F (2007) Introduction of computational chemistry, 2nd edn. Wiley, Chichester

79. Su ZW, Coppens P (1994) Acta Cryst A50:636–643

80. Haettig C, Hess BA (1994) Mol Phys 81:813–824

81. Haettig C (1996) Chem Phys Lett 260:341

Theor Chem Acc (2012) 131:1161
DOI 10.1007/s00214-012-1161-7

REGULAR ARTICLE

Optimization of the explicit polarization (X-Pol) potential using a hybrid density functional

Jaebeom Han · Donald G. Truhlar · Jiali Gao

Received: 26 September 2011 / Accepted: 19 November 2011 / Published online: 13 March 2012
© Springer-Verlag 2012

Abstract The explicit polarization (X-Pol) method is a self-consistent fragment-based electronic structure theory in which molecular orbitals are block-localized within fragments of a cluster, macromolecule, or condensed-phase system. To account for short-range exchange repulsion and long-range dispersion interactions, we have incorporated a pairwise, empirical potential, in the form of Lennard-Jones terms, into the X-Pol effective Hamiltonian. In the present study, the X-Pol potential is constructed using the B3LYP hybrid density functional with the 6-31G(d) basis set to treat interacting fragments, and the Lennard-Jones parameters have been optimized on a dataset consisting of 105 bimolecular complexes. It is shown that the X-Pol potential can be optimized to provide a good description of hydrogen bonding interactions; the root mean square deviation of the computed binding energies from full (i.e., nonfragmental) CCSD(T)/aug-cc-pVDZ results is 0.8 kcal/mol, and the calculated hydrogen bond distances have an average deviation of about 0.1 Å from those obtained by full B3LYP/aug-cc-pVDZ optimizations.

Keywords Explicit polarization · X-Pol · Quantum force field

Published as part of the special collection of articles: From quantum mechanics to force fields: new methodologies for the classical simulation of complex systems.

J. Han · D. G. Truhlar · J. Gao (✉)
Department of Chemistry and Supercomputing Institute, University of Minnesota, Minneapolis, MN 55455-0431, USA
e-mail: gao@jialigao.org

1 Introduction

The basic approach used in the most popular current parametrizations of molecular mechanics (MM) was established in the 1960s by Lifson, and this approach continues to play an essential role in providing force fields for dynamical simulations of macromolecular systems such as proteins and nucleic acids [1] as well as other nanomaterials. Despite its success, which was promoted by careful and laborious parameterization by many research groups over the past half century, there are also a number of well-known shortcomings, including redundancy of energy terms and parameters, the widespread use of harmonic approximations for bond stretching and angle bending, and the difficulty of treating electronic polarization and charge transfer (for a recent special issue on polarizable force fields, see Jorgensen [2]). Furthermore, molecular mechanics is not designed to treat chemical reactions and photochemical processes [3]. With continuing advances in computer architecture, it is natural to ask what type of force fields will be used for biomolecular and materials simulations in the future. To this end, we have introduced the explicit polarization (X-Pol) potential [4–8], which is an electronic structure method based on block localization of molecular orbitals [4, 5, 9]. The X-Pol method differs from the effective fragment potential (EFP) [10, 11] and SIBFA (sum of interactions between fragments computed ab initio) [12] potentials in that the latter models are derived by fitting results to ab initio results in terms of a multipole expansion of the electrostatics along with other energy terms. In the X-Pol method, a macromolecular system is partitioned into constituent blocks, also called fragments, each of which can be, for example, an individual solvent or solute molecule, an enzyme cofactor, a ligand or molecular fragment, or a peptide unit of a protein. The internal energies of the fragments are determined by an explicitly quantum mechanical method, and interfragment interactions are approximated in a way akin to a combined quantum mechanical and molecular mechanical (QM/MM) [3, 13–15] method. However, the electrostatic field in which each individual fragment is embedded is obtained from the corresponding instantaneous wave functions of all other fragments in the system, and the mutual electronic polarization among fragments is included self-consistently [4–8].

X-Pol can also be used as an electronic structure method such that any quantum chemical model, e.g., Hartree–Fock (HF) theory (or semiempirical models of HF), second-order Møller-Plesset perturbation theory (MP2), coupled cluster theory, or density functional theory (DFT), may be adopted to represent the individual fragment blocks. In this regard, one can treat all fragments by using the same method, or by mixing different electronic structure methods for different fragments (for example, MP2 for one fragment and DFT for all other fragments). Because a large system is partitioned into fragments, the X-Pol method can be made to scale well for fast calculations, and therefore, it can be used to establish a framework for the development of a next-generation force field [4] that goes beyond the conventional molecular mechanics by explicitly including a quantum mechanical treatment of electronic polarization and possibly charge transfer effects (which can be included, for example, by a recently proposed method [16] involving ensemble DFT). When X-Pol is used as a force field, we introduce a set of empirical terms to account for the missing exchange repulsion [17] and dispersion-like attractive, noncovalent interactions. Because these terms are empirical, they can increase the accuracy and, at the same time, reduce computational costs by using parameterization to compensate for errors introduced by using a low or modest level of electronic structure theory [8]. In the present study, we illustrate this by showing how we can use a modestly accurate density functional with a small basis set to treat the individual fragments in the X-Pol method. In particular, we employ the hybrid B3LYP model and a fairly small 6-31G(d) basis set, and we show that X-Pol with this choice can be parameterized to model hydrogen bonding interactions in good agreement with the results from full CCSD(T) calculations. Here, we emphasize that our goal is not to reproduce the geometries and energies at the B3LYP/6-31G(d) level that is used to represent the X-Pol fragment, but rather to obtain agreement with the higher-level CCSD(T) results by optimization of the parameters introduced in the X-Pol quantum force field [5, 8].

There are many other fragment-based molecular orbital methods [18]. For example, Zhang et al. [19, 20] developed a molecular fractionation with conjugated caps (MFCC) approach to treat proteins and protein–ligand interactions. In this method, the individual fragments are capped with a structure representative of the local structure of the original system, and the total energy is obtained by subtracting the energies of the common fragments used in the "caps". The method provides a good means to evaluate interfragment interactions and a straightforward procedure to incorporate the local electronic structure into a fragment-based molecular orbital approach [21, 22]. Another way of separating the total energy into fragmental contributions is the general interaction energy expansion approach described by Stoll and Preuss [23]. The key to achieve fast convergence in this method, in contrast to early schemes [24], is to optimize the monomer, dimer and many-body fragmental molecular orbitals in the presence of all other fragments, rather than using isolated gas-phase fragment terms. There are a number of applications of this strategy, including the fragment molecular orbital (FMO) method [25, 26] and the electrostatically embedded many-body (EE-MB) expansion method [27–30]. The SCF procedure used in the FMO model is identical to that developed in the X-Pol method [25, 26], whereas two-body and three-body exchange and charge transfer effects are included in the FMO2 and FMO3 implementations [23].

In Sect. 2, we briefly review the theoretical background of the X-Pol potential, and in Sect. 3, we present the computational details. In Sect. 4, we describe the optimization of parameters and compare the computed hydrogen bonding energies and geometries obtained from the X-Pol method with higher-level results. Finally, Sect. 5 summarizes the main findings from this work and presents concluding remarks.

2 Theoretical background

The X-Pol method has been described in detail elsewhere [4, 7, 8]. For completeness, we briefly describe the key aspects and approximations made in the X-Pol potential and the empirical parameters introduced to correct for these approximations. We note that the X-Pol method was developed based on block localization of the molecular wave function of the system, which includes a hierarchy of approximations [8]. There are numerous other methods based on localized molecular orbitals or molecular fragments. A recent review, which appeared online after the submission of this manuscript, contains an account of these methodologies [18].

First, we partition a macromolecular system into structural blocks, also called fragments. The molecular wave function, Φ, is approximated as a Hartree product of antisymmetric wave functions of the individual fragments, $\{\Psi^A\}$:

$$\Phi = \prod_{A=1}^{N} \Psi^A \tag{1}$$

where N is the number of fragments in the system, and Ψ^A is a Slater determinant of occupied molecular orbitals (MOs) that are constructed using an atomic orbital basis located on the atoms of fragment A. Thus, these MOs are block-localized by construction. In the present work, density functional theory is used to represent the molecular fragments, and the block-localized molecular orbitals (BLMO) are block-localized Kohn–Sham (BLKS) orbitals, in terms of which the electron density $\rho^A(\mathbf{r})$ of fragment A is given by

$$\rho^A(\mathbf{r}) = 2 \sum_i |\phi_i^A|^2 \tag{2}$$

where ϕ_i^A is the ith doubly occupied Kohn–Sham orbital of fragment A. In the present work, the molecular fragments are closed-shell molecules.

The X-Pol total energy of the system can be written as follows [4, 8]:

$$E_{\text{tot}} = \sum_{A=1}^{N} \left(E^A + \frac{1}{2} E_{\text{int}}^A \right) + E_{\text{XD}} \tag{3}$$

where E^A is the energy of fragment A with the wave function Ψ^A, which can be calculated at any given theoretical level, including Hartree–Fock (HF), density functional theory (DFT), or post-HF theories such as Møller-Plesset perturbation theory or coupled cluster theory, E_{int}^A is the Coulomb interaction energy between fragment A and other fragments, and E_{XD} accounts for the exchange-repulsion (X) and dispersion-correlation (D) interactions between the fragments. It should be pointed out that the wave function Ψ^A in Eq. 3 corresponds to that of fragment A polarized by the remaining fragments in the system, and it differs from the wave function of an isolated fragment in the gas phase (Ψ_0^A). The energy difference between the two states, Ψ^A and Ψ_0^A, is the energy penalty paid for distorting the fragmental wave function due to many-body polarization [13].

The use of the Hartree-product wave function in Eq. 1 implies that the short-range exchange repulsion and long-range and medium-range dispersion and dispersion-like interactions (for brevity, will just call these dispersion in the rest of the article) as well as charge transfer among fragments are neglected [4]. The exchange repulsion and dispersion energies can be determined in various ways, for example by antisymmetrizing the block-localized (i.e., fragmental) orbitals in Eq. 1 [9, 17, 31, 32] or by perturbation methods such as symmetry-adapted perturbation theory (SAPT) [33–35]. However, these methods are not suitable for the construction of a fast quantum mechanical force field for large systems due to their high computational cost as compared to the method adopted here, which is discussed next.

Because the exchange repulsion is short-ranged and approximately pairwise additive [17] and the dispersion

interactions can also be adequately modeled by pairwise potentials [4], such as those used in dispersion-corrected density functional theory (DFT-D) [36], we have used the Lennard-Jones potential to parametrically model the exchange-repulsion and dispersion interactions between each pair of fragments, A and B:

$$E_{\text{ErD}}^{AB} = 4 \sum_a^A \sum_b^B \varepsilon_{ab}^{AB} \left[\left(\frac{\sigma_{ab}^{AB}}{R_{ab}} \right)^{12} - \left(\frac{\sigma_{ab}^{AB}}{R_{ab}} \right)^6 \right] \tag{4}$$

where the A over the sum means that the sum is restricted to orbitals a on center A, and where ε_{ab}^{AB} and σ_{ab}^{AB} are parameters. These parameters are determined from atomic parameters by using standard combining rules: $\varepsilon_{ab}^{AB} = \sqrt{\varepsilon_a^A \varepsilon_b^B}$ and $\sigma_{ab}^{AB} = \sqrt{\sigma_a^A \sigma_b^B}$. The values of ε_a^A and σ_a^A depend on the atomic number of the atom and sometimes also on its hybridization. These parameters can be optimized for a particular electronic structure method used in the X-Pol potential, and the main objective of the present study is to illustrate the optimization of these parameters and the performance of the X-Pol potential with the B3LYP hybrid density functional and the modest 6-31G(d) basis set for calculating binding energies of bimolecular complexes. We will judge the accuracy by comparing to the results of higher-level CCSD(T) calculations.

For closed-shell fragments, the Kohn–Sham DFT energy of fragment A in the presence of the rest of the system is

$$E^A[\rho^A(\mathbf{r})] = \sum_i 2H_i^A + \sum_{i,j} 2J_{ij}^A + E_{xc}^A[\rho^A(\mathbf{r})] + E_{\text{nuc}}^A \tag{5}$$

where the superscript A labels the energies and densities for monomer fragment A, the indices i and j run through the doubly occupied, BLKS molecular orbitals of fragment A, H_i^A and J_{ij}^A are respectively the one-electron Hamiltonian integrals and the Coulomb integrals, $E_{xc}^A[\rho^A(\mathbf{r})]$ is the exchange–correlation functional, and E_{nuc}^A is the nuclear repulsion energy.

The Coulomb interaction energy, E_{int}^A, in Eq. 3 is given by:

$$E_{\text{int}}^A = -2 \sum_i \langle \psi_i^A | V^A(\mathbf{r}) | \psi_i^A \rangle + \sum_a^A Z_a^A V^A(\mathbf{R}_a^A) \tag{6}$$

where Z_a^A is the nuclear charge of atom a of fragment A, and $V^A(\mathbf{r})$ is the total external electrostatic potential (ESP) due to all other fragments in the system. The external ESP is defined as follows:

$$V^A(\mathbf{r}) = -\sum_{B \neq A} \left(\int \frac{\rho^B(\mathbf{r}')}{|\mathbf{r} - \mathbf{r}'|} d\mathbf{r}' + \sum_b^B \frac{Z_b^B}{|\mathbf{r} - \mathbf{R}_b^B|} \right) \tag{7}$$

where Z_b^B is the nuclear charge of atom b of fragment B located at $\mathbf{R}_b^B$. The potential $V^A(\mathbf{r})$ could be determined analytically and used to compute the two-electron integrals

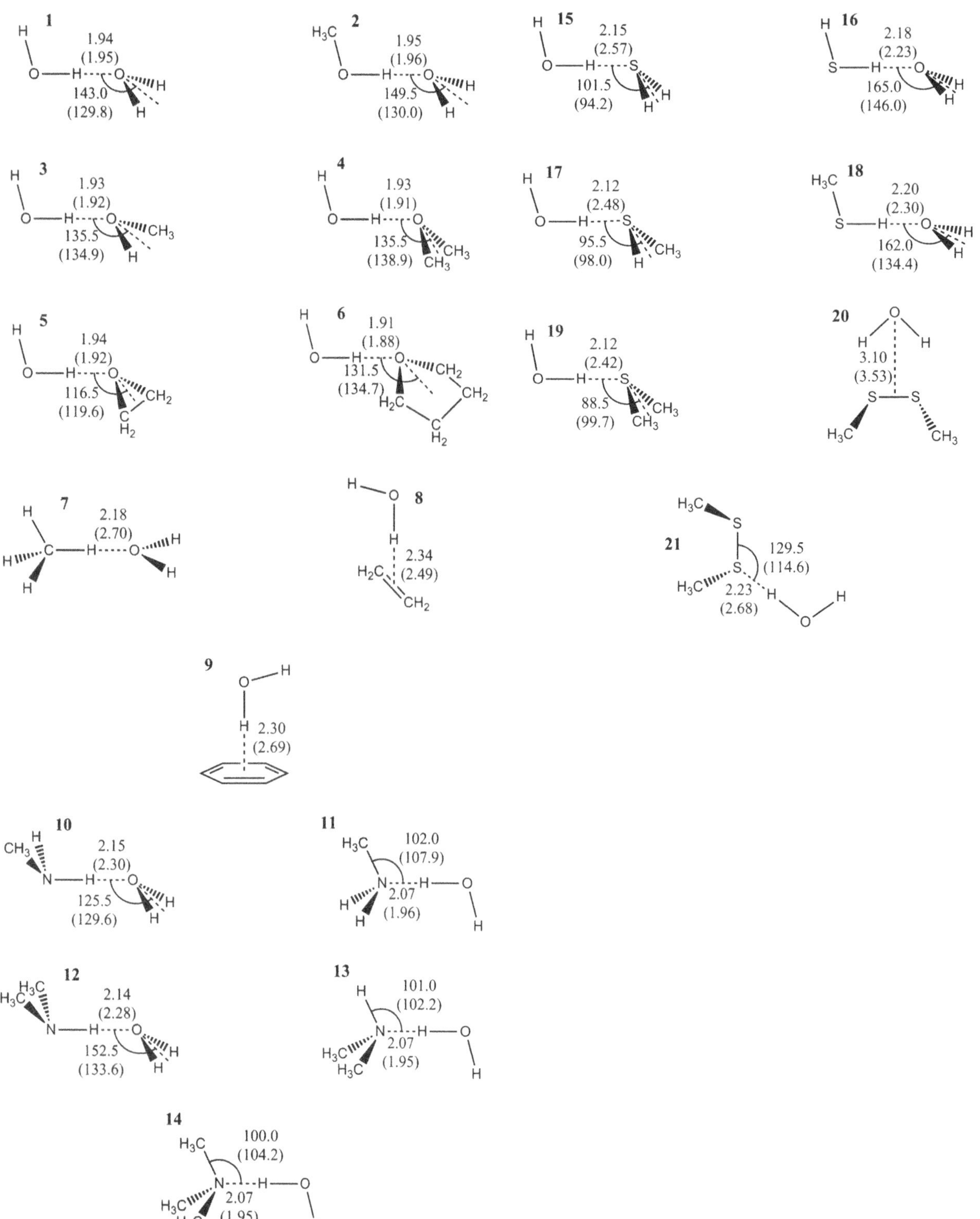

Fig. 1 Schematic illustration of the bimolecular complexes between water and small molecules. Optimized hydrogen bond distances and angles from the X-Pol potential, XP@B3LYP/6-31G(d), are first given, followed by values in parentheses by B3LYP/aug-cc-pVDZ. Distances are given in angstroms and angles in degrees

in Eq. 6 (the terms in the first summation), but this is time-consuming and not a useful choice for fast calculations on macromolecular systems. Alternatively, the ESP in Eq. 7 can be treated by a distributed multipole expansion, and the simplest approximation is to retain only the distributed monopole terms, possibly with scaling to make up for this approximation. This yields the following expression.

$$V^A(\mathbf{r}) = \lambda \sum_{B \neq A}^{B} \sum_b \frac{q_b^B}{|\mathbf{r} - \mathbf{R}_b^B|} \tag{8}$$

where λ is a scaling parameter, and q_b^B is the partial atomic charge on atom b of fragment B. The partial atomic charges can be determined in various ways, for example by fitting electrostatic potentials or by using a charge population analysis method [5]. We make the latter choice [8] for the present calculations, using Mulliken population charges from the BLKS orbitals, and the single parameter λ is set to unity.

In calculating the total energy of the system by Eq. 3, we use the double self-consistent-field (DSCF) method [6, 7]. With an initial guess of the one-electron density matrix for each fragment, the electronic structure calculations for each fragment are performed in the presence of the Mulliken charges of all the other fragments until the change in the total electronic energy or density matrix reaches a predefined tolerance. Although the X-Pol theory has been formulated variationally to allow efficient calculations of energy gradients [7], here we use the older, nonvariational sequential optimization energy formulation.

3 Computational details

For the bimolecular complexes, in both the "high-level" reference calculations and the X-Pol calculations, partial geometry optimizations were performed in which the monomer geometries are held fixed at the corresponding level of theory. Thus, in each bimolecular complex, the hydrogen bond distance and angle between the donor and acceptor molecules, as illustrated in Figs. 1, 2, 3, 4 and 5, are optimized. (Only one angle is involved because we adopt a high symmetry for each hydrogen bond, as illustrated). In all cases, the monomer geometries that were optimized at the corresponding level of theory were held fixed.

The reference geometries were calculated by full (i.e., nonfragmental) B3LYP [37–39] calculations with the aug-cc-pVDZ basis set. The reference energies were obtained by full CCSD(T) single-point calculations with the aug-cc-pVDZ basis set at the geometries of the complexes optimized using B3LYP/aug-cc-pVDZ.

The X-Pol calculations were carried out using the B3LYP/6-31G(d) method as the quantum mechanical level with the geometries optimized by the same level of X-Pol calculation. The hybrid B3LYP functional was chosen in the present study because it is a popular model that has been used widely; one can certainly select a more accurate and recent functional, but the goal here is not to compare the quality and performance of different functionals. We sometimes use the notation XP@B3LYP/6-31G(d) to specify such an X-Pol calculation.

The binding energy for a bimolecular complex, including the empirical Lennard-Jones terms to account for the exchange repulsion and dispersion contributions, is calculated by:

$$\Delta E_b(A \cdots B) = E(A) + E(B) - E_{\text{tot}}^{\text{X-Pol}}(A \cdots B) \tag{9}$$

where $E_{\text{tot}}^{\text{X-Pol}}(A \cdots B)$ is the X-Pol energy (Eq. 3) of the bimolecular complex in which each monomer, A or B, is treated as an individual fragment, and $E(A)$ and $E(B)$ are the B3LYP/6-31G(d) energies of the optimized monomer structures. All binding energies in the present article are zero-point-exclusive.

The Lennard-Jones parameters in Eq. 4 have been adjusted so that the XP@B3LYP/6-31G(d)-binding energies best reproduce the results calculated using CCSD(T)/aug-cc-pVDZ//B3LYP/aug-cc-pVDZ.

The full quantum mechanical calculations for all systems were performed using *Gaussian03* [40], whereas all X-Pol calculations were carried out using a local program that is coupled to a modified version of the GAMESS package [41].

4 Results and discussion

4.1 Optimization of the repulsion and dispersion interactions between fragments

We considered a total of 105 bimolecular complexes, each of which involves one water molecule and an organic or inorganic compound or ion; the organic compounds include ionic and neutral functional groups found in amino acids and nucleobases. Although experimental results for a number of hydrogen bonding complexes are available, we wish to examine a much larger dataset, and therefore, we used theoretical results as reference data, as explained in Sect. 3. The reference data for these complexes were used to optimize the Lennard-Jones parameters by an iterative procedure for the case where X-Pol fragments are treated by B3LYP/6-31G(d). In this process, we placed greater emphasis on the performance for binding energies than on hydrogen bond distances and angles. The resulting

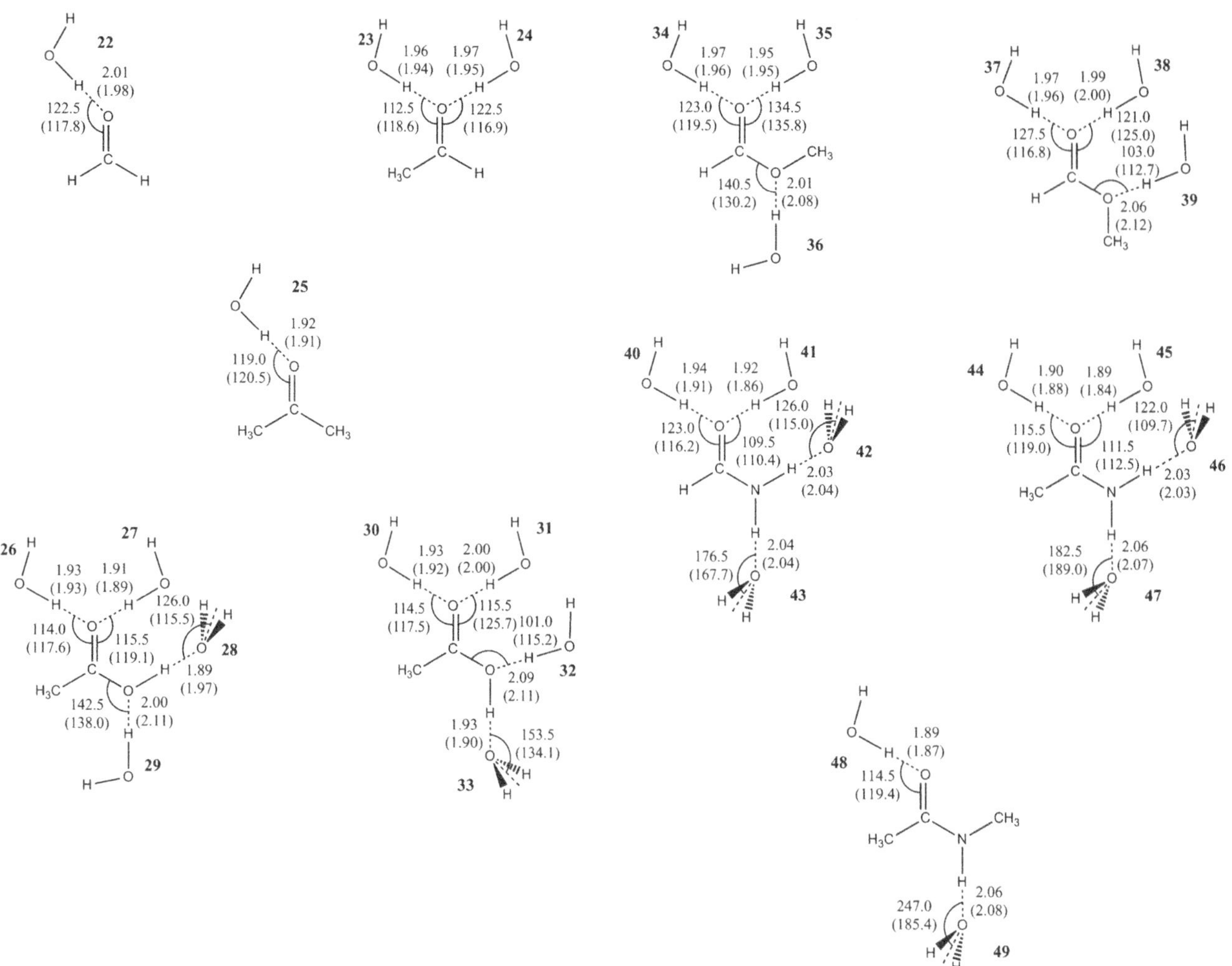

Fig. 2 Bimolecular complexes depicting the interactions between water and a series of carbonyl-containing compounds. Optimized hydrogen bond distances and angles from the X-Pol potential, XP@B3LYP/6-31G(d), are given first, followed by values in *parentheses* by B3LYP/aug-cc-pVDZ. Distances are given in angstroms and angles in degrees

parameters for H, C, O, N, and S atoms and for F⁻, Cl⁻, and Na⁺ ions are listed in Table 1.

Previously, we examined a small set of 14 bimolecular complexes and reported a set of parameters for several atoms [8], and the values listed in Table 1 for these atoms are very similar to those obtained in that work. In the present work, we introduced a new atom type for hydrogen attached to a sulfur atom (thiols and H_2S), and this atom type has a greater σ value than that used in other situations. Three atom types are assigned to oxygen, corresponding to an oxygen type in neutral functional groups and two types for anionic species. Previously, different Lennard-Jones parameters were used for sp^2 and sp^3 oxygen atoms [8], but a single oxygen type for both hybridizations is adequate here. The Lennard-Jones parameters for the carboxylate oxygen and neutral oxygen atoms are very similar; although it would be possible to use the same oxygen parameters in both cases, we kept the two atom types to increase flexibility. For nitrogen atoms, we distinguish atom types for neutral and protonated cases. For other elements, including carbon, a single set of parameters for each is sufficient for the present data set.

4.2 Energies and geometries of hydrogen bonded complexes

Figures 1, 2, 3, 4 and 5 depict the structural arrangements used in the present calculations, along with the optimized geometrical parameters from both the reference calculations and the X-Pol calculations. In many cases, more than one structure is considered for a given chemical species, corresponding to placing water molecules at different positions or in different orientations. Each structure is assigned a number for discussion purposes, and the computed binding energies are given in Tables 2, 3, 4, 5 and 6. The figures and tables are organized roughly according to functional groups.

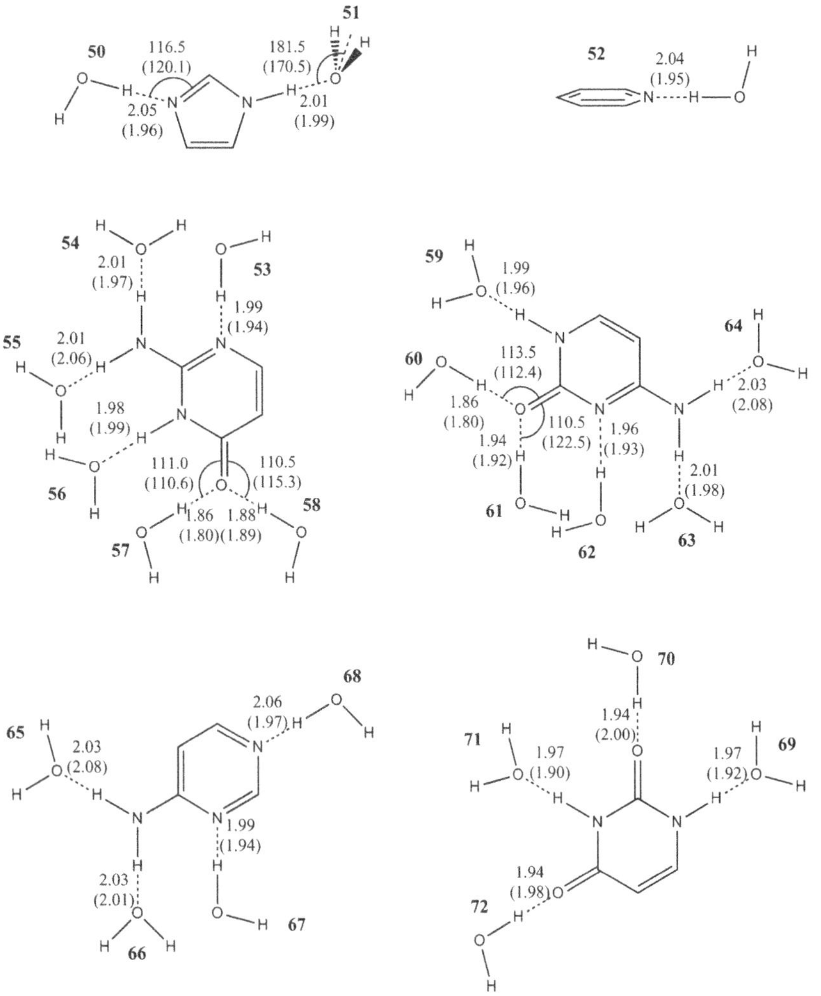

Fig. 3 Bimolecular complexes depicting the interactions between water and heterocyclic compounds. Optimized hydrogen bond distances and angles from the X-Pol potential, XP@B3LYP/6-31G(d), are given first, followed by values in *parentheses* by B3LYP/aug-cc-pVDZ. Distances are given in angstroms and angles in degrees

Both the B3LYP/aug-cc-pVDZ and CCSD(T)/aug-cc-pVDZ-binding energies are given in the tables for comparison; however, the B3LYP-binding energies are inaccurate due to a poor treatment of dispersion contributions, and only the CCSD(T) values should be considered as reference values.

4.2.1 Small molecules and simple functional groups

Figure 1 and Table 2 give the results for water complexes with small molecules. Six sp^3 oxygen structures are included in our study, including water, two structures of methanol, and three structures of ethers (structures **1–6**).

The interaction energy for a water dimer (**1**) is calculated to be 5.7 kcal/mol by the X-Pol method, which yields a hydrogen bond length and bond angle of 1.93 Å and 135.5° (Fig. 1). These may be compared with the corresponding reference values of 5.2 kcal/mol, 1.91 Å, and 138.9°. The best estimate of the water dimer interaction energy is 5.0 kcal/mol using CCSD(T) with extrapolation to a complete basis set [42, 43].

For the methanol–water complexes (**2** and **3**), both XP@B3LYP/6-31G(d) and CCSD(T) calculations predict that methanol is a better hydrogen bond acceptor (structure **3**) by 0.4–0.5 kcal/mol. As the number of alkyl groups on the oxygen increases in going from water to alcohol to ether, the calculated hydrogen bond strength is also enhanced, due to the electron donating effect of an alkyl group, to a final value of about 7 kcal/mol for the complex with tetrahydrofuran (**6**). The average unsigned errors in hydrogen bond lengths and angles are, respectively, 0.02 Å and 7° for the sp^3 oxygen-containing compounds.

The binding energies of methane (**7**) and benzene (**9**) with a single water molecule from the X-Pol optimizations are 0.3–0.4 kcal/mol greater than the CCSD(T) results, but the binding energy between ethane and water is 0.6 kcal/mol smaller (**8**).

We examined five complexes involving simple methyl amines (**10–14**). The primary and secondary amines are much better hydrogen bond acceptors than donors [44], both from the XP@B3LYP/6-31G(d) and CCSD(T)/aug-cc-pVDZ calculations. Although the X-Pol-binding energies for the donor complexes (**10** and **12**) are in good agreement

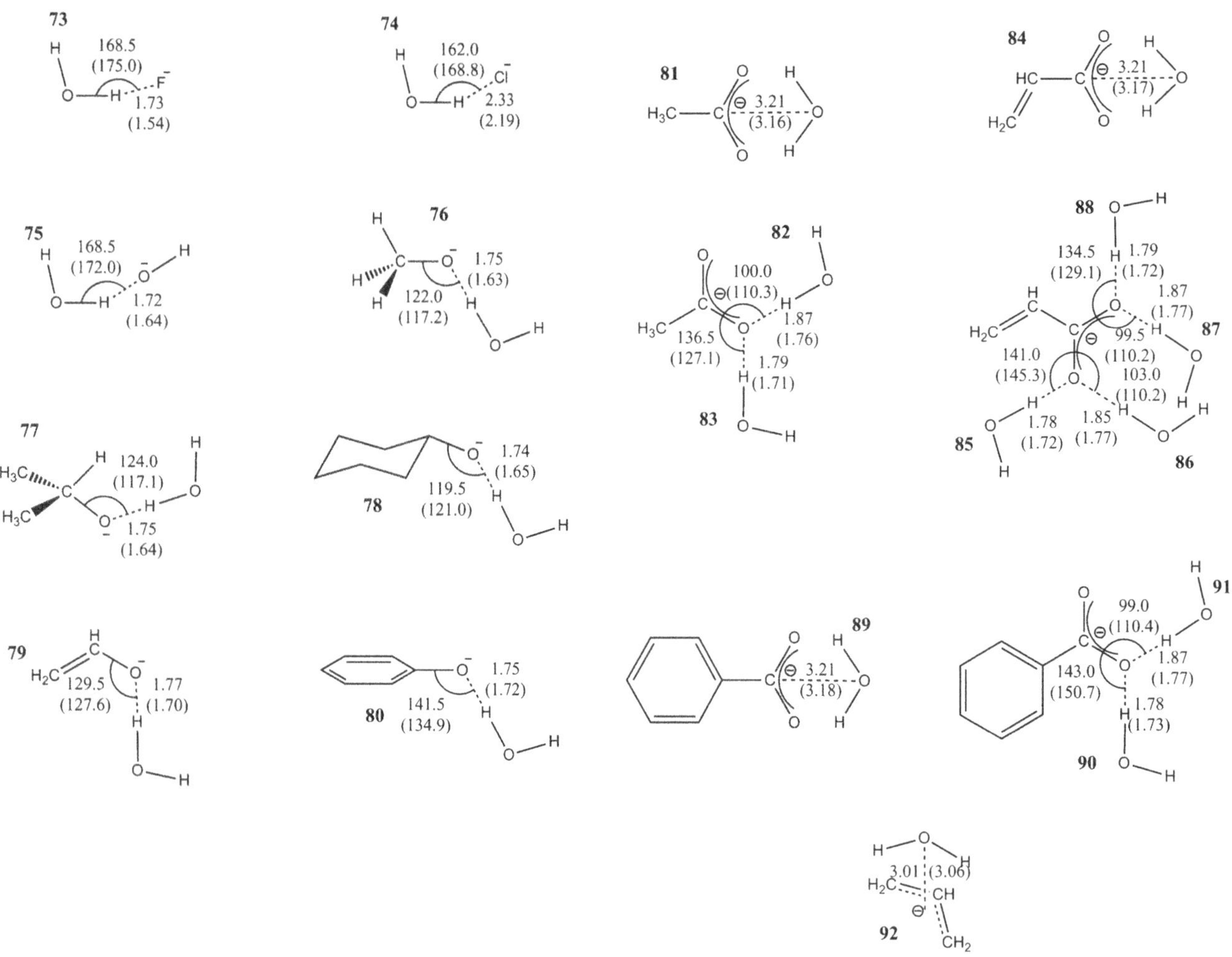

Fig. 4 Bimolecular complexes depicting the interactions between water and anionic species. Optimized hydrogen bond distances and angles from the X-Pol potential, XP@B3LYP/6-31G(d), are given first, followed by values in parentheses by B3LYP/aug-cc-pVDZ. Distances are given in angstroms and angles in degrees

with the reference data, the binding energies for the acceptor complexes are underestimated by 1.5–2.2 kcal/mol, and the deviation increases as the basicity of the amines increases with more methyl substitutions.

Figure 1 includes four sulfur compounds: hydrogen disulfide, methanethiol, dimethyl sulfide, and dimethyl disulfide (**15–21**). Similar to alkyl amine complexes, the binding energies for sulfur compounds are stronger when the sulfur atom acts as a hydrogen bond acceptor than a donor for H_2S and thiols; however, the difference is smaller than in the corresponding nitrogen compounds. These trends are correctly reproduced in the XP@B3LYP/6-31G(d) model in comparison with the reference data (Table 2). In fact, the X-Pol method performs very well, having an average unsigned error of less than 0.2 kcal/mol in binding energy. A somewhat less satisfactory finding is that an additional hydrogen atom type for H_2S and thiols needs to be introduced, whereas the hydrogen bond distance for the acceptor complexes is significantly shorter in

the X-Pol calculations than the values optimized using B3LYP/aug-cc-pVDZ.

4.2.2 Carbonyl-containing compounds

Figure 2 and Table 3 list results for carbonyl compounds, including aldehydes, ketones, carboxylic acids, esters, and amides. The X-Pol results for the aldehydes and acetone (**22–25**) are in reasonable agreement with the reference data, resulting in an average unsigned deviation in binding energy of 0.4 kcal/mol. It is especially encouraging that the X-Pol model correctly distinguishes the relative interaction energies between the two complexes of acetaldehyde with water, favoring the configuration with water oriented toward the methyl group. The optimized hydrogen bond geometries using XP@B3LYP/6-31G(d) are also in good agreement with the reference data.

For carboxylic acid systems, both the syn and anti conformations are considered (**26–33**). We note that the

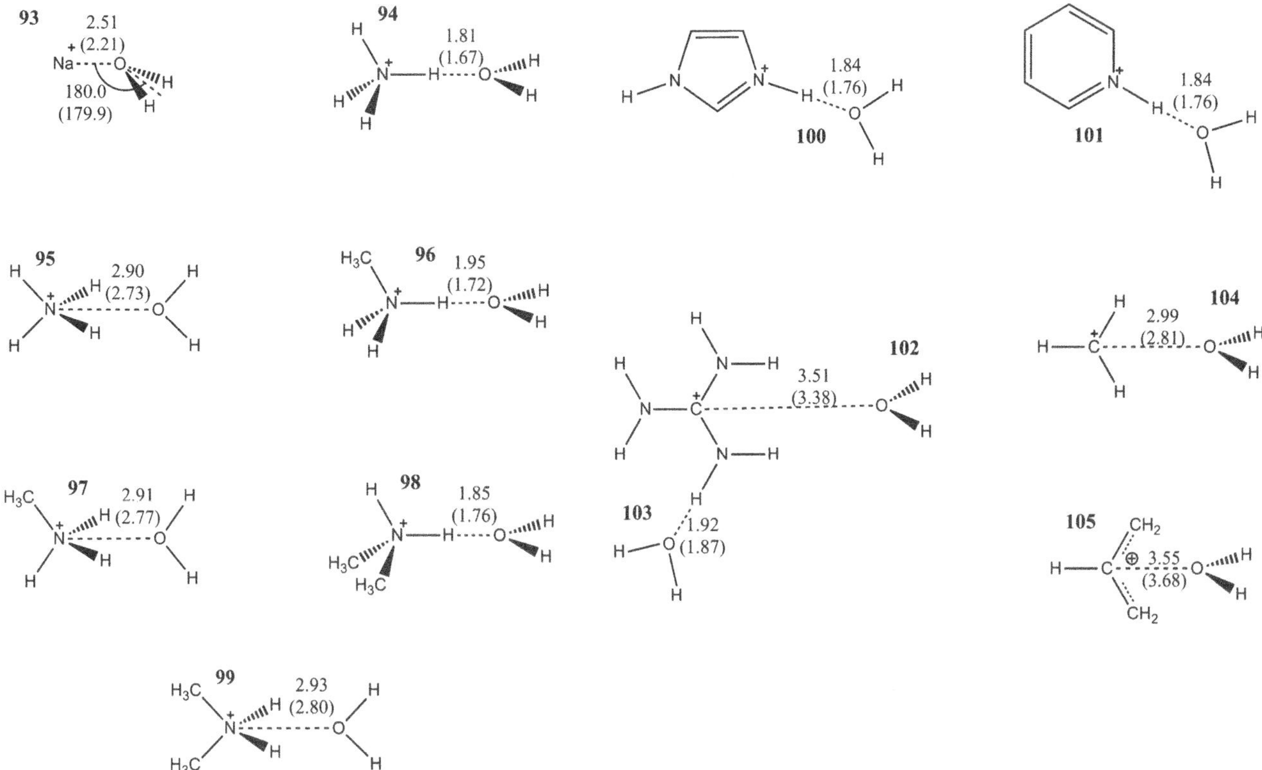

Fig. 5 Bimolecular complexes depicting the interactions between water and cationic species. Optimized hydrogen bond distances and angles from the X-Pol potential, XP@B3LYP/6-31G(d), are given first, followed by values in parentheses by B3LYP/aug-cc-pVDZ. Distances are given in angstroms and angles in degrees

Table 1 Optimized Lennard-Jones parameters in the X-Pol potential with B3LYP/6-31G(d)

Atom	σ (Å)	ε (kcal/mol)
H	1.31	0.04
H (–SH)	1.81	0.04
C	3.67	0.16
N (neutral)	3.60	0.20
N (cation)	3.47	0.20
O (both $sp2$ and $sp3$)	3.25	0.15
O^- (RO^-)	3.21	0.15
O^- (RCO_2^-)	3.24	0.15
S	3.11	0.56
Na^+	2.51	0.30
Cl^-	4.37	0.21
F^-	2.97	0.45

carboxylic acids are particularly good hydrogen bond donors, with computed binding energies of 7.6 and 8.4 kcal/mol for the syn (**28**) and anti (**33**) conformations of acetic acid using XP@B3LYP/6-31G(d), which may be compared with the reference values of 7.9 and 8.2 kcal/mol. The interaction energies on the carbonyl sites are of similar magnitude as those found in aldehyde and ketone complexes. The hydrogen bond accepting ability of the hydroxyl oxygen is relatively weak (3.3 kcal/mol) in the syn (**29**) configuration, while the structure in the anti conformer (**32**) enjoys a secondary hydrogen bonding interaction [45] to the carbonyl oxygen, increasing the XP@B3LYP-binding energy to 5.6 kcal/mol. The latter is 0.9 kcal/mol greater than the reference energy. The optimized geometrical parameters are also in excellent agreement between the two computational approaches. Analogously, both the syn and anti conformations for methyl formate are considered (**34–39**), and similar trends as the corresponding acids are found for these complexes. Overall, the average unsigned errors for all acid and ester complexes are just under 0.3 kcal/mol in binding energy and 0.1 Å in hydrogen bond distance.

The amides complexes are structures **40–49**. For the formamide-water complexes (**40–43**), the XP@B3LYP/6-31G(d) method predicts that the carbonyl group is a better hydrogen bond acceptor (**40** and **41**) than the amide group as a hydrogen bond donor (**42** and **43**) in agreement with the reference data, and the same trend is found in the N-methyl formamide and N-methyl acetamide complexes. However, for the donor complexes, XP@B3LYP/6-31G(d) yields a larger binding energy for **43** than complex **42**, due to the alignment of the carbonyl dipole in the direction of the N–H

Table 2 Binding energies of bimolecular complexes between water and simple functional groups containing oxygen, nitrogen, and sulfur atoms computed using the XP@B3LYP/6-31G(d), B3LYP/aug-cc-pVDZ, and CCSD(T)/aug-cc-pVDZ//B3LYP/aug-cc-pVDZ methods

Complex	XP@B3LYP/6-31G(d)	B3LYP/aug-cc-pVDZ	CCSD(T)/aug-cc-pVDZ
1	5.7	4.6	5.2
2	5.5	4.4	5.4
3	5.9	5.0	5.9
4	6.1	5.0	6.3
5	6.0	4.9	6.0
6	7.0	5.6	7.2
7	1.3	0.4	0.9
8	2.4	1.9	3.0
9	4.2	1.4	3.9
10	2.6	1.9	2.8
11	5.9	6.6	7.4
12	3.0	2.0	2.8
13	6.0	6.5	7.8
14	6.2	6.5	8.4
15	3.7	2.6	3.3
16	3.1	2.3	3.1
17	4.4	3.3	4.2
18	2.5	1.8	2.8
19	5.0	4.0	5.2
20	3.3	1.8	3.3
21	2.6	1.6	2.7

Energies are given in kilocalories per mole

Table 3 Binding energies of bimolecular complexes between water and carbonyl-containing compounds computed using the XP@B3LYP/6-31G(d), B3LYP/aug-cc-pVDZ, and CCSD(T)/aug-cc-pVDZ//B3LYP/aug-cc-pVDZ methods

Complex	XP@B3LYP/6-31G(d)	B3LYP/aug-cc-pVDZ	CCSD(T)/aug-cc-pVDZ
22	4.2	4.2	4.8
23	5.8	5.2	6.4
24	5.0	4.9	5.5
25	6.6	5.7	6.6
26	6.1	5.2	6.2
27	6.0	5.1	5.9
28	7.6	7.9	7.9
29	3.3	2.1	3.2
30	6.4	5.5	6.6
31	5.4	4.6	5.4
32	5.6	3.5	4.7
33	8.4	6.6	8.2
34	4.9	4.6	5.3
35	6.2	5.1	6.5
36	3.6	2.2	3.5
37	5.2	4.9	5.6
38	5.2	4.5	5.3
39	5.5	3.5	4.8
40	6.0	5.8	6.4
41	7.4	6.7	7.6
42	5.7	5.0	6.0
43	5.9	4.7	5.6
44	7.5	6.5	7.5
45	8.0	7.1	8.1
46	5.6	4.7	5.9
47	5.6	4.3	5.6
48	7.9	6.8	7.9
49	5.6	4.1	5.7

Energies are given in kilocalories per mole

bond, but the opposite is found in the reference data. In the full QM calculation, there is apparently an overlap interaction between water and the carbonyl group [46], suggested by the smaller hydrogen bond angles (**42**). This is absent in the present X-Pol method [47]. In all amide complexes, the optimized hydrogen bond lengths using XP@B3LYP/6-31G(d) are in excellent agreement with those at the B3LYP/aug-cc-pVDZ level.

4.2.3 Heterocyclic compounds

We considered a number of heterocyclic compounds, which are displayed in Fig. 3, and the corresponding interaction energies are given in Table 4. For both imidazole and pyridine complexes with water, the XP@B3LYP/6-31G(d) method yields weaker binding energies by about 1 kcal/mol than the corresponding reference data. In these cases, the hydrogen bond lengths from the XP@B3LYP/6-31G(d) optimization are about 0.1 Å longer than the B3LYP/aug-cc-pVDZ results. However, good agreement is obtained between the X-Pol- and CCSD(T)-binding energies for the remaining heterocyclic compounds, even

though we restricted the nitrogen atom type to just one for all neutral compounds.

Structures **53–72** represent hydrogen bonding interactions between a water molecule and the functionalities of nucleobases; cytosine and uracil are depicted in **59–64** and **69–72**, respectively, whereas only the six-member-ring portions of guanine (**53–58**) and adenine (**65–68**) are studied. In these complexes, each organic compound contains both hydrogen bond donor sites and acceptor sites, and the latter can be either an oxygen or a nitrogen atom. Thus, these species cover a large range of hydrogen bonding strengths. Although there are some variations, on average, the binding energies are about 1.3 kcal/mol (XP@B3LYP) and 1.4 kcal/mol (CCSD(T)) larger for structures that accept a hydrogen bond than for those that donate one to water. An exception is

Table 4 Binding energies for bimolecular complexes between water and heterocyclic compounds computed with the XP@B3LYP/6-31G(d), B3LYP/aug-cc-pVDZ, and CCSD(T)/aug-cc-pVDZ//B3LYP/aug-cc-pVDZ methods

Complex	XP@B3LYP/6-31G(d)	B3LYP/aug-cc-pVDZ	CCSD(T)/aug-cc-pVDZ
50	5.8	6.6	7.4
51	5.6	5.4	6.6
52	6.4	6.4	7.4
53	7.4	6.1	7.4
54	6.2	4.8	5.9
55	7.5	5.6	6.7
56	7.1	5.3	6.9
57	9.9	8.2	9.4
58	8.1	6.4	7.4
59	6.4	5.0	6.3
60	9.4	8.3	9.5
61	8.1	6.6	7.6
62	10.3	7.9	9.2
63	5.7	4.5	5.7
64	6.3	4.3	5.6
65	6.4	4.2	5.4
66	5.3	3.9	5.1
67	7.5	6.3	7.4
68	5.9	6.1	7.0
69	7.7	6.2	7.3
70	5.5	4.1	5.1
71	6.5	5.3	6.7
72	5.7	4.3	5.1

Energies are given in kilocalories per mole

found for the two carbonyl groups in uracil, which have binding energies of just over 5 kcal/mol, without which the above difference would be even greater (2 kcal/mol). The binding energies for several complexes are particularly strong, with values greater than 9 kcal/mol (**57**, **60**, and **62**); this can be attributed to contributions from secondary hydrogen bonding interactions [45]. The X-Pol method correctly reproduces these features, in good agreement with the reference binding energies; however, structure **60** is predicted to have a stronger hydrogen bond than **62** from CCSD(T) calculations, but the opposite is obtained using XP@B3LYP. For the whole set of 23 heterocyclic complexes, the mean unsigned error in binding energy is 0.5 kcal/mol, and the differences in hydrogen bond distance are all under 0.1 Å.

4.2.4 Ions

Complexes involving anions and cations, ranging from simple monatomic ions to delocalized organic species, are shown in Figs. 4 and 5, and the corresponding binding energies are given in Tables 5 and 6. The binding energies for the monatomic F^- (**73**) and Cl^- (**74**) ions are fitted in exact agreement with the reference data, but the hydrogen bond distances are 0.19 and 0.14 Å longer in the X-Pol model than the reference values.

For the oxyanions, we consider hydroxide ion, (**75**) alkoxide ions (**76–78**), and conjugated species (**79–80**). In these complexes, there is strong electronic overlap between the two fragments, particularly for the smaller ions. Thus, the block localization of the fragment orbitals in the X-Pol method tends to introduce greater errors as reflected in the hydroxide–water complex (**75**), for which the X-Pol-binding energy is 4.5 kcal/mol greater than in the CCSD(T) calculations. The agreement for the larger and delocalized oxyanions is much improved, with an average error of 1.3 kcal/mol.

A total of twelve carboxylate–water plus allyl anion–water complexes are shown in Fig. 4. We found that a different set of Lennard-Jones parameters than those used for the alkoxide anions has to be adopted for the carboxylate anions, perhaps due to the more electron-delocalized nature of the carboxylate group. The binding energy ranges from 14.7 to 18.3 kcal/mol from CCSD(T) calculations. The agreement between these results and the XP@B3LYP ones is generally good with a mean unsigned deviation in binding energy of 0.3 kcal/mol. The XP@B3LYP/6-31G(d) method correctly predicts that the bifurcated forms of the complexes with water (**81**, **84** and **89**) are the most stable in each case [48], and the differences from the least stable complexes are from 2.3 to 2.7 kcal/mol. This agrees with CCSD(T) calculations except that the least stable complexes are reversed for structures **85** and **88** in the XP@B3LYP/6-31G(d) method.

The computed binding energies for cation–water complexes are given in Table 6. For the ammonium ions, the single-site hydrogen bonding complex (**94**, **96** and **98**) yields stronger interactions than the symmetric two-site structure (**95**, **97** and **99**) both from the XP/B3LYP and CCSD(T) models. However, the reference calculations predict that the latter complexes are 0.7–0.9 kcal/mol more stable than the predictions of the X-Pol method. In the alkyl ammonium series, binding energies decrease progressively as the number of methyl substituents increases, primarily due to charge delocalization. For imidazolium and pyridinium ions, the agreement between XP@B3LYP and CCSD(T) is also reasonable, although the hydrogen bond distances from the X-Pol optimizations are 0.08 Å longer. Two structures are considered for the guanidium ion–water complex. In this case, the energy difference between the two structures is predicted to be smaller than that from CCSD(T) calculations, but the average of the two binding energies is

Table 5 Binding energies for bimolecular complexes between water and anions computed using the XP@B3LYP/6-31G(d), B3LYP/aug-cc-pVDZ, and CCSD(T)/aug-cc-pVDZ//B3LYP/aug-cc-pVDZ methods

Complex	XP@B3LYP/6-31G(d)	B3LYP/aug-cc-pVDZ	CCSD(T)/aug-cc-pVDZ
73	24.9	24.8	24.9
74	14.1	13.8	14.1
75	25.6	21.2	21.1
76	19.4	21.1	21.7
77	19.2	19.5	20.6
78	19.1	19.3	20.5
79	16.4	15.8	16.4
80	16.3	14.0	14.7
81	18.0	17.1	18.3
82	17.8	15.3	18.0
83	15.3	17.0	16.2
84	17.7	16.4	17.7
85	15.4	14.2	15.4
86	17.4	16.4	17.4
87	17.4	16.4	17.4
88	15.0	14.7	15.7
89	17.3	15.7	17.2
90	15.0	13.3	14.7
91	17.1	15.7	16.8
92	15.2	13.9	15.9

Energies are given in kilocalories per mole

Table 6 Binding energies for bimolecular complexes between water and cations computed using the XP@B3LYP/6-31G(d), B3LYP/aug-cc-pVDZ, and CCSD(T)/aug-cc-pVDZ//B3LYP/aug-cc-pVDZ methods. Energies are given in kilocalories per mole

Complex	XP@B3LYP/6-31G(d)	B3LYP/aug-cc-pVDZ	CCSD(T)/aug-cc-pVDZ
93	22.0	23.9	22.0
94	20.0	20.4	20.1
95	15.4	15.8	16.3
96	18.4	17.9	18.3
97	14.5	14.3	15.2
98	17.0	16.3	17.2
99	13.7	13.2	14.4
100	16.2	14.8	15.6
101	16.6	14.9	16.0
102	16.5	15.8	17.4
103	14.3	12.3	13.3
104	14.7	12.8	14.7
105	16.9	13.9	14.0

consistent with the ab initio data. Finally, two carbocations are considered, both of which are found to be adequately modeled by the present X-Pol potential.

4.3 Overall assessment

The performance of the present XP@B3LYP/6-31G(d) method, based on comparisons to the results of CCSD(T)/aug-cc-pVDZ//B3LYP/aug-cc-pVDZ calculations, is shown in Fig. 6. Overall, the root mean square deviation (RMSD) in binding energy between the XP@B3LYP/6-31G(d) predictions and the CCSD(T)/aug-ccc-pVDZ//B3LYP/aug-cc-pVDZ reference data for all 105 bimolecular complexes, covering a range of binding energies of more than 20 kcal/mol, is 0.8 kcal/mol. For a similar set of bimolecular systems, combined QM/MM calculations using the AM1 [13, 49] and HF/3-21G [50] methods along with the three-point charge TIP3P model for water yielded RMSDs of 1.2 and 0.5 kcal/mol, respectively. In those studies, the optimization target was obtained from HF/6-31 + G(d) calculations. In other studies, Riccardi et al. [51] calculated the binding energies for a series of bimolecular complexes of water and organic compounds representing amino acid side chains using the SCC-DFTB/CHARMM potential and obtained an RMSD of 1.2 kcal/mol with respect to B3LYP/6-311 ++G(d,p)//B3LYP/6-31 +G(d) dataset. Freindorf et al. [52] obtained an RMSD of 1.5 kcal/mol for small organic molecule/water complexes using the B3LYP/6-31 +G(d)/AMBER potential. The present X-Pol potential, making use of B3LYP/6-31G(d) for each fragment, yields slightly better agreement with the target dataset than several combined QM/MM methods that employed the same strategy of optimizing the van der Waals parameters. Although the origin of the good

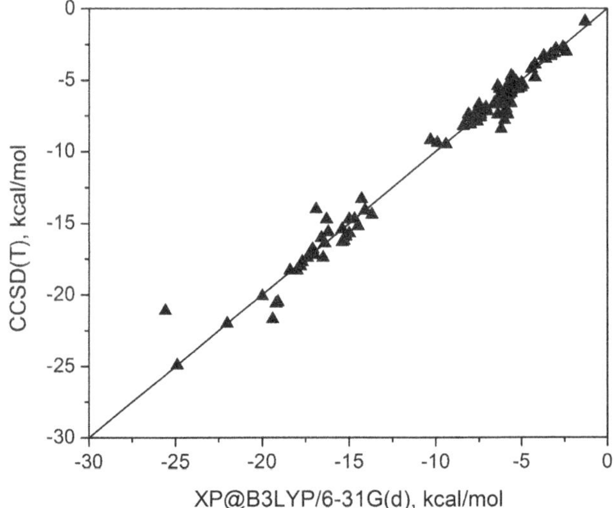

Fig. 6 Comparison of the computed hydrogen bond interaction energies obtained using the XP@B3LYP/6-31G(d) and CCSD(T)/aug-cc-pVDZ//B3LYP/aug-cc-pVDZ methods. The geometries used in the X-Pol calculations were obtained at the same level of theory, whereas those used in the coupled cluster energy evaluations were optimized with B3LYP/aug-cc-pVDZ

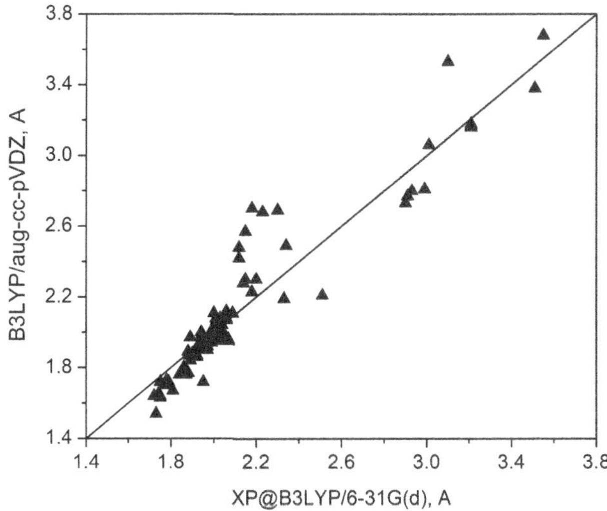

Fig. 7 Comparison of the optimized hydrogen bond distances using the XP@B3LYP/6-31G(d) and B3LYP/aug-cc-pVDZ methods

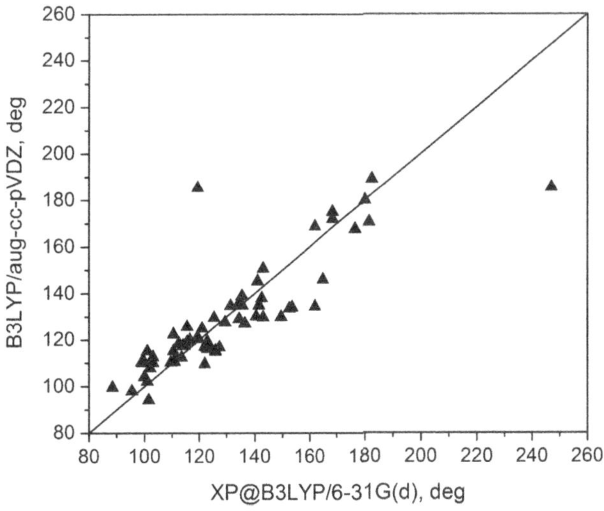

Fig. 8 Comparison of the optimized hydrogen bond angles using the XP@B3LYP/6-31G(d) and B3LYP/aug-cc-pVDZ methods

performance of the X-Pol method needs to be more carefully investigated, a main difference from these QM/MM approaches is that the mutual electronic polarization effects between the two monomers in each complex are included in X-Pol, whereas fixed charge MM force fields are used in the QM/MM calculations.

The RMSD of the optimized hydrogen bond distance between the XP@B3LYP/6-31G(d) optimization and that of the B3LYP/aug-cc-pVDZ method is 0.13 Å (Fig. 7), which is slightly greater than a value of 0.08 Å for a much smaller set of 14 structures in a previous study [8]. This may also be compared with the AI-3/MM [50], SCC-DFTB/MM [51], and B3LYP/6-31 +G(d)/AMBER methods [52], which have RMSDs of 0.07 Å, 0.08 Å, and 0.11 Å relative to the respective datasets. The RMSD errors for hydrogen bond angles are 12° when comparing XP@B3LYP/ 6-31G(d) to B3LYP/aug-cc-pVDZ optimizations (Fig. 8). Although the errors in the optimized hydrogen bond angles are relatively large, the potential energy surfaces for these interactions are typically flat and they do not affect the binding energies significantly. Overall, the present XP@B3LYP/6-31G(d) model yields reasonable hydrogen bond geometries for a variety of organic functional groups interacting with a water molecule.

5 Concluding remarks

The X-Pol potential uses an electronic structure method to model the mutual polarization effects between interacting fragments. It employs an effective Hamiltonian to model energy contributions from interactions between fragments beyond the electrostatic ones that are included self-consistently in the quantum mechanical fragment calculations.

These additional interactions are dominated by exchange repulsion and dispersion energies [8, 17], which are approximated by the Lennard-Jones model [4]. The parameters of the Lennard-Jones function can be optimized to reproduce experimental data or accurate results from high-level calculations; in the present study, the CCSD(T)// aug-cc-pVDZ-binding energies and the optimized hydrogen bond distances and angles using B3LYP/aug-cc-pVDZ are chosen as the optimization target. Implicitly, the X-Pol results obtained with the optimized Lennard-Jones parameters for different atom types account for the energy component due to charge transfer in a way analogous to molecular mechanics force fields. If the specific description of charge transfer effects is important for a given problem, the fragment partitions need to be assigned in such a way that the electron donor and acceptor groups are both included in the same fragment. Alternatively, resonance charge delocalization effects can be modeled by the multiconfigurational, generalized X-Pol (GX-Pol) theory highlighted recently [47, 53] or by using ensemble DFT [16]. Here, however, we tested the simpler approach in which charge transfer is only implicit. The goal is to develop and test a computational approach involving a modest computational cost that can be applied for fast calculations on large systems.

The explicit polarization (X-Pol) method is based on block localization of molecular orbitals within each fragment. The fragments can be assigned, for example, as individual molecules such as solvent molecules or as or amino acid residues in a protein. If desired, important portions of the system can be treated as single large fragments, for example, one may take an entire active site of an enzyme as a single fragment. A key feature of the X-Pol method is that the block-localized orbitals of each fragment are optimized in the presence of the instantaneous electric field due to all other fragments, and the mutual polarization among all fragments is determined using self-consistent field methods [6, 7]. When the electronic integrals between different fragments are approximated by an external potential expansion, the computational costs can be greatly reduced [4, 8]. Furthermore, the fragment block-localization scheme naturally leads to linear scaling in electronic structural calculations on large systems. The X-Pol approach provides a theoretical framework for developing next-generation force fields for macromolecular simulations using an explicitly quantum mechanical electronic structure theory [54].

The present calculations employ the B3LYP hybrid density functional, and the Lennard-Jones parameters are optimized to higher-level reference data for a dataset containing 105 bimolecular, trimolecular, tetramolecular, pentamolecular, and heptamolecular complexes between one or more water molecules and an organic or inorganic compound or ion, representing the functional groups of amino acids and nucleobases. We found that the average deviation between the binding energies calculated by the XP@B3LYP/6-31G(d) and the CCSD(T)/aug-cc-pVDZ// B3LYP/aug-cc-pVDZ methods is about 0.8 kcal/mol, whereas the deviation in hydrogen bond distance is about 0.1 Å. It will be interesting to further test this kind of model through condensed-phase simulations, including the computation of liquid properties and solvation free energies.

Acknowledgments We thank Dr. Yen-lin Lin for assistance. This work has been supported by the National Institutes of Health (RC1-GM091445 and GM46736) and the National Science Foundation (CHE09-56776 and CHE09–57162).

References

1. Levitt M (2001) Nat Struct Biol 8:392
2. Jorgensen WL (2007) J Chem Theory Comput 3:1877
3. Gao J (1996) Acc Chem Res 29:298
4. Gao J (1997) J Phys Chem B 101:657
5. Gao J (1998) J Chem Phys 109:2346
6. Xie W, Gao J (2007) J Chem Theory Comput 3:1890
7. Xie W, Song L, Truhlar DG, Gao J (2008) J Chem Phys 128:234108/1
8. Song L, Han J, Lin YL, Xie W, Gao J (2009) J Phys Chem A 113:11656
9. Mo Y, Gao J, Peyerimhoff SD (2000) J Chem Phys 112:5530
10. Day PN, Jensen JH, Gordon MS, Webb SP, Stevens WJ, Krauss M, Garmer D, Basch H, Cohen D (1996) J Chem Phys 105:1968
11. Gordon MS, Slipchenko L, Li H, Jensen JH (2007) Ann Rep Comput Chem 3:177
12. Gresh N, Cisneros GA, Darden TA, Piquemal J-P (2007) J Chem Theory Comput 3:1960
13. Gao J, Xia X (1992) Science 258:631
14. Lin H, Truhlar Donald G (2007) Theor Chem Acc 117:185
15. Senn HM, Thiel W (2009) Angew Chem Int Ed 48:1198
16. Isegawa M, Gao J, Truhlar DG (2011) J Chem Phys 135 (in press)
17. Cembran A, Bao P, Wang Y, Song L, Truhlar DG, Gao J (2010) J Chem Theory Comput 6:2469
18. Gordon MS, Fedorov DG, Pruitt SR, Slipchenko LV (2012) Chem Rev 112:632
19. Zhang DW, Xiang Y, Zhang JZH (2003) J Phys Chem B 107:12039
20. Xiang Y, Zhang DW, Zhang JZH (2004) J Comput Chem 25:1431
21. Duan LL, Mei Y, Zhang DW, Zhang QG, Zhang JZH (2010) J Am Chem Soc 132:11159
22. Tong Y, Mei Y, Li YL, Ji CG, Zhang JZH (2010) J Am Chem Soc 132:5137
23. Stoll H, Preuss H (1977) Theor Chem Acc 46:12
24. Hankins D, Moskowitz JW, Stillinger FH (1970) J Chem Phys 53:4544
25. Kitaura K, Ikeo E, Asada T, Nakano T, Uebayasi M (1999) Chem Phys Lett 313:701
26. Fedorov DG, Kitaura K (2007) J Phys Chem A 111:6904
27. Truhlar DG, Dahlke EE (2007) J Chem Theory Comput 3:1342
28. Truhlar DG, Dahlke EE, Leverentz HR (2008) J Chem Theory Comput 4:33
29. Tempkin JOB, Leverentz HR, Wang B, Truhlar DG (2011) J Phys Chem Lett 2:2141
30. Leverentz HR, Truhlar DG (2009) J Chem Theory Comput 5:1573
31. Mo Y, Peyerimhoff SD (1998) J Chem Phys 109:1687
32. Mo Y, Gao J (2000) J Comput Chem 21:1458
33. Jacobson LD, Herbert JM (2011) J Chem Phys 134:094118
34. Jeziorski B, Moszynski R, Szalewicz K (1994) Chem Rev (Washington, DC) 94:1887
35. Misquitta AJ, Podeszwa R, Jeziorski B, Szalewicz K (2005) J Chem Phys 123:214103
36. Grimme S (2006) J Comput Chem 27:1787
37. Lee C, Yang W, Parr RG (1988) Phys Rev B 37:785
38. Becke AD (1993) J Chem Phys 98:5648
39. Stephens PJ, Devlin FJ, Chabalowski CF, Frisch MJ (1994) J Phys Chem 98:11623
40. Frisch MJ, Trucks GW, Schlegel HB, Scuseria GE, Robb MA, Cheeseman JR, Montgomery J, J. A., Vreven T, Kudin KN, Burant JC, Millam JM, Iyengar SS, Tomasi J, Barone V, Mennucci B, Cossi M, Scalmani G, Rega N, Petersson GA, Nakatsuji H, Hada M, Ehara M, Toyota K, Fukuda R, Hasegawa J, Ishida M, Nakajima T, Honda Y, Kitao O, Nakai H, Klene M, Li X, Knox JE, Hratchian HP, Cross JB, Bakken V, Adamo C, Jaramillo J, Gomperts R, Stratmann RE, Yazyev O, Austin AJ, Cammi R, Pomelli C, Ochterski JW, Ayala PY, Morokuma K, Voth GA, Salvador P, Dannenberg JJ, Zakrzewski VG, Dapprich S, Daniels AD, Strain MC, Farkas O, Malick DK, Rabuck AD, Raghavachari K, Foresman JB, Ortiz JV, Cui Q, Baboul AG, Clifford S, Cioslowski J, Stefanov BB, Liu G, Liashenko A, Piskorz P, Komaromi I, Martin RL, Fox DJ, Keith T, Al-Laham MA, Peng CY, Nanayakkara A, Challacombe M, Gill PMW, Johnson B, Chen W, Wong MW, Gonzalez C, Pople JA In, Gaussian, Inc, Wallingford, CT
41. Schmidt MW, Baldridge KK, Boatz JA, Elbert ST, Gordon MS, Jensen JH, Koseki S, Matsunaga N, Nguyen KA, Su SJ, Windus TL, Dupuis M, Montgomery JS (1993) J Comput Chem 14:1347
42. Bryantsev VS, Diallo MS, van Duin ACT, Goddard WAI (2009) J Chem Theory Comput 5:1016
43. Zhang P, Fiedler L, Leverentz HR, Truhlar DG, Gao JL (2011) J Chem Theory Comput 7:857
44. Gao J, Xia X, George TF (1993) J Phys Chem 97:9241
45. Jorgensen WL, Pranata J (1990) J Am Chem Soc 112:2008
46. Mo Y, Schleyer PvR WuW, Lin M, Zhang Q, Gao J (2003) J Phys Chem A 107:10011

47. Mo YR, Bao P, Gao JL (2011) Phys Chem Chem Phys 13:6760
48. Gao J, Garner DS, Jorgensen WL (1986) J Am Chem Soc 108:4784
49. Gao J (1994) ACS Symp Ser 569:8
50. Freindorf M, Gao J (1996) J Comput Chem 17:386
51. Riccardi D, Li G, Cui Q (2004) J Phys Chem B 108:6467
52. Freindorf M, Shao YH, Furlani TR, Kong J (2005) J Comput Chem 26:1270
53. Gao J, Cembran A, Mo Y (2010) J Chem Theory Comput 6:2402
54. Xie W, Orozco M, Truhlar DG, Gao J (2009) J Chem Theory Comput 5:459

Theor Chem Acc (2012) 131:1136
DOI 10.1007/s00214-012-1136-8

REGULAR ARTICLE

Analytic gradient and molecular dynamics simulations using the fragment molecular orbital method combined with effective potentials

Takeshi Nagata · Dmitri G. Fedorov · Kazuo Kitaura

Received: 9 May 2011 / Accepted: 29 August 2011 / Published online: 25 February 2012
© Springer-Verlag 2012

Abstract The completely analytic energy gradients are derived and implemented for the two-body fragment molecular orbital (FMO2) method combined with the model core potentials (MCP) and effective fragment potentials (EFP). The many-body terms in EFP require solving coupled-perturbed Hartree-Fock equations, which are derived and implemented. The molecular dynamics (MD) simulations are performed using the FMO2/MCP method for the capped alanine decamer and with the FMO2/EFP method for the zwitterionic conformer of glycine tetramer immersed in the water layer of 6.0 Å (135 water molecules). The results of the MD simulations using the FMO2/EFP and FMO2/MCP gradients show that the total energy is conserved at the time steps less than 1 fs.

Keywords FMO · EFP · MCP · MD · FMO/EFP · FMO/ MCP · Polypeptide · Zwitterion · NVT · Nose-Hoover · Large molecule · Protein · Solvent · QM/MM · Relativistic effect · Analytic gradient · SCZV

Published as part of the special collection of articles. From quantum mechanics to force fields: new methodologies for the classical simulation of complex systems.

T. Nagata (✉) · D. G. Fedorov · K. Kitaura
NRI, National Institute of Advanced Industrial Science and Technology (AIST), 1-1-1 Umezono, Tsukuba, Ibaraki 305-8568, Japan
e-mail: takeshi.nagata@aist.go.jp

T. Nagata · K. Kitaura
Graduate School of Pharmaceutical Sciences, Kyoto University, 46-29 Yoshidashimo Adachi, Sakyo-ku, Kyoto 606-8501, Japan

1 Introduction

Molecular dynamics (MD) of large systems such as proteins and DNA has become a widely used tool [1–7]. Ab initio MD [8, 9] is applicable to chemical reactions, but it is challenging for biological molecules, because the computational cost of ab initio methods scales at least as N^3 (N: system size) for a single step in MD. The reduction of scaling is a major problem in computational chemistry. For this purpose, a variety of quantum-mechanical methods has been developed based on fragmentation of large systems [10–25], whose relation has been introduced by Nagata et al. [26] and discussed in more detail in the recent review [27].

The fragment molecular orbital (FMO) method [28–31] is one such approach; it closely reproduces conventional ab initio properties while showing a nearly linear scaling. FMO has been extended to various kinds of wave functions, and its accuracy has been verified by comparing the FMO and conventional ab initio properties for Møller-Plesset perturbation theory (MP) [32–34], coupled cluster [35], density functional theory (DFT) [36, 37], multiconfigurational self-consistent field [38], configuration interaction [39, 40], time-dependent DFT [41–43], and open shell methods [44]. Continuum solvation models [45] can also be used with FMO [46–48].

The analytic energy gradient (the first derivative with respect to a nuclear coordinate) originally developed by Kitaura et al. [49] neglected the terms arising from the electrostatic potentials (ESP) coupled to the fragment electron densities, which can be shown to be related to the occupied-virtual orbital response terms, obtained by solving the coupled-perturbed Hartree-Fock (CPHF) equations of the entire system. Recently, Nagata et al. have formulated the CPHF equations necessary to obtain these terms in gas phase two-body FMO (FMO2) both at the restricted

Hartree-Fock (RHF) [26] and MP2 levels [50]. Their self-consistent Z-vector (SCZV) method [26] adds only a fraction to the total cost because it is successfully formulated for the CPHF equations of fragments, avoiding CPHF for the entire systems, although these fragment CPHF equations are coupled and require a self-consistent solution. It has been found that the response contribution due to the electrostatic potentials to the energy gradient is significant for polarizable systems and large basis sets. Using SCZV, the FMO2 energy gradients in gas phase are completely analytic, which ensures that MD simulations can be carried out without the accumulation of errors as the time evolves.

Komeiji et al. have implemented FMO/MD using the original incomplete energy gradient and shown that the loss of the energy conservation can take place [51]. In the following years, Komeiji et al. have suggested a number of improvements for FMO/MD [52–54], reviewed in 2009 [55]; Ishimoto et al. have developed alternative formulations [56, 57] and Fujita et al. have proposed path integral FMO/MD [58] and FMO/MD with periodic boundary conditions [59]. FMO/MD has been applied to a large number of systems, so far limited to small solute molecules in water, including the sampling of configurations for a consequent excited state calculations [60] and chemical reactions [61–64]. Recently, an important missing term in the gradient has been identified and implemented by Nagata et al. [65], which appears across the covalent bonds connecting fragments in proteins and other systems.

One well-established and efficient way to treat solvent explicitly is given by the effective fragment potential (EFP) method, which is derived from ab initio calculations [66, 67]. EFP can be thought of as a new generation of force fields in molecular mechanics (MM), which explicitly considers the Coulomb interaction and the polarization interaction of the solvent molecules represented in the Stone's multipole expansion and induced dipoles [68], respectively, while quantum-mechanical effects such as the electron exchange-repulsion and charge transfer are incorporated via effective one-electron operators, whose parameters are fitted to molecular orbital (MO) calculations. The first implementation of EFP using fitted parameters is denoted by EFP1 [66], which is typically used for water. Another version of EFP for the treatment of the electron-exchange and the charge-transfer contributions, which can consider any solvent, is denoted by EFP2 [69–72]. Development and applications of EFP are quite extensive [73–78]. Nagata et al. [79, 80] have interfaced FMO with EFP (FMO/EFP1) for the energy and an incomplete analytic gradient. In the application of FMO2/EFP1 to hydrated chignolin [81], the first structure in the experimental nuclear magnetic resonance (NMR) set of measured data showed the 0.819 Å root mean square deviation to the FMO optimized structure [80].

The model core potential (MCP) [82–84] is an extension to the effective core potentials [85, 86] with the proper consideration of the nodal structure of valence orbitals. MCP can be generated for heavy atoms using high level relativistic calculations [87]. A combined FMO and MCP method (FMO/MCP) has been developed by Ishikawa et al. and applied to hydrated Hg^{2+} and the Pt-containing cisplatin-DNA complex [88].

In this study, we develop completely analytic FMO2/EFP1 and FMO2/MCP gradients, of which the former presents particular difficulties because of its treatment of the many-body polarization. Secondly, we demonstrate the quality of the developed gradients by performing FMO/MD simulations for covalently connected fragments and thus extend the applicability of FMO/MD, previously limited mainly to molecular clusters.

2 Mathematical formulation

2.1 Energy gradient expression

FMO2 calculations [28, 31, 89] proceed as follows. At the first stage, individual fragment (=monomer) calculations are done using ab initio QM methods in the presence of the electrostatic potentials due to the remaining fragments. Because the monomer electron densities are mutually polarized via the electrostatic potentials, the calculations are repeated self-consistently. The procedure for determining the monomer densities and the electrostatic potentials at the monomer stage is referred to as the self-consistent charge (SCC). After convergence, dimers (=pairs of fragments) are calculated in the presence of the electrostatic potentials due to the monomer densities, which are frozen at this point. Finally, thus obtained monomer and dimer energies are summed up giving the total energy.

We briefly summarize the FMO2 energy expression [28, 30, 31, 90] and the corresponding gradient. The FMO2 total energy is given by

$$E = \sum_I^N E_I' + \sum_{I>J}^N (E_{IJ}' - E_I' - E_J') + \sum_{I>J}^N \mathrm{Tr}(\Delta \mathbf{D}^{IJ} \mathbf{V}^{IJ}), \quad (1)$$

where E_X' is the internal fragment energy of fragment X ($X = I$ or IJ, for monomers and dimers, respectively). $\mathbf{V}^{IJ}$ is the matrix of the electrostatic potential (ESP) for dimer IJ due to the electron densities and nuclei of the remaining fragments, i.e.,

$$V_{\mu\nu}^{IJ} = \sum_{K \neq IJ}^N \left(u_{\mu\nu}^K + v_{\mu\nu}^K \right). \quad (2)$$

The one-electron and two-electron integrals in $V_{\mu\nu}^{IJ}$ are, respectively,

$$u_{\mu v}^{K} = \sum_{A \in K} \left\langle \mu \left| \frac{-Z_A}{\|\mathbf{r} - \mathbf{R}_A\|} \right| v \right\rangle, \tag{3}$$

$$v_{\mu v}^{K} = \sum_{\lambda \sigma \in K} D_{\lambda \sigma}^{K} (\mu v | \lambda \sigma), \tag{4}$$

where $D_{\lambda\sigma}^{K}$ is the density matrix element of fragment K and $(\mu v|\lambda\sigma)$ is the two-electron integral in the AO basis. The dimer density matrix difference $\Delta \mathbf{D}^{IJ}$ in Eq. 1 is defined by

$$\Delta \mathbf{D}^{IJ} = \mathbf{D}^{IJ} - (\mathbf{D}^{I} \oplus \mathbf{D}^{J}). \tag{5}$$

The internal fragment energy E'_X in Eq. 1 has the following form:

$$E'_X = \sum_{\mu v \in X} D_{\mu v}^{X} h_{\mu v}^{X} + \frac{1}{2} \sum_{\mu v \lambda \sigma \in X} \left[D_{\mu v}^{X} D_{\lambda \sigma}^{X} - \frac{1}{2} D_{\mu\lambda}^{X} D_{v\sigma}^{X} \right] (\mu v | \lambda \sigma)$$
$$+ \sum_{\mu v \in X} D_{\mu v}^{X} P_{\mu v}^{X} + E_X^{\mathrm{NR}}, \tag{6}$$

where $h_{\mu v}^{X}$ is the X-mer one-electron Hamiltonian, $P_{\mu v}^{X}$ is the hybrid orbital projection matrix [65], and E_X^{NR} is the nuclear repulsion energy of X. In this study, the Roman induces $ijkl$ and Greek induces $\mu v\lambda\sigma$ denote the MO integrals and AO integrals, respectively.

The differentiation of E'_X with respect to a nuclear coordinate a leads to

$$\frac{\partial E'_X}{\partial a} = \sum_{\mu v \in X} D_{\mu v}^{X} \frac{\partial h_{\mu v}^{X}}{\partial a} + \frac{1}{2} \sum_{\mu v \lambda \sigma \in X} \left[D_{\mu v}^{X} D_{\lambda \sigma}^{X} - \frac{1}{2} D_{\mu\lambda}^{X} D_{v\sigma}^{X} \right]$$
$$\times \frac{\partial (\mu v | \lambda \sigma)}{\partial a} + \sum_{\mu v \in X} D_{\mu v}^{X} \frac{\partial P_{\mu v}^{X}}{\partial a}$$
$$- 2 \sum_{i,j \in X}^{\mathrm{occ}} S_{ji}^{a,X} F_{ji}^{\prime X} - \overline{U}^{a,X,X} + \frac{\partial E_X^{\mathrm{NR}}}{\partial a}, \tag{7}$$

where the internal fragment Fock matrix element $F_{ij}^{\prime X}$ is given in terms of MOs:

$$F_{ij}^{\prime X} = h_{ij}^{X} + \sum_{k \in X}^{\mathrm{occ}} [2(ij|kk) - (ik|jk)] + P_{ij}^{X}, \tag{8}$$

and the overlap derivative is defined by

$$S_{ij}^{a,X} = \sum_{\mu v \in X} C_{\mu i}^{X*} \frac{\partial S_{\mu v}^{X}}{\partial a} C_{vj}^{X}. \tag{9}$$

The response term in Eq. 7 is defined as [91]

$$\overline{U}^{a,X,Y} = 4 \sum_{i \in X}^{\mathrm{occ}} \sum_{r \in X}^{\mathrm{vir}} U_{ri}^{a,X} V_{ri}^{Y}, \tag{10}$$

where the MO coefficient derivative is

$$\frac{\partial C_{\mu i}^{X}}{\partial a} = \sum_{m \in X}^{\mathrm{occ+vir}} U_{mi}^{a,X} C_{\mu m}^{X}, \tag{11}$$

$U_{mi}^{a,X}$ is the orbital response of fragment X.

The differentiation of the ESP energy contribution in Eq. 1 with respect to a nuclear coordinate a leads to

$$\frac{\partial}{\partial a} \mathrm{Tr}(\Delta \mathbf{D}^{IJ} \mathbf{V}^{IJ}) = \sum_{\mu v \in IJ} \Delta D_{\mu v}^{IJ} \sum_{K \neq IJ}^{N} \left[\frac{\partial u_{\mu v}^{K}}{\partial a} + \sum_{\lambda \sigma \in K} D_{\lambda\sigma}^{K} \frac{\partial (\mu v | \lambda \sigma)}{\partial a} \right]$$
$$- 2 \sum_{\mu v \in IJ} W_{\mu v}^{IJ} \frac{\partial S_{\mu v}^{IJ}}{\partial a} + 2 \sum_{\mu v \in I} W_{\mu v}^{I} \frac{\partial S_{\mu v}^{I}}{\partial a}$$
$$+ 2 \sum_{\mu v \in J} W_{\mu v}^{J} \frac{\partial S_{\mu v}^{J}}{\partial a} - 2 \sum_{K \neq IJ}^{N} \sum_{\mu v \in K} \Delta X_{\mu v}^{K(IJ)} S_{\mu v}^{a,K}$$
$$+ \overline{U}^{a,IJ,IJ} - \overline{U}^{a,I,IJ} - \overline{U}^{a,J,IJ}$$
$$+ 4 \sum_{K \neq IJ}^{N} \sum_{\mu v \in IJ} \sum_{r \in K}^{\mathrm{vir}} \sum_{i \in K}^{\mathrm{occ}} U_{ri}^{a,K} \Delta D_{\mu v}^{IJ} (\mu v | ri), \tag{12}$$

where

$$W_{\mu v}^{X} = \frac{1}{4} \sum_{\lambda \sigma \in X} D_{\mu\lambda}^{X} V_{\lambda\sigma}^{IJ} D_{\sigma v}^{X}, \tag{13}$$

and

$$\Delta X_{\mu v}^{K(IJ)} = \frac{1}{4} \sum_{\lambda \sigma \in K} D_{\mu\lambda}^{K} \left[\sum_{\zeta \eta \in IJ} \Delta D_{\zeta\eta}^{IJ} (\zeta\eta | \lambda\sigma) \right] D_{\sigma v}^{K}. \tag{14}$$

Collecting the response terms in Eqs. 7 and 12, the total contribution of response terms to the energy gradient can be given by the sum of $\overline{U}^{a}$ and $\Re^{a}$, where

$$\overline{U}^{a} = - \sum_{I}^{N} \overline{U}^{a,I,I} - \sum_{I>J}^{N} \left(\overline{U}^{a,IJ,IJ} - \overline{U}^{a,I,I} - \overline{U}^{a,J,J} \right)$$
$$+ \sum_{I>J}^{N} \left(\overline{U}^{a,IJ,IJ} - \overline{U}^{a,I,IJ} - \overline{U}^{a,J,IJ} \right), \tag{15}$$

and

$$\Re^{a} = 4 \sum_{K}^{N} \sum_{(I>J) \neq K}^{N} \sum_{\mu v \in IJ} \sum_{r \in K}^{\mathrm{vir}} \sum_{i \in K}^{\mathrm{occ}} U_{ri}^{a,K} \Delta D_{\mu v}^{IJ} (\mu v | ri). \tag{16}$$

Note that $\overline{U}^{a}$ cancels out unless some but not all ESP are computed with approximations (e.g., point charges) [26]. In this study, we do not employ them and use only the electrostatic dimer (ES-DIM) approximation, thereby one should compute just the $\Re^{a}$ term.

One purpose of this study is to derive the completely analytic energy gradient for the FMO2/EFP1 method and the FMO2/MCP method by considering the response contributions, Eq. 16. The FMO/EFP1 gradient has already been derived without the response contribution in Ref. [80]. To obtain the analytic FMO2/MCP gradient, one has to add the MCP derivative terms and the response terms. Since the fragment energies are variationally determined with those effective potentials in the fragment

Hamiltonians, there are no direct contributions to Eq. 16 for those methods. The additional contributions should appear only in the coupled-perturbed Hartree-Fock (CPHF) equations. We should note that Eq. 16 arises because the monomer densities are fixed in the dimer calculations.

2.2 CPHF equations for the FMO gradient

The FMO energy gradient expression is given in the previous subsection. Here, the set of CPHF equations is introduced for obtaining the orbital response contributions, Eq. 16. In the dimer RHF calculations, the dimer densities are computed without mutual self-consistency to monomer densities. Therefore, the response contribution, Eq. 16, emerges in the FMO energy gradient, and only the monomer responses $U_{ri}^{a,K}$ contribute to the gradient.

The CPHF equations in FMO are given by [26]

$$\mathbf{A}\mathbf{U}^a = \mathbf{B}_0^a, \tag{17}$$

where the fragment diagonal block of matrix $\mathbf{A}$ and the corresponding fragment off-diagonal block are, respectively,

$$A_{ij,kl}^{I,I} = \delta_{ik}\delta_{jl}\left(\epsilon_j^I - \epsilon_i^I\right) - [4(ij|kl) - (ik|jl) - (il|jk)], \tag{18}$$

and

$$A_{ij,kl}^{I,K} = -4(ij|kl). \tag{19}$$

Note that Eq. 17 has the dimension of the entire molecular system. In Eqs. 18 and 19, the former corresponds to the orbital Hessian of the target fragment I and the latter comes from the ESP acting upon I. The $ij \in I$ element of the derivative integrals $\mathbf{B}_0^a$ is

$$B_{0,ij}^{a,I} = F_{ij}^{a,I} - S_{ij}^{a,I}\epsilon_j^I - \sum_{kl\in I}^{\text{occ}} S_{kl}^{a,I}[2(ij|kl) - (ik|jl)]$$
$$- \sum_{K\neq I}\sum_{kl\in K}^{\text{occ}} 2S_{kl}^{a,K}(ij|kl). \tag{20}$$

where the Fock matrix derivative $F_{ij}^{a,I} = F_{ij}^{'a,I} + V_{ij}^{a,I}$ includes the derivative of the electrostatic potential in the MO basis $V_{ij}^{a,I}$ (more detailed definitions can be found in [26]). The computation of $\mathbf{B}_0^a$ is straightforward [26]. We should note that the set of CPHF equations, Eq. 17, is derived from the perturbation of monomer RHF equations.

Substitution of Eq. 17 into the response contribution $\Re^a$ yields

$$\Re^a = \sum_K^N \sum_{r\in K}^{\text{vir}} \sum_{i\in K}^{\text{occ}} U_{ri}^{a,K} L_{ri}^K \tag{21}$$
$$= \mathbf{L}^T\mathbf{U}^a = \mathbf{L}^T\mathbf{A}^{-1}\mathbf{B}_0^a = \mathbf{Z}^T\mathbf{B}_0^a$$

where the Lagrangian is defined by

$$L_{ri}^K = 4 \sum_{(I > J)\neq K}^N \sum_{\mu\nu\in IJ} \Delta D_{\mu\nu}^{IJ}(\mu\nu|ri), \tag{22}$$

and the corresponding Z-vector equations with the dimension of the entire system should be solved:

$$\mathbf{A}\mathbf{Z} = \mathbf{L}. \tag{23}$$

In practice, Eq. 23 is solved by the SCZV procedure [26]. Next, the contributions due to the effective potentials are derived to be added into Eqs. 20, 18, and 19.

2.3 Effective potentials

This section considers the model core potential and the effective fragment potential (EFP1 is used throughout in this study). These effective potentials can be added into the Hamiltonian of fragment X. The particle treatment of some core electrons is replaced by MCP, and N_v represents the number of explicitly treated valence electrons. The FMO Hamiltonian of fragment X [89] is replaced by the corresponding MCP Hamiltonian [88] as follows:

$$\widehat{H}^{\text{MCP},X} = \sum_{i\in X}^{N_v}\left[\widehat{h}_i^{\text{eff},X} + \widehat{h}_i^{\text{MCP},X} + \widehat{V}^{\text{eff},X}(\mathbf{r}_i) + \sum_{j(\in X)<i}^{N_v}\frac{1}{|\mathbf{r}_i - \mathbf{r}_j|}\right]$$
$$+ E_X^{\text{NR,eff}}, \tag{24}$$

where the effective one-electron Hamiltonian of fragment X, the one-electron potential inherent in MCP, and the effective nuclear repulsion energy are, respectively,

$$\widehat{h}_i^{\text{eff},X} = -\frac{1}{2}\nabla_i^2 - \sum_{L\in X}\frac{Z_L^{\text{eff}}}{|\mathbf{r}_i - \mathbf{R}_L|} \tag{25}$$

$$\widehat{h}_i^{\text{MCP},X} = -\sum_{L\in X}\frac{Z_L^{\text{eff}}}{|\mathbf{r}_i - \mathbf{R}_L|}\left[\sum_k^{n_{L,\alpha}} A_{L,k}\exp(-\alpha_{L,k}|\mathbf{r}_i - \mathbf{R}_L|^2)\right.$$
$$\left. + |\mathbf{r}_i - \mathbf{R}_L|\sum_k^{n_{L,\alpha}} B_{L,k}\exp(-\beta_{L,k}|\mathbf{r}_i - \mathbf{R}_L|^2)\right]$$
$$+ \sum_{L\in X}\sum_c^{N_{L,c}} B_{L,c}|\psi_{L,c}\rangle\langle\psi_{L,c}| \tag{26}$$

$$E_X^{\text{NR,eff}} = \sum_{B\in X}\sum_{A(\in X)>B}\frac{Z_A^{\text{eff}}Z_B^{\text{eff}}}{R_{AB}}, \tag{27}$$

and $\widehat{V}_i^{\text{eff},X}(\mathbf{r}_i)$ corresponds to the electrostatic potential in Eq. 2 in which the nuclear charge Z_A is replaced by the effective nuclear charge, $Z_A^{\text{eff}} = Z_A - N_{A,c}$ and $N_{A,c}$ is the number of core electrons on atom A. L runs over atoms. The parameters $A_{L,k}$, $\alpha_{L,k}$, $B_{L,k}$, and $\beta_{L,k}$ are fitted from all-electron calculations when MCP is generated. The last term on the right hand side of Eq. 26 is the projection operator that shifts the core orbitals $|\psi_{L,c}\rangle$.

In the EFP method, the effective potential for electron i is composed of the Coulomb, polarization, and the remainder (the electron exchange-repulsion and charge transfer) terms:

$$\widehat{V}_i^{\text{EFP},X} = \widehat{V}_i^{\text{coul},X} + \widehat{V}_i^{\text{pol},X} + \widehat{V}_i^{\text{rem},X}. \tag{28}$$

The Coulomb potential acting on X is represented using the Stone's multipole expansion [68]:

$$\widehat{V}_i^{\text{coul},X} = \sum_C^{N_{\text{Coul}}} \left(-\frac{q_C}{r_{iC}} - \sum_a^{x,y,z} \mu_a^C \widehat{F}_a(\mathbf{r}_{iC}) - \frac{1}{3} \sum_{ab}^{x,y,z} \Theta_{ab}^C \widehat{F}_{ab}'(\mathbf{r}_{iC}) \right.$$
$$\left. - \frac{1}{15} \sum_{abc}^{x,y,z} \Omega_{abc}^C \widehat{F}_{abc}''(\mathbf{r}_{iC}) \cdots \right), \tag{29}$$

where index C runs over the N_{Coul} expansion points of the whole solvent system (usually atoms and bond midpoints), and $r_{iC} = |\mathbf{r}_i - \mathbf{R}_C|$ is the distance between electron in X and point C. μ_a^C, Θ_{ab}^C and Ω_{abc}^C are the dipole, quadrupole, and octupole moments at C, respectively. $\widehat{F}_a(\mathbf{r}_{iC}) = -(a - a_C)/r_{iC}^3$, $\widehat{F}_{ab}'(\mathbf{r}_{iC})$, and $\widehat{F}_{abc}''(\mathbf{r}_{iC})$ give the electric field due to the QM charge, the field gradient, and the field second derivative operators, respectively. To improve the point multipole model accounting for the overlapping electron densities, the first term on the right hand side of Eq. 29 can be replaced by

$$-\frac{q_C}{r_{iC}} \rightarrow (1 - \beta_C \exp[-\alpha_C r_{iC}^2]) \left[-\frac{q_C}{r_{iC}} \right], \tag{30}$$

where α_C and β_C are fitted parameters [66].

The remainder potential for electron i due to EFP fragments is a one-electron operator,

$$\widehat{V}_i^{\text{rem},X} = \sum_m^{N_{\text{rem}}} \widehat{V}(m, i), \tag{31}$$

where m runs over the EFP remainder expansion points, and N_{rem} is the total number of them. For water acting as solvent, the ab initio electron exchange-repulsion plus charge-transfer contributions are fitted to Gaussian functions of $\widehat{V}^{\text{rem}}(m, i)$ for water dimer [92].

The effective potentials have one-body terms, whose gradients are derived below. The polarization (induced dipole) potential contribution $\widehat{V}_i^{\text{pol},X}$ in the EFP method needs a different treatment as a many-body contribution, introduced separately.

2.4 One-body terms

As mentioned above, the FMO/EFP1 gradient except the response contribution has been derived [80] and the FMO/MCP derivative contributions are straightforward to derive and implement. However, for these methods, an additional contribution should be introduced through the CPHF equations. Because they are derived from the HF equations, this subsection considers the contribution of a one-body potential to the Fock matrix.

Generally, an effective potential $\widehat{O}_1$ in the one-electron operator form combined with the electron density $D_{\mu\nu}^X$ gives rise to the following contribution to the total energy:

$$O^X(\mathbf{D}^X) = \sum_{\mu\nu \in X} D_{\mu\nu}^X \langle \mu | \widehat{O}_1 | \nu \rangle. \tag{32}$$

In FMO, the CPHF equations are obtained from monomer HF equations [26], thereby the contribution to the monomer Fock matrix in the MO basis is given by

$$O_{ij}^I = \langle i | \widehat{O}_1 | j \rangle. \tag{33}$$

The differentiation of O_{ij}^I with respect to a nuclear coordinate a yields

$$\frac{\partial O_{ij}^I}{\partial a} = \sum_{m \in I}^{\text{occ+vir}} \left(U_{mi}^{a,I} \langle m | \widehat{O}_1 | j \rangle + U_{mj}^{a,I} \langle i | \widehat{O}_1 | m \rangle \right)$$
$$+ \sum_{\mu\nu \in X} C_{\mu i}^{I*} C_{\nu j}^I \frac{\partial \langle \mu | \widehat{O}_1 | \nu \rangle}{\partial a}. \tag{34}$$

The corresponding derivative terms contribute to the gradient directly (e.g., via the one-electron contribution to the Lagrangian and the AO-based integral derivatives similarly to the standard ab initio gradients; we do not repeat these well known formulas, see Ref. [93]). Here, we only concentrate on the modifications of the CPHF-related terms.

The first parenthesized term on the right hand side of Eq. 34 becomes a part of the Fock matrix element, leading to the orbital energies $\epsilon_j^I - \epsilon_i^I$ in Eq. 18, the second term is added into $F_{ij}^{a,I}$ in Eq. 20, and its computation is straightforward unless $\widehat{O}_1$ contains the electron density. The MCP potential $\widehat{O}_1 = \widehat{h}_1^{\text{MCP},X}$, the Coulomb potential $\widehat{O}_1 = \widehat{V}_1^{\text{coul},X}$, and the electron-exchange plus charge-transfer potential $\widehat{O}_1 = \widehat{V}_1^{\text{rem},X}$ in EFP1 have this form (see Eqs. 26, 29–31), which is straightforward to implement. The many-body contribution contains the electron density in the operator, which is discussed in the following section.

2.5 Many-body terms

The addition of the many-body contributions to the response term in the CPHF equations is much more complicated than the one-body potentials to the Fock operator, because they usually contain the product of the electron densities.

The total EFP polarization energy contribution in the FMO2/EFP method is given by

$$E^{\text{pol}} = \sum_{I}^{N} E_I^{\text{pol}} + \sum_{I>J}^{N} (E_{IJ}^{\text{pol}} - E_I^{\text{pol}} - E_J^{\text{pol}}) + E^{\text{pol,efp}}, \quad (35)$$

where the X-mer polarization energy contribution E_X^{pol} and the EFP-EFP polarization energy $E^{\text{pol,efp}}$ are, respectively,

$$E_X^{\text{pol}} = -\frac{1}{2} \left[\mathbf{E}^X \right]^T \mathbf{p}^X - E^{\text{pol,efp}} = -\frac{1}{2} \left[\mathbf{p}^X \right]^T \mathbf{E}^X - E^{\text{pol,efp}}$$

$$= -\frac{1}{2} \left[\mathbf{E}^X \right]^T \mathbf{\Xi}^{-1} \mathbf{E}^X - E^{\text{pol,efp}}, \quad (36)$$

$$E^{\text{pol,efp}} = -\frac{1}{2} \left[\mathbf{E}^{\text{efp}} \right]^T \mathbf{p}^{\text{efp}} = -\frac{1}{2} \left[\mathbf{E}^{\text{efp}} \right]^T \mathbf{\Xi}^{-1} \mathbf{E}^{\text{efp}}. \quad (37)$$

Equations 36 and 37 are defined in Ref. [80]. The superscript T stands for the transpose. The electrostatic field vector $\mathbf{E}^X$ and the induced dipole vector $\mathbf{p}^X$ are, respectively,

$$\mathbf{E}^X = \mathbf{E}^{X,\text{el}}(\mathbf{D}^X) + \mathbf{E}^{X,\text{nuc}} + \mathbf{E}^{\text{efp}}, \quad (38)$$

$$\mathbf{p}^X = \mathbf{\Xi}^{-1} \mathbf{E}^X. \quad (39)$$

These vectors consist of $3N^{\text{pol}}$ elements, where N^{pol} is the number of the polarizability points. In the case of a water molecule, the centroids of two OH bond orbitals and two lone-pair orbitals are chosen as the polarizability points. $\mathbf{E}^{X,\text{nuc}}$ and $\mathbf{E}^{\text{efp}}$ are the electrostatic fields due to a nuclei of X and the remaining EFPs. The ii diagonal block matrix of $\mathbf{\Xi}$ (i is a polarizability point) is defined by

$$\mathbf{\Xi}_{ii} = \begin{pmatrix} \alpha_{i,xx} & \alpha_{i,xy} & \alpha_{i,xz} \\ \alpha_{i,yx} & \alpha_{i,yy} & \alpha_{i,yz} \\ \alpha_{i,zx} & \alpha_{i,zy} & \alpha_{i,zz} \end{pmatrix}^{-1}, \quad (40)$$

$\mathbf{\Xi}_{ij} = \mathbf{0}$ for polarizability points i and j belonging to the same molecule. For different EFP molecules,

$$\mathbf{\Xi}_{ij} = -\begin{pmatrix} \frac{1}{r_{ij}^3} - \frac{3x_{ij}x_{ij}}{r_{ij}^5} & -\frac{3x_{ij}y_{ij}}{r_{ij}^5} & -\frac{3x_{ij}z_{ij}}{r_{ij}^5} \\ -\frac{3y_{ij}x_{ij}}{r_{ij}^5} & \frac{1}{r_{ij}^3} - \frac{3y_{ij}y_{ij}}{r_{ij}^5} & -\frac{3y_{ij}z_{ij}}{r_{ij}^5} \\ -\frac{3z_{ij}x_{ij}}{r_{ij}^5} & -\frac{3z_{ij}y_{ij}}{r_{ij}^5} & \frac{1}{r_{ij}^3} - \frac{3z_{ij}z_{ij}}{r_{ij}^5} \end{pmatrix}, \quad (41)$$

where the differences of the coordinates are defined by $x_{ij} = (x_i - x_j)$, $y_{ij} = (y_i - y_j)$, $z_{ij} = (z_i - z_j)$ and r_{ij} is the distance between polarizability points i and j.

Equation 40 represents the inverse of the polarizability tensor at a polarizability point i. The electronic contribution in Eq. 38 is described by

$$\mathbf{E}^{X,\text{el}}(\mathbf{D}^X) = \sum_{\mu\nu \in X} D_{\mu\nu}^X \langle \mu | \widehat{\mathbf{E}}^{X,\text{el}} | \nu \rangle, \quad (42)$$

where $\widehat{\mathbf{E}}^{X,\text{el}}$ is the electrostatic field operator.

The EFP polarization contribution to the MO Fock matrix of monomer fragment I has a somewhat different form from the polarization energy E_I^{pol} because of its many-body feature:

$$W_{ij}^I = \langle i | \widehat{V}_1^{\text{pol},I} | j \rangle, \quad (43)$$

where

$$\widehat{V}_1^{\text{pol},I} = -\frac{1}{2} (\mathbf{p}^I + \widetilde{\mathbf{p}}^I)^T \widehat{\mathbf{E}}^{I,\text{el}} \quad (44)$$

$$\widetilde{\mathbf{p}}^I = \left(\mathbf{\Xi}^{-1} \right)^T \mathbf{E}^I. \quad (45)$$

W_{ij}^I can be easily derived from the variation of $E_I^{\text{pol}} = -\frac{1}{2}[\mathbf{E}^I]^T \mathbf{\Xi}^{-1} \mathbf{E}^I - E^{\text{pol,efp}}$ with respect to the MO coefficients, and this expression is common both in QM/EFP (index I is not used) and FMO/EFP (for fragment I), because the variation of the EFP polarization contribution $E^{\text{pol,efp}}$ is zero. Note that the induced dipoles in Eqs. 39 and 45 depend on the electron density $\mathbf{D}^X$ through the electric field $\mathbf{E}^X$, thus the operators in Eq. 43 are fundamentally different from that in Eq. 33.

The polarization contribution to the derivatives of the MO Fock matrix elements in the CPHF equations is obtained by the differentiation of W_{ij}^I with respect to a nuclear coordinate a:

$$\frac{\partial W_{ij}^I}{\partial a} = -\frac{1}{2} (\mathbf{p}^I + \widetilde{\mathbf{p}}^I)^T \left[\sum_{m \in I}^{\text{occ+vir}} \left(U_{mi}^{a,I} \langle m | \widehat{\mathbf{E}}^{I,\text{el}} | j \rangle + U_{mj}^{a,I} \langle i | \widehat{\mathbf{E}}^{I,\text{el}} | m \rangle \right) \right.$$
$$\left. + \sum_{\mu\nu \in I} C_{\mu i}^{I*} C_{\nu j}^I \frac{\partial \langle \mu | \widehat{\mathbf{E}}^{I,\text{el}} | \nu \rangle}{\partial a} \right] - \frac{1}{2} \frac{\partial (\mathbf{p}^I + \widetilde{\mathbf{p}}^I)^T}{\partial a} \langle i | \widehat{\mathbf{E}}^{I,\text{el}} | j \rangle. \quad (46)$$

Similarly to the one-body contributions (see Eq. 34 and its discussion below), these terms contribute to the gradient directly (for EFP, one can also see [80]), in addition to the CPHF terms. Below, we focus exclusively on the latter. In Eq. 46, the orbital response contributions are absorbed as a part of MO Fock matrix in the CPHF equation, leading to $\epsilon_j^I - \epsilon_i^I$ and the second term describes the atomic orbital (AO) derivative terms and the Hellman-Feynman term, corresponding to the contribution to the Fock derivative term, $F_{ij}^{a,I}$. Their implementation is straightforward.

The last term on the right hand side of Eq. 46 can be further transformed as follows:

$$-\frac{1}{2} \frac{\partial (\mathbf{p}^I + \widetilde{\mathbf{p}}^I)^T}{\partial a} \langle i | \widehat{\mathbf{E}}^{I,\text{el}} | j \rangle$$

$$= -\frac{1}{2} \left(\frac{\partial \mathbf{\Xi}^{-1}}{\partial a} \mathbf{E}^I + \mathbf{\Xi}^{-1} \frac{\partial \mathbf{E}^I}{\partial a} + \frac{\partial (\mathbf{\Xi}^{-1})^T}{\partial a} \mathbf{E}^I \right.$$

$$\left. + \left(\mathbf{\Xi}^{-1} \right)^T \frac{\partial \mathbf{E}^I}{\partial a} \right)^T \langle i | \widehat{\mathbf{E}}^{I,\text{el}} | j \rangle$$

$$= -\frac{1}{2} \left(-(\mathbf{p}^I)^T \frac{\partial \mathbf{\Xi}^T}{\partial a} \left(\mathbf{\Xi}^{-1} \right)^T + \frac{\partial (\mathbf{E}^I)^T}{\partial a} \left(\mathbf{\Xi}^{-1} \right)^T \right.$$

$$\left. - (\widetilde{\mathbf{p}}^I)^T \frac{\partial \mathbf{\Xi}}{\partial a} \mathbf{\Xi}^{-1} + \frac{\partial (\mathbf{E}^I)^T}{\partial a} \mathbf{\Xi}^{-1} \right) \langle i | \widehat{\mathbf{E}}^{I,\text{el}} | j \rangle, \quad (47)$$

where the relation

$$\frac{\partial \mathbf{\Xi}^{-1}}{\partial a} = -\mathbf{\Xi}^{-1}\frac{\partial \mathbf{\Xi}}{\partial a}\mathbf{\Xi}^{-1}, \tag{48}$$

is used. The substitution of Eq. 47 into Eq. 46 leads to

$$\begin{aligned}
\frac{\partial W_{ij}^{I}}{\partial a} = &-\frac{1}{2}(\mathbf{p}^{I} + \widetilde{\mathbf{p}}^{I})^{T}\left[\sum_{m\in I}^{\mathrm{occ+vir}}\left(U_{mi}^{a,I}\langle m|\widehat{\mathbf{E}}^{I,\mathrm{el}}|j\rangle\right.\right.\\
&\left.+ U_{mj}^{a,I}\langle i|\widehat{\mathbf{E}}^{I,\mathrm{el}}|m\rangle\right) + \sum_{\mu\nu\in I}C_{\mu i}^{I*}C_{\nu j}^{I}\frac{\partial\langle\mu|\widehat{\mathbf{E}}^{I,\mathrm{el}}|\nu\rangle}{\partial a}\Bigg]\\
&+\frac{1}{2}\left(\left(\mathbf{p}^{I}\right)^{T}\frac{\partial\mathbf{\Xi}^{T}}{\partial a}\left(\mathbf{\Xi}^{-1}\right)^{T} - \frac{\partial\left(\mathbf{E}^{I}\right)^{T}}{\partial a}\left(\mathbf{\Xi}^{-1}\right)^{T}\right.\\
&\left.+ \left(\widetilde{\mathbf{p}}^{I}\right)^{T}\frac{\partial\mathbf{\Xi}}{\partial a}\mathbf{\Xi}^{-1} - \frac{\partial\left(\mathbf{E}^{I}\right)^{T}}{\partial a}\mathbf{\Xi}^{-1}\right)\langle i|\widehat{\mathbf{E}}^{I,\mathrm{el}}|j\rangle\\
= &-\frac{1}{2}(\mathbf{p}^{I} + \widetilde{\mathbf{p}}^{I})^{T}\left[\sum_{m\in I}^{\mathrm{occ+vir}}\left(U_{mi}^{a,I}\langle m|\widehat{\mathbf{E}}^{I,\mathrm{el}}|j\rangle\right.\right.\\
&\left.+ U_{mj}^{a,I}\langle i|\widehat{\mathbf{E}}^{I,\mathrm{el}}|m\rangle\right) + \sum_{\mu\nu\in X}C_{\mu i}^{I*}C_{\nu j}^{I}\frac{\partial\langle\mu|\widehat{\mathbf{E}}^{I,\mathrm{el}}|\nu\rangle}{\partial a}\Bigg]\\
&-\frac{1}{2}\frac{\partial\left(\mathbf{E}^{I}\right)^{T}}{\partial a}\left(\mathbf{p}^{I,ij} + \widetilde{\mathbf{p}}^{I,ij}\right)\\
&+\frac{1}{2}\left(\left(\widetilde{\mathbf{p}}^{I,ij}\right)^{T}\frac{\partial\mathbf{\Xi}}{\partial a}\mathbf{p}^{I} + \left(\widetilde{\mathbf{p}}^{I}\right)^{T}\frac{\partial\mathbf{\Xi}}{\partial a}\mathbf{p}^{I,ij}\right),
\end{aligned} \tag{49}$$

where

$$\mathbf{p}^{I,ij} = \mathbf{\Xi}^{-1}\langle i|\widehat{\mathbf{E}}^{I,\mathrm{el}}|j\rangle \tag{50}$$

$$\widetilde{\mathbf{p}}^{I,ij} = \left(\mathbf{\Xi}^{-1}\right)^{T}\langle i|\widehat{\mathbf{E}}^{I,\mathrm{el}}|j\rangle. \tag{51}$$

In the second half of Eq. 49, the $\mathbf{E}^{I,\mathrm{el}}(\mathbf{D}^{I})$ derivative term extracted from the second term leads to

$$\begin{aligned}
&-\frac{1}{2}\frac{\partial\left[\mathbf{E}^{I,\mathrm{el}}(\mathbf{D}^{I})\right]^{T}}{\partial a}\left(\mathbf{p}^{I,ij} + \widetilde{\mathbf{p}}^{I,ij}\right)\\
= &-\frac{1}{2}\left[\sum_{l\in I}^{\mathrm{occ}}\sum_{m\in I}^{\mathrm{occ+vir}}2\left(U_{ml}^{a,I}\langle m|\widehat{\mathbf{E}}^{I,\mathrm{el}}|l\rangle + U_{ml}^{a,I}\langle l|\widehat{\mathbf{E}}^{I,\mathrm{el}}|m\rangle\right)\right.\\
&\left.+ \sum_{\mu\nu\in I}D_{\mu\nu}^{I}\frac{\partial\langle\mu|\widehat{\mathbf{E}}^{I,\mathrm{el}}|\nu\rangle}{\partial a}\right]^{T}\left(\mathbf{p}^{I,ij} + \widetilde{\mathbf{p}}^{I,ij}\right)\\
= &-\frac{1}{2}\sum_{l\in I}^{\mathrm{occ}}\sum_{k\in I}^{\mathrm{vir}}4U_{kl}^{a,I}\left(\langle k|\widehat{\mathbf{E}}^{I,\mathrm{el}}|l\rangle\right)^{T}\left(\mathbf{p}^{I,ij} + \widetilde{\mathbf{p}}^{I,ij}\right)\\
&-\frac{1}{2}\sum_{kl\in I}^{\mathrm{occ}}\left(-2S_{kl}^{a,I}\right)\left(\langle k|\widehat{\mathbf{E}}^{I,\mathrm{el}}|l\rangle\right)^{T}\left(\mathbf{p}^{I,ij} + \widetilde{\mathbf{p}}^{I,ij}\right)\\
&-\frac{1}{2}\left[\sum_{\mu\nu\in X}D_{\mu\nu}^{I}\frac{\partial\langle\mu|\widehat{\mathbf{E}}^{I,\mathrm{el}}|\nu\rangle}{\partial a}\right]^{T}\left(\mathbf{p}^{I,ij} + \widetilde{\mathbf{p}}^{I,ij}\right).
\end{aligned} \tag{52}$$

Substitution of Eq. 52 into Eq. 49 yields

$$\begin{aligned}
\frac{\partial W_{ij}^{I}}{\partial a} = &-\frac{1}{2}(\mathbf{p}^{I} + \widetilde{\mathbf{p}}^{I})^{T}\left[\sum_{m\in I}^{\mathrm{occ+vir}}\left(U_{mi}^{a,I}\langle m|\widehat{\mathbf{E}}^{I,\mathrm{el}}|j\rangle\right.\right.\\
&\left.+ U_{mj}^{a,I}\langle i|\widehat{\mathbf{E}}^{I,\mathrm{el}}|m\rangle\right) + \sum_{\mu\nu\in I}C_{\mu i}^{I*}C_{\nu j}^{I}\frac{\partial\langle\mu|\widehat{\mathbf{E}}^{I,\mathrm{el}}|\nu\rangle}{\partial a}\Bigg]\\
&- 2\sum_{l\in I}^{\mathrm{occ}}\sum_{k\in I}^{\mathrm{vir}}U_{kl}^{a,I}\left(\langle k|\widehat{\mathbf{E}}^{I,\mathrm{el}}|l\rangle\right)^{T}\left(\mathbf{p}^{I,ij} + \widetilde{\mathbf{p}}^{I,ij}\right)\\
&+ \sum_{kl\in I}^{\mathrm{occ}}S_{kl}^{a,I}\left(\langle k|\widehat{\mathbf{E}}^{I,\mathrm{el}}|l\rangle\right)^{T}\left(\mathbf{p}^{I,ij} + \widetilde{\mathbf{p}}^{I,ij}\right)\\
&- \frac{1}{2}\left(\sum_{\mu\nu\in X}D_{\mu\nu}^{I}\frac{\partial\langle\mu|\widehat{\mathbf{E}}^{I,\mathrm{el}}|\nu\rangle}{\partial a} + \frac{\partial\left[\mathbf{E}^{I,\mathrm{nuc}} + \mathbf{E}^{\mathrm{efp}}\right]}{\partial a}\right)^{T}\\
&\times\left(\mathbf{p}^{I,ij} + \widetilde{\mathbf{p}}^{I,ij}\right)\\
&+ \frac{1}{2}\left(\left(\widetilde{\mathbf{p}}^{I,ij}\right)^{T}\frac{\partial\mathbf{\Xi}}{\partial a}\mathbf{p}^{I} + \left(\widetilde{\mathbf{p}}^{I}\right)^{T}\frac{\partial\mathbf{\Xi}}{\partial a}\mathbf{p}^{I,ij}\right).
\end{aligned} \tag{53}$$

The terms in Eq. 53 can be classified into the contributions to $B_{0,ij}^{a,I}$ and $F_{ij}^{a,I}$:

$$\begin{aligned}
B_{0,ij}^{a,I,\mathrm{pol}} = &\sum_{kl\in I}^{\mathrm{occ}}S_{kl}^{a,I}\left(\langle k|\widehat{\mathbf{E}}^{I,\mathrm{el}}|l\rangle\right)^{T}\left(\mathbf{p}^{I,ij} + \widetilde{\mathbf{p}}^{I,ij}\right)\\
&- \frac{1}{2}\left(\sum_{\mu\nu\in I}D_{\mu\nu}^{I}\frac{\partial\langle\mu|\widehat{\mathbf{E}}^{I,\mathrm{el}}|\nu\rangle}{\partial a}\right)^{T}\left(\mathbf{p}^{I,ij} + \widetilde{\mathbf{p}}^{I,ij}\right)\\
&- \frac{1}{2}\frac{\partial\left[\mathbf{E}^{I,\mathrm{nuc}} + \mathbf{E}^{\mathrm{efp}}\right]^{T}}{\partial a}\left(\mathbf{p}^{I,ij} + \widetilde{\mathbf{p}}^{I,ij}\right)\\
&+ \frac{1}{2}\left(\left(\widetilde{\mathbf{p}}^{I,ij}\right)^{T}\frac{\partial\mathbf{\Xi}}{\partial a}\mathbf{p}^{I} + \left(\widetilde{\mathbf{p}}^{I}\right)^{T}\frac{\partial\mathbf{\Xi}}{\partial a}\mathbf{p}^{I,ij}\right),
\end{aligned} \tag{54}$$

$$F_{ij}^{a,I,\mathrm{pol}} = -\frac{1}{2}(\mathbf{p}^{I} + \widetilde{\mathbf{p}}^{I})^{T}\sum_{\mu\nu\in X}C_{\mu i}^{I*}C_{\nu j}^{I}\frac{\partial\langle\mu|\widehat{\mathbf{E}}^{I,\mathrm{el}}|\nu\rangle}{\partial a}, \tag{55}$$

and the contribution to the matrix $\mathbf{A}$:

$$\begin{aligned}
A_{ij,kl}^{I,I,\mathrm{pol}} &= 2\left(\langle k|\widehat{\mathbf{E}}^{I,\mathrm{el}}|l\rangle\right)^{T}\left(\mathbf{p}^{I,ij} + \widetilde{\mathbf{p}}^{I,ij}\right)\\
&= 2\left(\mathbf{p}^{I,kl} + \widetilde{\mathbf{p}}^{I,kl}\right)^{T}\langle i|\widehat{\mathbf{E}}^{I,\mathrm{el}}|j\rangle.
\end{aligned} \tag{56}$$

Note that the sign of the first term in Eq. 56 is opposite in the above contribution to the $\mathbf{A}$ matrix compared to the correction term in Eq. 53.

Practically, the product of $A_{ij,kl}^{I,I,\mathrm{pol}}$ and the Z-vector Z_{kl}^{I} is computed in the SCZV procedure (also see Eq. 23) [26]:

$$\begin{aligned}
\sum_{k\in I}^{\mathrm{vir}}\sum_{l\in I}^{\mathrm{occ}}A_{ij,kl}^{I,I,\mathrm{pol}}Z_{kl}^{I} &= 2\sum_{k\in I}^{\mathrm{vir}}\sum_{l\in I}^{\mathrm{occ}}\left(\mathbf{p}^{I,kl} + \widetilde{\mathbf{p}}^{I,kl}\right)^{T}\langle i|\widehat{\mathbf{E}}^{I,\mathrm{el}}|j\rangle Z_{kl}^{I}\\
&= 2\sum_{k\in I}^{\mathrm{vir}}\sum_{l\in I}^{\mathrm{occ}}\left(\left\{\mathbf{\Xi}^{-1} + \left[\mathbf{\Xi}^{-1}\right]^{T}\right\}\langle k|\widehat{\mathbf{E}}^{I,\mathrm{el}}|l\rangle\right)^{T}\\
&\quad\times\langle i|\widehat{\mathbf{E}}^{I,\mathrm{el}}|j\rangle Z_{kl}^{I}
\end{aligned}$$

$$=2\left(\left\{\Xi^{-1}+\left[\Xi^{-1}\right]^{T}\right\}\mathbf{E}^{I,\mathrm{el}}(\overline{\mathbf{Z}}^{I})\right)^{T}\langle i|\widehat{\mathbf{E}}^{I,\mathrm{el}}|j\rangle$$

$$=2\left(\mathbf{p}^{I,\mathbf{Z}^{I}}+\widetilde{\mathbf{p}}^{I,\mathbf{Z}^{I}}\right)^{T}\langle i|\widehat{\mathbf{E}}^{I,\mathrm{el}}|j\rangle, \tag{57}$$

where the following definitions

$$\mathbf{E}^{I,\mathrm{el}}(\overline{\mathbf{Z}}^{I})=\sum_{\mu\nu\in I}\langle\mu|\widehat{\mathbf{E}}^{I,\mathrm{el}}|\nu\rangle\overline{Z}_{\mu\nu}^{I} \tag{58}$$

$$\overline{Z}_{\mu\nu}^{I}=\sum_{k\in I}^{\mathrm{vir}}\sum_{l\in I}^{\mathrm{occ}}\frac{C_{\mu k}^{I*}Z_{kl}^{I}C_{\nu l}^{I}+C_{\nu k}^{I*}Z_{kl}^{I}C_{\mu l}^{I}}{2}, \tag{59}$$

are used for the derivation of Eq. 57. The induced dipoles due to the symmetrized Z-vector, $\mathbf{p}^{I,\mathbf{Z}^{I}}$ and $\widetilde{\mathbf{p}}^{I,\mathbf{Z}^{I}}$ should be solved by the following equations on each SCZV iteration (because of the $\overline{\mathbf{Z}}^{I}$ dependence)

$$\Xi\mathbf{p}^{I,\mathbf{Z}^{I}}=\mathbf{E}^{I,\mathrm{el}}(\overline{\mathbf{Z}}^{I}) \tag{60}$$

$$\Xi^{T}\widetilde{\mathbf{p}}^{I,\mathbf{Z}^{I}}=\mathbf{E}^{I,\mathrm{el}}(\overline{\mathbf{Z}}^{I}). \tag{61}$$

The SCZV equations considering the effective potentials explicitly do not introduce a considerable extra cost. The calculation of the induced dipoles due to the Z-vector in Eqs. 60 and 61 corresponds to that of the induced dipoles due to the density matrix in the SCC calculation [80].

3 Computational details

The developed FMO2/EFP1 and FMO2/MCP gradients were implemented into GAMESS [94, 95]. To verify that the analytic FMO2/EFP1 and FMO2/MCP energy gradients are complete, short MD simulations of 50 fs were performed with the NVE ensemble. The accuracy of FMO2/MD based on the analytic gradient is investigated. The zwitterionic form of glycine tetramer is computed in gas phase and immersed in a water layer of 6.0 Å (135 water molecules) using VEGA [96]; alanine dacamer capped with acetyl and –NHCH$_3$ groups, (ALA)$_{10}$, is computed in gas phase. For all polypeptide molecules, the one residue per fragment partition is employed.

To determine the initial structure for the demonstrative short MD simulation, the geometry optimization [97] was performed for the trans-structure of hydrated tetraglycine at the FMO2-RHF/EFP1/cc-pVDZ level. The same optimization calculation was carried out at the conventional MO-RHF/EFP1/cc-pVDZ level as well. A solute part of the optimized structures were used for the respective short MD simulations in gas phase. Using the optimized structure of hydrated tetraglycine, the FMO2/EFP1 MD simulation proceeded at the same level with the time step of 0.2 fs in the NVT ensemble at 315 K (using Nose-

Hoover thermostats [98]) until the temperature is stable, followed by the further MD simulation of 593 fs under the same conditions. The relaxed structure (Fig. 1a) from this MD simulation was used in the consequent shorter FMO2/EFP1 MD simulations. One can see from Fig. 1a that the EFP water molecules are still bound with each other and charged groups of the solute after the MD simulation. These structures were used for the short MD simulations in the conventional MO method and the FMO2/EFP1 method with the incomplete gradients for comparison. All MD simulations, whose results are compared to each other, always had the same initial structure and atomic velocities.

For (ALA)$_{10}$, the core 1s electrons of C, N, and O were treated with MCP. In the covalent bond detachment in FMO [89], the bond detached atom in each detached bond is put into two fragments (redundantly) and MCPs were used only in the main fragment containing this bond detached atom, while the other fragment, where this atom is only used to describe the detached bond, did not have the MCP on this atom. To obtain the initial structure for the MD simulation, the geometry was optimized at the FMO2-RHF/MCP level with the MCP-DZP basis set (see Fig. 1b) [83]. This is also used as the initial geometry of the MD simulation with the incomplete FMO2/MCP gradients. We should note that the MD simulations with the corresponding conventional MO gradients (using the geometry optimized at the corresponding level) and the incomplete FMO2 gradients were executed for comparison.

For all the FMO calculations, the electrostatic dimer (ES-DIM) approximation was applied with the default value, $L_{\mathrm{ES\text{-}DIM}}=2.0$ and the other electrostatic approximations are switched off [90, 91]. In the ES-DIM approximation, only the response term of Eq. 16 can be considered [26]. There are two types of EFP1 parameter sets for the electron-exchange plus charge-transfer potential: one is the RHF parameter set and another is the DFT parameter set [99]. This study used the former throughout [92]. The velocity-Verlet algorithm was used for time integration throughout. All gradients developed in this work are parallelized with the generalized distributed data interface (GDDI) [100].

4 Results and discussion

4.1 Gradient accuracy

To verify the accuracy of the analytic FMO2/EFP1 and FMO2/MCP gradients, we compared them with the corresponding numeric gradients. Numeric FMO2/EFP1-RHF and FMO2/MCP-RHF energy gradients were computed

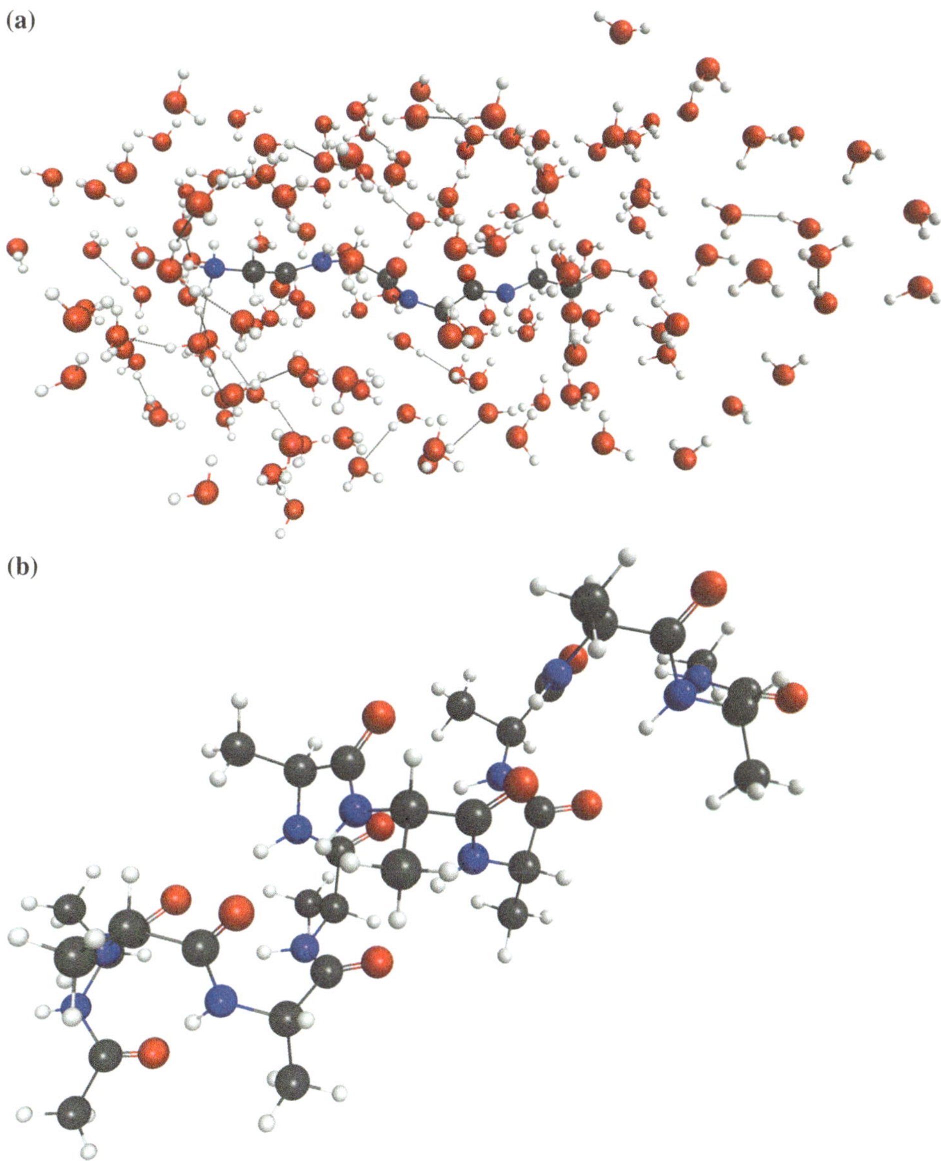

Fig. 1 a Structure of zwitterionic tetraglycine in the water layer of 6.0 Å relaxed with the MD simulation in the NVT ensemble (at 315 K) at the FMO-RHF/EFP/cc-pVDZ level and **b** structure of alanine decamer (ALA)$_{10}$ optimized at the FMO-RHF/MCP/MCP-DZP level

with double differencing and a coordinate step of 0.0005 Å. The initial structures of hydrated tetraglycine and (ALA)$_{10}$ for the short MD simulations were used for the respective gradient calculations.

Table 1 displays the errors of analytic FMO2/MCP gradients relative to the corresponding numeric gradients for (ALA)$_{10}$. The bond detached atoms (BDA) and bond attached atoms (BAA) at the two ends of detached bonds form a representative set for the accuracy tests. As shown in Table 1, all of the deviations are sufficiently small.

Next, the analytic and numeric FMO2/EFP1 gradient calculations were performed for hydrated tetraglycine with the geometry obtained after the NVT MD simulation of 593 fs. For the solute, the RMSD between the analytic and numeric gradients is 8.1×10^{-7} a.u. and the maximum absolute deviation is 5.9×10^{-6} a.u.

4.2 Zwitterionic tetraglycine

There are some arguments on the structural stability of peptides in water solvent both empirically and theoretically [101–105]. It is important to sample the structures and to check the free energy difference between the zwitterionic and neutral forms. When a 0 K geometry optimization is performed, the thermal movement of solvent is not accounted for, overestimating the hydrogen bonding. Also, MD sampling is capable of overcoming local minima and converging toward the global one.

Before discussing the results of the MD simulations, we first check the energy conservation of zwitterionic tetraglycine in gas phase. For systems with the fragmentation across the covalent bonds such as tetraglycine, the FMO2/MD simulation using the incomplete gradients does not

Table 1 Deviations of the analytic and numeric gradients (a.u.) for a representative set of bond detached atoms (BDA) and bond attached atoms (BAA) in (ALA)$_{10}$ at the FMO-RHF/MCP-DZP level

	x	y	z
C1(BDA)	−0.000007	0.000000	0.000000
C1(BAA)	−0.000004	0.000003	0.000004
C2(BDA)	−0.000001	0.000006	0.000007
C2(BAA)	0.000005	−0.000002	0.000008
C3(BDA)	0.000012	0.000005	0.000000
C3(BAA)	0.000000	0.000007	0.000000
C4(BDA)	−0.000002	0.000009	−0.000004
C4(BAA)	−0.000009	0.000011	−0.000002
C5(BDA)	−0.000002	0.000001	0.000007
C5(BAA)	−0.000007	0.000004	0.000008
C6(BDA)	0.000007	0.000001	0.000006
C6(BAA)	0.000002	−0.000001	0.000004
C7(BDA)	0.000003	0.000010	−0.000003
C7(BAA)	0.000009	0.000008	−0.000006
C8(BDA)	−0.000018	0.000005	0.000000
C8(BAA)	−0.000003	0.000002	0.000000
C9(BDA)	0.000003	0.000001	0.000011
C9(BAA)	0.000003	0.000003	0.000002

conserve the energy [51]. Figure 2a displays a double logarithmic plot of the root mean square deviation (RMSD) of the total energies relative to the energy of 10th step against time step Δt for zwitterionic tetraglycine in gas phase. For the FMO2/MD simulations with the completely analytic gradients, the slope of 2.0 indicates that the energy is sufficiently conserved even at the smaller time step such as 0.1 fs and the gradient is analytic because the RMSD value is proportional to Δt^2 in the velocity-Velret method [51]. This tendency is in good agreement with that in the conventional MO-MD simulations, while the FMO2/MD simulations with the incomplete gradients give a wrong description at the shorter time steps. The corresponding numeric values in Table 2 reinforce these findings. The results imply that FMO2/MD with the one residue per fragment partition is sufficiently accurate and shows perfect energy conservation. The energy conservation at 0.1 fs also means that it is possible to pursue the course of chemical reactions.

The initial structure of tetraglycine in gas phase for the MD simulations was obtained from the FMO2 geometry optimization for hydrated tetraglycine. Since the geometry optimization is done at 0 K, the structure is different from that determined at 315 K, which is typical for biological molecules in solution. Thus, the structure is more realistic when obtained by the MD simulation in the NVT ensemble. This study employed the Nose-Hoover thermostats in the NVT ensemble [98], and we performed the MD

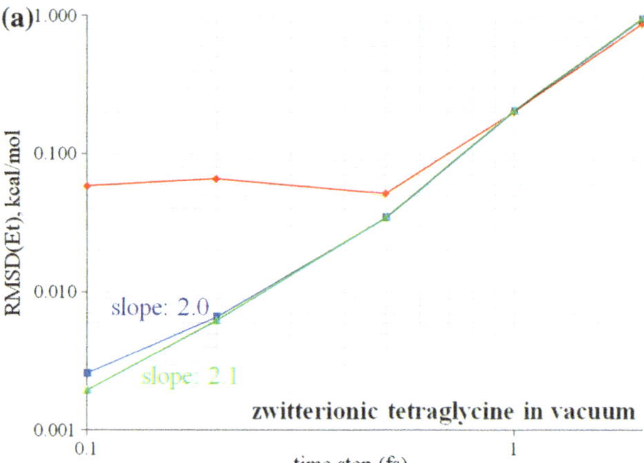

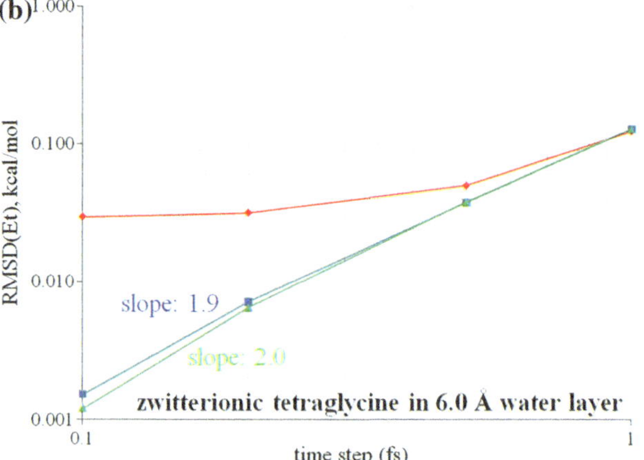

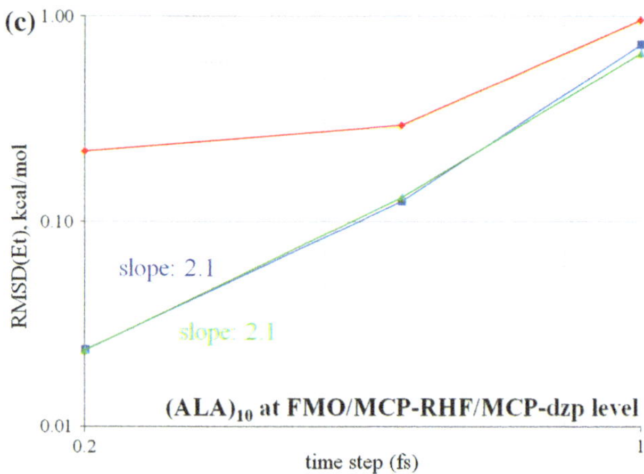

Fig. 2 Double logarithmic plots of the RMSD of the total energies relative to the energy of the 10th step against time step Δt in the short MD simulations of 50 fs in the NVE ensemble. *Triangle* markers are used for MO-MD, *square* for FMO/MD with the analytic gradients, and *diamond* for FMO/MD with the incompletely analytic gradients

simulation at 315 K for the hydrated tetraglycine until the total energy and temperature are stable and a stationary structure is taken for the short MD simulations of 50 fs for time steps of 0.1, 0.2, 0.5, and 1.0 fs. Figure 2b displays that the slopes in the MO- and FMO2-based MD/EFP1

Table 2 Root mean square deviations of the total energies relative to the 10th value (in kcal/mol) in the MD simulations of 50 fs at several time steps

Time step	Ab initio MO	FMO[a]	FMO[b]
Zwitterionic tetraglycine in gas phase			
0.1 fs	0.0020	0.0026	0.0592
0.2 fs	0.0063	0.0067	0.0667
0.5 fs	0.0352	0.0348	0.0524
1.0 fs	0.2068	0.2055	0.2030
2.0 fs	0.9608	0.9484	0.8769
Zwitterionic tetraglycine in 6.0 Å water layer			
0.1 fs	0.0012	0.0015	0.0298
0.2 fs	0.0065	0.0071	0.0319
0.5 fs	0.0384	0.0377	0.0501
1.0 fs	0.1293	0.1294	0.1243
Capped alanine decamer: $(ALA)_{10}$			
0.2 fs	0.0237	0.0239	0.2204
0.5 fs	0.1324	0.1256	0.2944
1.0 fs	0.6642	0.7295	0.9633

[a] Analytic gradient in this work

[b] Original incomplete gradient

Table 3 Time-Averages and RMSDs of the kinetic energy, potential energy, total energy (in kcal/mol) and temperature (in K)

	Kinetic energy	Potential energy	Total energy	Temperature
Hydrated tetraglycine at time step of 0.1 fs in the NVE ensemble				
MO/MD				
Average	276.737	-5.67877066×10^5	-5.67600324×10^5	
RMSD	7.058	7.058		
FMO/MD[a]				
Average	276.651	-5.67877335×10^5	-5.67600681×10^5	
RMSD	7.090	7.091		
FMO/MD[b]				
Average	276.637	-5.67877344×10^5	-5.67600704×10^5	
RMSD	7.092	7.105		
Hydrated tetraglycine at time step of 0.2 fs in the NTV ensemble (315K)				
FMO/MD[a]				
Average	281.536	-5.67695274×10^5	-5.67413738×10^5	313.891
RMSD	8.107	9.028	4.190	9.036
$(ALA)_{10}$ at time step of 0.2 fs in the NVE ensemble				
MO/MD				
Average	49.198	-3.11678589×10^5	-3.11629391×10^5	
RMSD	5.706	5.715		
FMO/MD[a]				
Average	48.984	-3.11672215×10^5	-3.11623232×10^5	
RMSD	6.029	6.039		
FMO/MD[b]				
Average	48.979	-3.11672195×10^5	-3.11623216×10^5	
RMSD	6.046	6.117		

[a] Analytic gradient in this work

[b] Original incomplete gradient

simulations are 1.9 and 2.0, showing that accurate long MD simulations are possible without accumulating the errors due to the incomplete gradients.

Table 3 shows the time-averaged kinetic energy, potential energy, and total energy calculated for the hydrated tetraglycine in the MD simulations of 50 fs for the time step of 0.1 fs. The FMO2/MD energies are in good agreement with MO-MD. The FMO2/MD simulation with the incomplete gradients also give reasonably accurate energies, because there is little accumulation of errors in

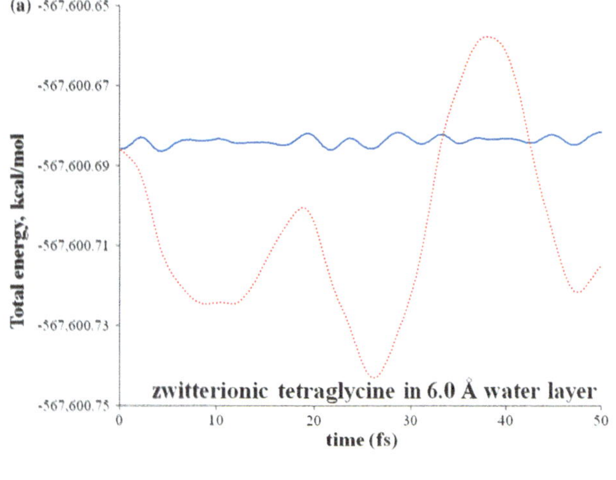

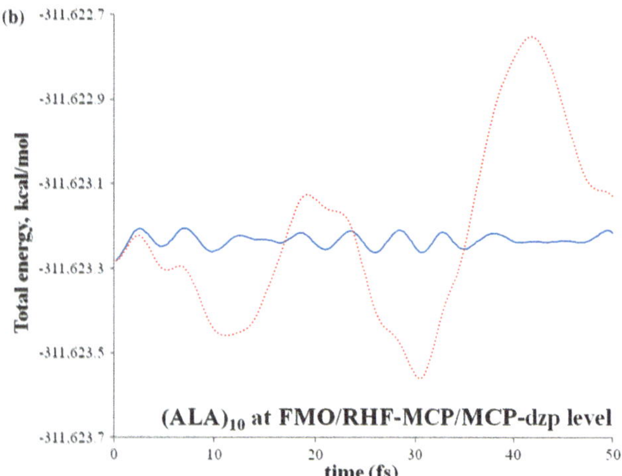

Fig. 3 Plots of the total energies against the time in the short FMO/MD simulations with the NVE ensemble: **a** for the zwitterionic tetraglycine in the water layer of 6.0 Å for the time step of 0.1 fs and **b** (ALA)$_{10}$ at the time step of 0.2 fs. *Solid line* is used for the analytic energy gradient. *Dashed line* is used for the incompletely analytic energy gradient

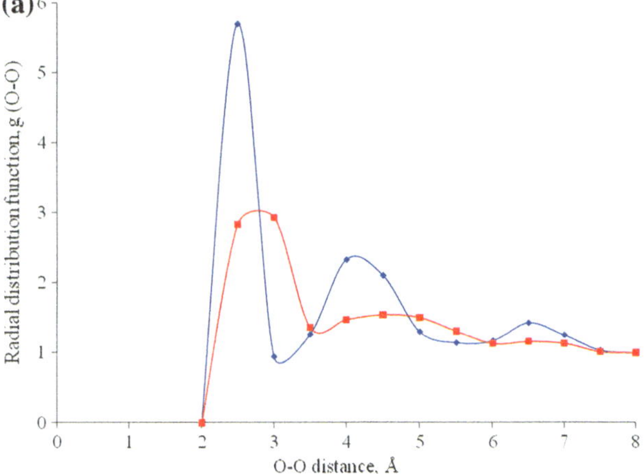

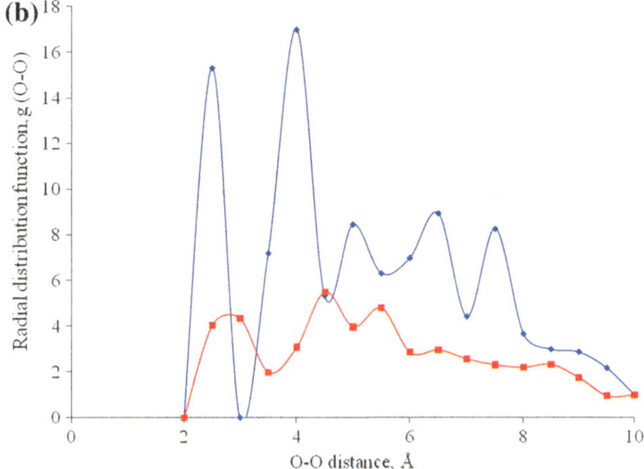

Fig. 4 Radial distribution functions of **a** O–O distances in the EFP water molecules of hydrated tetraglycine and **b** distances between oxygens of the COO$^-$ group and of the EFP water molecules of hydrated tetraglycine. *Diamond* markers show the data for the structure optimized at the FMO-RHF/EFP/cc-pVDZ level. *Square* markers show the data for the time-averaged structures in the MD simulation of 592 fs

such a short MD simulation. The plot of the total energy against the time in the MD simulation is also another way of checking the energy conservation. In Fig. 3a, one can see that the total energies in the FMO2/MD simulation with the analytic gradients are conserved, while those with the incomplete gradients fluctuate. The preliminary MD simulation of 593 fs (see Table 3) is close to 315 K, meaning that the NVT MD simulations are properly conducted.

Figure 4a shows the radial distribution functions (RDF) against the O–O distances of the water molecules in the structures of hydrated tetraglycine optimized at the FMO2-RHF/EFP1/cc-pVDZ level versus those obtained in the NVT MD simulation of 593 fs for the time step of 0.2 fs. The fluctuations of O–O distances are larger for the latter, as one can easily expect, and the FMO2/EFP1-MD simulations properly describe the fluctuation of the water molecules. For the MD simulations, the peak in the first solvation shell reasonably corresponds to the typical O–O

distance in water at 315 K, while the O–O distance optimized in the FMO method is somewhat shorter. Figure 4b plots the RDFs between the oxygen atoms of the COO$^-$ group and oxygen atoms of the water molecules obtained at the same levels of theory as in Fig. 4a. For the RDF obtained in the MD simulation, the peak in the first solvation shell also shifts to the larger O–O distance compared with that for the FMO2 geometry optimization. In addition, for the latter, there is a vacancy between the first shell and the second shell, while for the former, the water molecules can be found there because of the fluctuation and the second shell shifts to the larger O–O distances.

4.3 Alanine decamer

The FMO/MCP method can applied to systems containing heavy metals such as transition metals, for which the relativistic effects are significant. Another use of the

FMO/MCP method is to reduce the basis set superposition errors (BSSE) [88], because the core orbitals, which is one of main causes of BSSE, are replaced by the potentials. The α-helix conformer of $(ALA)_{10}$ has several hydrogen bonds, so it is a good test system for the FMO2/MCP method. The intramolecular hydrogen bonds hold together the structures of polypeptides, proteins, and DNA.

In Fig. 2c, the slope of 2.1 in the FMO2/MD simulations shows that the FMO2/MD simulations at short time steps are accurate even with the one residue per fragment partition, and the slope is in good agreement with the conventional MO-MD simulations. Another energy conservation test in Fig. 3b shows that the total energy is well conserved for this system, when the analytic energy gradient is used in simulations. So far $(ALA)_{10}$ appears to be the largest system fragmented across covalent bonds as treated in the FMO2/MD simulations, and the applications to larger systems such as proteins and DNA are now possible.

We measured the total wall clock times for the FMO2/MCP and MO/MCP MD simulations of 50 fs at the time step of 0.2 fs using the Soroban cluster: 192 cores of Xeon and compared them and the number of GDDI groups was set to 5 for the former [100]. The total wall clock times for the former and the latter were 3763.9 min and 3721.7 min, respectively. The latter calculation is faster, because the system is too small to treat in the FMO method and the MO method in GAMESS is well parallelized. The computation time in the FMO gradient calculation is discussed in Ref. [26], which directly affects that in the MD simulation. It is concluded that FMO/MD is preferable for large systems with many fragments.

5 Conclusions

The analytic two-body FMO2/EFP1 and FMO2/MCP energy gradients have been successfully derived and implemented in this study. The results of FMO2/MD simulations using these gradients are in good agreement with those of the corresponding conventional MO-MD simulations. Our study has demonstrated that FMO2/MD can be performed for systems with covalently connected fragments, which very considerably expands its application field previously limited to molecular clusters.

By verifying the accuracy in MD simulations, we have been able to use an efficient partition of one residue per fragment, while earlier studies [51] suggested that even larger partitions of two residues per fragment do not have sufficient accuracy. It is clear that the ability to perform molecular dynamics using FMO will largely extend its usefulness, for instance, to evaluate the entropic contribution in the protein-ligand binding, so far often neglected in the applications of FMO or evaluated using force fields.

Acknowledgments This work has been supported by the Next Generation Super Computing Project, Nanoscience Program (MEXT, Japan), and Computational Materials Science Initiative (CMSI, Japan). We thank Dr. Yuto Komeiji, Prof. Mark Gordon and Kurt Brorsen for fruitful discussions.

References

1. Pearlman DA, Case DA, Caldwell JW, Ross WS, Cheatham TE, Debolt S, Ferguson D, Seibel G, Kollman P (1995) Comput Phys Commun 91:1
2. Kollman P, Massova I, Reyes C, Kuhn B, Huo SH, Chong L, Lee M, Lee T, Duan Y, Wang W, Donini O, Cieplak P, Srinivasan J, Case DA, Cheatham TE (2000) Acc Chem Res 33:889
3. Bashford D, Case DA (2000) Annu Rev Phys Chem 51:129
4. Mackerell AD (2004) J Comput Chem 25:1584
5. Wang W, Donini O, Reyes CM, Kollman PA (2001) Annu Rev Biophys Biomol Struct 30:211
6. Warshel A (2003) Annu Rev Biophys Biomol Struct 32:425
7. Kollman P (1993) Chem Rev 93:2395
8. Car R, Parrinello M (1985) Phys Rev Lett 55:2471
9. Schlegel HB, Millam JM, Iyengar SS, Voth GA, Daniels AD, Scuseria GE, Frisch MJ (2001) J Chem Phys 114:9758
10. Morokuma K (1971) J Chem Phys 55:1236
11. Ohno K, Inokuchi H (1972) Theor Chim Acta (Berl.) 26:331
12. Otto P, Ladik J (1975) Chem Phys 8:192
13. Barandiaran Z, Seijo L (1988) J Chem Phys 89:5739
14. Gao JL (1997) J Phys Chem B 101:657
15. Xie W, Orozco M, Truhlar DG, Gao J (2009) J Chem Theory Comput 5:459
16. Leverentz HR, Truhlar DG (2009) J Chem Theory Comput 5:1573
17. Gordon MS, Mullin JM, Pruitt SR, Roskop LB, Slipchenko LV, Boatz JA (2009) J Phys Chem B 113:9646
18. Huang L, Massa L, Karle I, Karle J (2009) Proc Natl Acad Sci USA 106:3664
19. Tong Y, Mei Y, Zhang JZH, Duan LL, Zhang QG (2009) J Theor Comput Chem 8:1265
20. Söderhjelm P, Kongsted J, Ryde U (2010) J Chem Theory Comput 6:1726
21. Mata RA, Stoll H, Cabral BJC (2009) J Chem Theory Comput 5:1829
22. Yeole SD, Gadre S (2010) J Chem Phys 132:094102
23. Makowski M, Korchowiec J, Gu FL, Aoki Y (2010) J Comput Chem 31:1733
24. Kobayashi M, Kunisada T, Akama T, Sakura D, Nakai H (2011) J Chem Phys 134:034105
25. He X, Merz KM (2010) J Chem Theory Comput 6:405
26. Nagata T, Brorsen K, Fedorov DG, Kitaura K, Gordon MS (2011) J Chem Phys 134:124115
27. Gordon MS, Fedorov DG, Pruitt SR, Slipchenko LV (2012) Chem Rev 112:632
28. Kitaura K, Ikeo E, Asada T, Nakano T, Uebayasi M (1999) Chem Phys Lett 313:701
29. Fedorov DG, Kitaura K (2004) J Chem Phys 120:6832
30. Fedorov DG, Kitaura K (2007) J Phys Chem A 111:6904
31. Fedorov, DG, Kitaura, K (eds) (2009) The fragment molecular orbital method: practical applications to large molecular systems. CRC Press, Boca Raton
32. Fedorov DG, Kitaura K (2004) J Chem Phys 121:2483
33. Mochizuki Y, Nakano T, Koikegami S, Tanimori S, Abe Y, Nagashima U, Kitaura K (2004) Theor Chem Acc 112:442
34. Mochizuki Y, Yamashita K, Fukuzawa K, Takematsu K, Watanabe H, Taguchi N, Okiyama Y, Tsuboi M, Nakano T, Tanaka S (2010) Chem Phys Lett 493:346

35. Fedorov DG, Kitaura K (2005) J Chem Phys 123:134103/1
36. Sugiki SI, Kurita N, Sengoku Y, Sekino H (2003) Chem Phys Lett 382:611
37. Fedorov DG, Kitaura K (2004) Chem Phys Lett 389:129
38. Fedorov DG, Kitaura K (2005) J Chem Phys 122:054108/1
39. Mochizuki Y, Koikegami S, Amari S, Segawa K, Kitaura K, Nakano T (2005) Chem Phys Lett 406:283
40. Mochizuki Y, Tanaka K, Yamashita K, Ishikawa T, Nakano T, Amari S, Segawa K, Murase T, Tokiwa H, Sakurai M (2007) Theor Chem Acc 117:541
41. Chiba M, Fedorov DG, Kitaura K (2007) Chem Phys Lett 444:346
42. Chiba M, Fedorov DG, Kitaura K (2007) J Chem Phys 127:104108
43. Chiba M, Fedorov DG, Kitaura K (2008) J Comput Chem 29:2667
44. Pruitt SR, Fedorov DG, Kitaura K, Gordon MS (2010) J Chem Theory Comput 6:1
45. Tomasi J, Mennucci B, Cammi R (2005) Chem Rev 105:2999
46. Fedorov DG, Kitaura K, Li H, Jensen JH, Gordon MS (2006) J Comput Chem 27:976
47. Li H, Fedorov DG, Nagata T, Kitaura K, Jensen JH, Gordon MS (2010) J Comput Chem 31:778
48. Watanabe H, Okiyama Y, Nakano T, Tanaka S (2010) Chem Phys Lett 500:116
49. Kitaura K, Sugiki SI, Nakano T, Komeiji Y, Uebayasi M (2001) Chem Phys Lett 336:163
50. Nagata T, Fedorov DG, Ishimura K, Kitaura K (2011) J Chem Phys 135:044110
51. Komeiji Y, Nakano T, Fukuzawa K, Ueno Y, Inadomi Y, Nemoto T, Uebayasi M, Fedorov DG, Kitaura K (2003) Chem Phys Lett 372:342
52. Komeiji Y, Ishikawa T, Mochizuki Y, Yamataka H, Nakano T (2009) J Comput Chem 30:40
53. Komeiji Y, Mochizuki Y, Nakano T (2010) Chem Phys Lett 484:380
54. Mochizuki Y, Nakano T, Komeiji Y, Yamashita K, Okiyama Y, Yoshikawa H, Yamataka H (2011) Chem Phys Lett 504:95
55. Komeiji Y, Mochizuki Y, Nakano T, Fedorov DG (2009) J Mol Str (THEOCHEM) 898:2
56. Ishimoto T, Tokiwa H, Teramae H, Nagashima U (2004) Chem Phys Lett 387:460
57. Ishimoto T, Tokiwa H, Teramae H, Nagashima U (2005) J Chem Phys 122:094905
58. Fujita T, Watanabe H, Tanaka S (2009) J Phys Soc Jpn 78:104723
59. Fujita T, Nakano T, Tanaka S (2011) Chem Phys Lett 506:112
60. Mochizuki Y, Komeiji Y, Ishikawa T, Nakano T, Yamataka H (2007) Chem Phys Lett 437:66
61. Sato M, Yamataka H, Komeiji Y, Mochizuki Y, Ishikawa T, Nakano T (2008) J Am Chem Soc 130:2396
62. Fujiwara T, Mochizuki Y, Komeiji Y, Okiyama Y, Mori H, Nakano T, Miyoshi E (2010) Chem Phys Lett 490:41
63. Fujiwara T, Mori H, Mochizuki Y, Tatewaki H, Miyoshi E (2010) J Mol Str (THEOCHEM) 949:28
64. Sato M, Yamataka H, Komeiji Y, Mochizuki Y, Nakano T (2010) Chem Eur J 16:6430
65. Nagata T, Fedorov DG, Kitaura K (2010) Chem Phys Lett 492:302
66. Day NP, Jensen HJ, Gordon SM, Webb PS (1996) J Chem Phys 105:1968
67. Gordon MS, Freitag MA, Bandyopadhyay P, Jensen JH, Kairys V, Stevens WJ (2001) J Phys Chem A 105:293
68. Stone AJ (1996) The theory of intermolecular forces. Oxford University Press, New York
69. Jensen JH, Gordon MS (1998) J Chem Phys 108:4772
70. Jensen JH (2001) J Chem Phys 114:8775
71. Adamovic I, Gordon MS (2005) Mol Phys 103:379
72. Li H, Gordon MS (2006) J Chem Phys 124:214108
73. Aguilar CM, Rocha WR (2011) J Phys Chem B 115:2030
74. Minezawa N, De Silva N, Zahariev F, Gordon MS (2011) J Chem Phys 134:124115
75. Kosenkov D, Slipchenko LV (2011) J Phys Chem A 115:392
76. Ghosh D, Kosenkov D, Vanovschi V, Williams CF, Herbert JM, Gordon MS, Schmidt MW, Slipchenko LV, Krylov AI (2010) J Phys Chem A 114:12739
77. Slipchenko LV (2010) J Phys Chem A 114:8824
78. Arora P, Slipchenko LV, Webb SP, DeFusco A, Gordon MS (2010) J Phys Chem A 114:6742
79. Nagata T, Fedorov DG, Kitaura K, Gordon MS (2009) J Chem Phys 131:024101
80. Nagata T, Fedorov DG, Sawada T, Kitaura K, Gordon MS (2011) J Chem Phys 134:034110
81. Honda S, Yamasaki K, Sawada Y, Morii H (2004) Structure 12:1507
82. Huzinaga S (1995) Can J Chem 73:619
83. Miyoshi E, Mori H, Hirayama R, Osanai Y, Noro T, Honda H, Klobukowski M (2005) J Chem Phys 122:074104
84. Klobukowski M, Huzinaga S, Sakai Y (1999) In: Leszczynski J (eds) Computational chemistry: reviews of current trends, vol 3, World Scientific, Singapore, p 49
85. Frenking G, Antes I, Böhme M, Dapprich S, Ehlers AW, Jonas V, Neuhaus MOA, Stegmann R, Veldkamp A, Vyboishchikov SF (1996) In: Lipkowitz KB, Boyd DB (eds) Reviews in computational chemistry, vol 8, VCH Publishers, New York, p 63
86. Cundari TR, Benson MT, Lutz ML, Sommerer SO (1996) In: Lipkowitz KB, Boyd DB (eds) Reviews in computational chemistry, vol 8, VCH Publishers, New York, p 145
87. Zeng T, Fedorov DG, Klobukowski M (2010) J Chem Phys 133:114107
88. Ishikawa T, Mochizuki Y, Nakano T, Amari S, Mori H, Honda H, Fujita T, Tokiwa H, Tanaka S, Komeiji Y, Fukuzawa K, Tanaka K, Miyoshi E (2006) Chem Phys Lett 427:159
89. Nagata T, Fedorov DG, Kitaura K (2011) In: Zalesny R, Papadopoulos MG, Mezey PG, Leszczynski J (eds) Linear-scaling techniques in computational chemistry and physics, Springer, Berlin, pp 17–64
90. Nakano T, Kaminuma T, Sato T, Fukuzawa K, Akiyama Y, Uebayasi M, Kitaura K (2002) Chem Phys Lett 351:475
91. Nagata T, Fedorov DG, Kitaura K (2009) Chem Phys Lett 475:124
92. Chem W, Gordon MS (1996) J Chem Phys 105:11081
93. Yamaguchi Y, Schaefer HF III, Osamura Y, Goddard J (1994) A new dimension to quantum chemistry: analytical derivative methods in ab initio molecular electronic structure theory. Oxford University Press, New York CAN 123:66488 65-3 General Physical Chemistry USA Book written in English
94. Schmidt MW, Baldridge KK, Boatz JA, Elbert ST, Gordon MS, Jensen JJ, Koseki S, Matsunaga N, Nguyen KA, Su S, Windus TL, Dupuis M, Montgomery JA (1993) J Comput Chem 14:1347
95. Gordon MS, Schmidt MW (2005) Theory and applications of computational chemistry, the first forty years. Elsevier, Amsterdam
96. Pedretti A, Villa L, Vistoli G (2002) J Mol Graph Model 21:47
97. Fedorov DG, Ishida T, Uebayasi M, Kitaura K (2007) J Phys Chem A 111:2722
98. Nose S (1991) Prog Theor Phys Suppl 103:1
99. Adamovic I, Freitag MA, Gordon MS (2003) J Chem Phys 118:6725
100. Fedorov DG, Olson RM, Kitaura K, Gordon MS, Koseki S (2004) J Comput Chem 25:872

101. Jensen JH, Gordon MS (1995) J Am Chem Soc 117:8159
102. Aikens CM, Gordon MS (2006) J Am Chem Soc 128:12835
103. Wada G, Tamura E, Okina M, Nakamura M (1982) Bull Chem Soc Jpn 55:3064
104. Mullin JM, Gordon MS (2009) J Phys Chem B 113:8657
105. Mullin JM, Gordon MS (2009) J Phys Chem B 113:14413

Theor Chem Acc (2012) 131:1145
DOI 10.1007/s00214-012-1145-7

REGULAR ARTICLE

Density-functional expansion methods: grand challenges

Timothy J. Giese · Darrin M. York

Received: 13 May 2011 / Accepted: 9 September 2011 / Published online: 21 February 2012
© Springer-Verlag 2012

Abstract We discuss the source of errors in semiempirical density-functional expansion (VE) methods. In particular, we show that VE methods are capable of well reproducing their standard Kohn-Sham density-functional method counterparts, but suffer from large errors upon using one or more of these approximations: the limited size of the atomic orbital basis, the Slater monopole auxiliary basis description of the response density, and the one- and two-body treatment of the core-Hamiltonian matrix elements. In the process of discussing these approximations and highlighting their symptoms, we introduce a new model that supplements the second-order density-functional tight-binding model with a self-consistent charge-dependent chemical potential equalization correction; we review our recently reported method for generalizing the auxiliary basis description of the atomic orbital response density; and we decompose the first-order potential into a summation of additive atomic components and many-body corrections, and from this examination, we provide new insights and preliminary results that motivate and inspire new approximate treatments of the core-Hamiltonian.

Keywords Tight-binding models · Density-functional theory · Electronic structure

Published as part of the special collection of articles: From quantum mechanics to force fields: new methodologies for the classical simulation of complex systems.

T. J. Giese · D. M. York (✉)
Department of Chemistry and Chemical Biology and BioMaPS Institute for Quantitative Biology, Rutgers University, Piscataway, NJ 08854-8087, USA
e-mail: york@biomaps.rutgers.edu

1 Introduction

A great deal of effort, spanning several decades, has been devoted to the design, implementation, and parametrization of semiempirical quantum models to make them practical, efficient, and accessible to a wide variety of problems [1–22]. As a consequence of the community's success, it can be tempting to dismiss the approximations used in these models as being technical curiosities or to simply underestimate their severity; it is only upon a careful examination of the magnitude of the corrections required to make these methods successful that one gains an appreciation for their ingenuity, and a somewhat worrying sense of amazement that they should even work at all. Although their approximations may be severe, there are only two metrics by which semiempirical models are ultimately judged: they must be *fast* and *accurate*. It is rare for the approximations to be criticized for their severity because the efficiency of the models is founded upon them, and the models can yet be persuaded to reproduce good results. It was for these reasons, after all, that the original developers of the semiempirical models rationalized their use, and it is easy to justify their continued use in the newer models if they are but minor modifications of those original works. This is not to say that the semiempirical development community has stagnated in groupthink, we mean only to highlight the realities that drive the direction of model development. The designer of a next-generation semiempirical model will do so under the influence of one of two prejudices: (1) current models are "fine as they are" and need only minor modifications to their parameters or functional forms; or (2) the parameters of the original semiempirical models have now been sufficiently trained [19], and the parametric freedom of the *ad hoc* functions used to replace the model's missing physics has now been

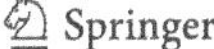 Springer

sufficiently extended [15], that a further significant advance will need to address the fundamental approximations of the method in some inherently new way [23–25]. The former option is attractive because it promises to offer instant gratification and can be more easily demonstrated to be fast and accurate. The latter option is higher risk and potentially requires the development of new mathematical and technical advances to convince people that it is worthwhile.

The above discussion used the words "semiempirical models" and "approximation" vaguely, so it is useful to continue by discussing the effects of a specific approximation used by a specific model. Consider the self-consistent-charge density-functional tight-binding (SCC-DFTB) model [22]: its "core-Hamiltonian" matrix, that is, the matrix of electron kinetic energies and the first-order density-response interactions with the neutral-atom reference, is constructed by an approximation whereby the one-center atomic orbital (AO) blocks presume that the system is but a single atom (itself), and the two-center AO blocks are computed without regard to the presence of any third atom, even if that atom lies directly between the two. Although we feel it is fair to claim that this approximation is severe by using no more than common sense, we have recently examined its use within the framework of Kohn-Sham potential energy expansion (VE) models to quantify its effect [26]. We performed second-order VE calculations with the PBE exchange-correlation (xc) functional [27] and an all-electron 6-31G* AO basis, but which was otherwise free of approximations and parameters, and showed that this reproduces standard PBE/6-31G* results almost exactly; however, when we then applied the above discussed approximation, it caused bond lengths to decrease by about 0.45 Å and angles to have errors in excess of 10°. These poor geometries resulted in dipole moment errors in excess of 50%, and the water dimer was found to be overbound by approximately two orders of magnitude. Although this set of approximations has severe consequences if left unchecked, SCC-DFTB mitigates the adverse effects with surprising success primarily by repelling the atoms with a pairwise additive correction to the energy.

Our goal is not to single-out and criticize SCC-DFTB for the above-mentioned approximation, which has been noted many times in the past [22, 28–30], and other semiempirical models have used conceptually similar approximations. Modified neglect of differential overlap (MNDO) models [11, 13] constructs their core-Hamiltonian matrices using a concept similar to the one described above, even though the theoretical foundations for these methods are quite different [28]. MNDO methods compute the two-center core-Hamiltonian AO blocks from scaled overlap integrals instead of looking up their values from precomputed tables, but like SCC-DFTB, explicit contributions from other atoms are neglected. MNDO's

treatment of the one-center core-Hamiltonian AO blocks is also similar to SCC-DFTB in that it considers the other atoms only insofar as it tries to eliminate the influence of neutral atoms in the resulting "Fock" matrix. And like SCC-DFTB, MNDO relies on the use of pairwise repulsive potentials to retain good geometries. Of course, there are important differences between these methods, such as their treatment of electrostatics and AO overlaps, and these differences should lead one to suspect that each method has varying degrees of success at modeling various properties. Indeed, articles that compare SCC-DFTB and MNDO-like methods suggest that this is true [31], but the differences are such that it is difficult to conclude that one is an overall better method than the other. When we read articles that compare these methods [31], and we ponder their similarities and differences [28], we may verily ask: Could it be that MNDO and SCC-DFTB models are of comparable overall accuracy because their core-Hamiltonians employ conceptually similar approximations, and that these approximations are mitigated in both methods by pairwise additive repulsive energy corrections? Is this approximation a *glass ceiling* that will prevent new models from making significant advances in overall accuracy?

Much of our work has focused on the exploration of new methods and the development of mathematical tools which, together, motivate the design of next-generation VE models [26, 32–37]. These techniques are meant to improve current models by reintroducing the physics that are otherwise only implicitly accounted for. In the process of demonstrating that these techniques add missing physics, we isolate and judge the severity of the underlying approximations used within existing models. To this end, we often find it difficult, and sometimes misleading, to judge the severity of specific approximations by interpreting the failures of parametrized models, because multiple approximations can be parametrized to cancel in unpredictable ways. To minimize the obfuscation of our comparisons, we begin with a parameter-free ab initio-like VE model that reproduces density-functional theory (DFT) results very well and then introduce various approximations to identify where and quantify by how much the model breaks down. The insights that are gained from these explorations, in turn, provide the concepts and motivation for new approximations. Using this strategy, we have concluded [26] that a second-order VE method reproduces standard DFT results extremely well if the model is not encumbered by the following approximations: (1) the limited size of the AO basis, (2) the Slater monopole auxiliary basis representation of the response density, and (3) the one- and two-body treatment of the first-order matrix elements (as discussed above). There are other, more benign approximations, but overcoming the most severe approximations with computational efficiency is *the grand challenge* for Kohn-Sham potential energy expansion models.

Over the years, we have attempted to address these approximations, and we will review and extend some of those works here so as to unify their motivation within the perspective of this grand challenge. The purpose of this work is threefold: (1) We present a new model that supplements second-order SCC-DFTB (DFTB2) with a self-consistent *charge-dependent* dipole polarization correction (DFTB2+CPEQ). This model is based on our earlier work [36, 37] (post-self-consistent field MNDO/d+CPE), but is made self-consistent, and is applied to a new Hamiltonian. In addition, we go beyond the analysis presented in our previous works by making explicit comparison with a *charge-independent* model (DFTB2+CPE0), to ascertain the benefits of the charge dependence. (2) We review our recent work that generalizes the auxiliary basis representation of the response density beyond Slater monopoles [32]. (3) We present new insights into the atomic decomposition of the nonadditive VE reference potential energy and its functional derivatives and discuss how this relates to the modeling of the core-Hamiltonian. Furthermore, we demonstrate how our observations lead to a new mathematical justification for the approximations used within spin-polarized SCC-DFTB models [38].

2 Methods

2.1 Expansion of the Kohn-Sham potential energy

Expansion models are derived from Kohn-Sham DFT through Taylor series expansion [29] of the potential energy [39] $V[\rho, \omega]$ in density response $\delta\rho(\mathbf{r}) = \rho(\mathbf{r}) - \rho_{\text{ref}}(\mathbf{r})$ and spin density $\omega(\mathbf{r}) = \rho^{\alpha}(\mathbf{r}) - \rho^{\beta}(\mathbf{r})$ about the reference density composed from the spin-averaged sum of isolated neutral atoms $\rho_{\text{ref}}(\mathbf{r}) = \sum_a \rho_a(\mathbf{r})$ and $\omega_{\text{ref}}(\mathbf{r}) = 0$. The VE energy $E_{\text{VE}} \equiv E_{\text{VE}}[\rho_{\text{ref}} + \delta\rho, \omega]$ is

$$E_{\text{VE}} = \sum_{ij} P_{ij} T_{ij} + V^{(0)}[\rho_{\text{ref}}] + V^{(1)}[\delta\rho] + W[\delta\rho, \omega], \quad (1)$$

where $\mathbf{P}$ is the AO basis single-particle density matrix; $\mathbf{T}$ is the kinetic energy matrix;

$$V^{(0)}[\rho_{\text{ref}}] = E_{\text{xc}}[\rho_{\text{ref}}, 0] + \sum_{b > a} \frac{Z_a Z_b}{R_{ab}}$$
$$+ J[\rho_{\text{ref}}] - \sum_{ab} \int \frac{Z_a \rho_b(\mathbf{r})}{|\mathbf{r} - \mathbf{R}_a|} d^3 r \quad (2)$$

is the reference potential energy; Z_a is a nuclear charge; $\mathbf{R}_a$ is its atomic position;

$$J[\rho] = \frac{1}{2} \int \int \frac{\rho(\mathbf{r})\rho(\mathbf{r}')}{|\mathbf{r} - \mathbf{r}'|} d^3 r d^3 r' \quad (3)$$

is the Coulomb functional; $E_{\text{xc}}[\rho, \omega]$ is the xc-functional; $V^{(1)}[\delta\rho] = \sum_{ij} (P_{ij} - P_{ij}^{(0)}) V_{ij}^{(1)}$, where

$$V_{ij}^{(1)} = \int \frac{\delta V[\rho, \omega]}{\delta\rho(\mathbf{r})}\bigg|_0 \chi_i(\mathbf{r})\chi_j(\mathbf{r}) d^3 r, \quad (4)$$

is the first-order potential energy; $\mathbf{P}^{(0)}$ is the reference density matrix, which reproduces $\rho_{\text{ref}}(\mathbf{r})$ from the AO basis $\{\chi\}$; the "0" subscript outside functional derivatives indicate its evaluation about the spin-averaged reference; and $W[\delta\rho, \omega]$ is a functional that describes all other response interactions. If the Taylor series expansion was carried out to infinite order, then $W[\delta, \omega]$ would be the second- and all higher-order energy corrections; however, we specifically write $W[\delta, \omega]$ as a generic functional to allow us the notational freedom to insert other empirical correction terms that depend on the response density in some complicated way. We will explore the use of empirical correction terms in later sections and continue here by considering the case of a second-order expansion; that is, $W[\delta\rho, \omega] \rightarrow V^{(2)}[\delta\rho, \omega]$, where

$$V^{(2)}[\delta\rho, \omega] = J[\delta\rho] + \frac{1}{2} \int \int \frac{\delta^2 E_{\text{xc}}[\rho, \omega]}{\delta\rho(\mathbf{r})\delta\rho(\mathbf{r}')}\bigg|_0$$
$$\times \delta\rho(\mathbf{r})\delta\rho(\mathbf{r}') d^3 r d^3 r'$$
$$+ \frac{1}{2} \int \int \frac{\delta^2 E_{\text{xc}}[\rho, \omega]}{\delta\omega(\mathbf{r})\delta\omega(\mathbf{r}')}\bigg|_0$$
$$\times \omega(\mathbf{r})\omega(\mathbf{r}') d^3 r d^3 r'. \quad (5)$$

If $E_{\text{xc}}[\rho, \omega]$ is a local functional of the density, that is, if it does not require the calculation of nonlocal integrals, then the last two terms in Eq. 5 simplify [40],

$$\frac{\delta^2 E_{\text{xc}}[\rho, \omega]}{\delta\rho(\mathbf{r})\delta\rho(\mathbf{r}')}\bigg|_0 = \delta(\mathbf{r} - \mathbf{r}')\frac{\delta^2 E_{\text{xc}}[\rho, \omega]}{\delta\rho(\mathbf{r})\delta\rho(\mathbf{r})}\bigg|_0; \quad (6)$$

and the functional derivatives can be expressed as local functions of space

$$E_{\text{VE}} = \sum_{ij} P_{ij} T_{ij} + V^{(0)}[\rho_{\text{ref}}] + \int v_{\text{ref}}^{(1)}(\mathbf{r})\delta\rho(\mathbf{r}) d^3 r$$
$$+ J[\delta\rho] + \frac{1}{2} \int v_{\text{ref}}^{(2)}(\mathbf{r})\delta\rho(\mathbf{r})^2 d^3 r$$
$$+ \frac{1}{2} \int w_{\text{ref}}^{(2)}(\mathbf{r})\omega(\mathbf{r})^2 d^3 r; \quad (7)$$

where

$$v_{\text{ref}}^{(1)}(\mathbf{r}) = -\sum_a \frac{Z_a}{|\mathbf{r} - \mathbf{R}_a|} + \int \frac{\rho_{\text{ref}}(\mathbf{r}')}{|\mathbf{r} - \mathbf{r}'|} d^3 r'$$
$$+ \frac{\delta E_{\text{PBE}}[\rho, \omega]}{\delta\rho(\mathbf{r})}\bigg|_0, \quad (8)$$

and $v_{\text{ref}}^{(2)}(\mathbf{r})$ and $w_{\text{ref}}^{(2)}(\mathbf{r})$ are the second-functional derivatives of the local xc-functional. Various models can be constructed by applying approximations to the above equations, and these models are described in the next sections.

2.2 Second-order SCC-DFTB

The DFTB2 model [22] differs from Eqs. 1–7 in the following ways:

$$V_{ij}^{(1)} \approx \begin{cases} \int v_a^{(1)}(\mathbf{r})\chi_i(\mathbf{r})\chi_j(\mathbf{r})d^3r, & \text{if } i,j \in a \\ \int v_{ab}^{(1)}(\mathbf{r})\chi_i(\mathbf{r})\chi_j(\mathbf{r})d^3r, & i \in a, j \in b, \end{cases} \tag{9}$$

where $v_a^{(1)}(\mathbf{r}) = \delta V[\delta\rho, \omega]/\delta\rho|_{\rho(\mathbf{r})=\rho_a(\mathbf{r})}$ is the first-order potential of atom a and $v_{ab}^{(1)}(\mathbf{r})$ is the potential of the ab-dimer; the second-order energy is evaluated using a Slater monopole auxiliary basis

$$\delta\rho(\mathbf{r}) = -\sum_a q^a \varphi^a(\mathbf{r} - \mathbf{R}_a), \tag{10}$$

where

$$q^a = -\sum_i \sum_{j \in a}(P_{ij} - P_{ij}^{(0)})\frac{1}{2}S_{ij}$$

$$\qquad -\sum_{i \in a}\sum_j (P_{ij} - P_{ij}^{(0)})\frac{1}{2}S_{ij} \tag{11}$$

is the Mulliken-partitioned partial electric charge of atom a, and $\mathbf{S}$ is the overlap matrix. The second-order energy $V^{(2)}$ is Coulomb-approximated; the auxiliary basis representation transforms the functional into a function of partial charges and atomic positions $V^{(2)}[\delta\rho, \omega] \rightarrow J(\mathbf{q}, \mathbf{R})$, where

$$J(\mathbf{q}, \mathbf{R}) = \sum_{ab} q^a q^b \frac{1}{2}\int\int \frac{\varphi^a(\mathbf{r})\varphi^b(\mathbf{r}')}{|\mathbf{r} - \mathbf{r}'|}d^3r d^3r'. \tag{12}$$

Although this only explicitly models the electrostatics, the second-order xc-energy is implicitly accounted for by choosing the Slater exponents so that the Coulomb self-energies of the atoms reproduce their experimental chemical hardness [30, 41]. The matrix elements (Eq. 9) are evaluated using the atomic densities and minimal valence AOs resulting from isolated atom calculations in the presence of a radial confinement potential, and the xc-potentials are evaluated with the PBE functional [27]. To compensate for the above approximations, the reference potential energy is replaced by a sum of pairwise additive corrections

$$V^{(0)}[\rho_{\text{ref}}] \rightarrow \sum_{b > a} f_{ab}(R_{ab}), \tag{13}$$

which are parametrized to reproduce molecular geometries and bond energies.

The DFTB2 energy is minimized via a standard self-consistent field (SCF) procedure, and the expression for the Fock matrix $F_{ij} = \delta E/\delta P_{ij}$ is

$$F_{ij} = T_{ij} + V_{ij}^{(1)} - \frac{1}{2}S_{ij}\left(\frac{\partial J(\mathbf{q}, \mathbf{R})}{\partial q^a} + \frac{\partial J(\mathbf{q}, \mathbf{R})}{\partial q^b}\right), \tag{14}$$

where $i \in a$ and $j \in b$.

2.3 DFTB2 with self-consistent polarization corrections

This section describes a new method that is similar to Refs. [42–44]; in that, it is founded upon our earlier works [36, 37]. The difference between the present method and Refs. [42–44] is: our new method is charge-dependent; but unlike Refs. [36, 37], the present work is self-consistent and applied to a different Hamiltonian. Our motivation for introducing this method here is to address the first grand challenge: The limited size of the AO basis.

Like all minimal valence basis models, DFTB2 lacks the sufficient AO degrees of freedom to adequately respond to external perturbations, and this manifests itself in its underprediction of dipole polarizabilities [45]. Obviously, one could increase the size of the AO basis in an attempt to alleviate the underlying cause of the problem [46]; however, the benefits of doing so would not be fully realized if the electrostatics were limited to monopolar interactions. In lieu of developing an entirely new model, the dipole polarizabilities of DFTB2 can be improved by including a chemical potential equalization (CPE) auxiliary response density [47], represented by atom-centered dipole functions, whose response to external perturbations is not constrained by the limited size of the AO basis.

The CPE response density is the namesake of *the CPE principle*, sometimes referred to as the electronegativity equalization principle [48–51], and is derived from DFT by Taylor expanding the density functional in density response [52, 53]. In this context, the DFTB2 AO density $\rho_{\text{AO}}(\mathbf{r})$ is the reference density in the CPE Taylor expansion, and the CPE density $\delta\widetilde{\rho}(\mathbf{r})$ is the response about this reference; that is,

$$E[\rho_{\text{AO}} + \delta\widetilde{\rho}] = E_{\text{VE}}[\rho_{\text{AO}}]$$

$$\qquad + \int \frac{\delta E[\rho]}{\delta\rho(\mathbf{r})}\bigg|_{\rho_{\text{AO}}} \delta\widetilde{\rho}(\mathbf{r})d^3r$$

$$\qquad + \frac{1}{2}\int\int \frac{\delta^2 E[\rho]}{\delta\rho(\mathbf{r})\delta\rho(\mathbf{r}')}\bigg|_{\rho_{\text{AO}}}$$

$$\qquad \times \delta\widetilde{\rho}(\mathbf{r})\delta\widetilde{\rho}(\mathbf{r}')d^3r d^3r', \tag{15}$$

and the CPE correction to the VE energy is

$$E_{\text{CPE}}[\rho_{\text{AO}}, \delta\widetilde{\rho}] \equiv E[\rho_{\text{AO}} + \delta\widetilde{\rho}] - E_{\text{VE}}[\rho_{\text{AO}}]. \tag{16}$$

Although they are both Taylor series expansions, Eq. 15 is fundamentally different from Eqs. 1, 15 expands the total density functional, including the electron kinetic energy, as opposed to the Kohn-Sham potential energy functional 39. This distinction is important because it has serious implications on the ability of the model to describe covalent bonding. Equation (15) is not a reasonable model for describing covalent bonds, which are difficult to describe outside of the linear combination of atomic orbitals (LCAO) approach, because the kinetic energy is not well modeled through the density alone [54]. Fortunately, we can rely on DFTB2 for that purpose and use the CPE correction to account for the missing response due to those interactions that are sufficiently separated that they can be accurately modeled by electrostatics only. This is, in fact, what the polarizabilities ultimately *are*: measures of response due to distant electrostatic perturbations. In particular, the isotropic dipole polarizability of a molecule is $\alpha = (\alpha_{xx} + \alpha_{yy} + \alpha_{zz})/3$, where

$$\alpha_{xy} = \frac{\partial^2 E}{\partial E_x \partial E_y} = -\frac{\partial q_x^{(1)}}{\partial E_y}, \tag{17}$$

and

$$q_x^{(1)} = \int x \left[\sum_a \frac{Z_a}{|\mathbf{r} - \mathbf{R}_a|} - \rho_{AO}(\mathbf{r}) - \delta\widetilde{\rho}(\mathbf{r}) \right] d^3 r \tag{18}$$

and E_x are the electric dipole moment and field strength in the x-direction, respectively.

For the reasons discussed above, we approximate $E[\rho] \approx J[\rho]$ and damp the short-range Coulomb interactions between the CPE density and the DFTB2 AO density, because the short-range interactions of the density are better modeled by the underlying DFTB2 model. Furthermore, being that only the well-separated Coulomb interactions are treated, we ignore the underlying neutral-atom DFTB2 reference; only the DFTB2 partial atomic charges (and any applied electric field) invoke a CPE response $\rho_{AO}(\mathbf{r}) \approx -\sum_a q^a \varphi^a(\mathbf{r})$. The CPE response is represented by atom-centered primitive Gaussian-dipole functions $\{\widetilde{\varphi}\}$

$$\delta\widetilde{\rho}(\mathbf{r}) = -\sum_a \sum_\mu d_\mu^a \widetilde{\varphi}_\mu^a(\mathbf{r} - \mathbf{R}_a)$$

$$= -\sum_a \sum_{\mu=-1}^{1} d_\mu^a C_{1,\mu}(\nabla_a) \left(\frac{\zeta_a}{\pi} \right)^{3/2} e^{-\zeta_a |\mathbf{r} - \mathbf{R}_a|^2}, \tag{19}$$

where $C_{l\mu}(\nabla_a)$ is the spherical tensor gradient operator [55] acting on $\mathbf{R}_a$, and $\mathbf{d}$ is the vector of atomic dipole moments. The basis representation of $\delta\widetilde{\rho}(\mathbf{r})$ allows us to rewrite the CPE correction energy in a simple algebraic form $E_{CPE}[\rho_{AO}, \delta\widetilde{\rho}] \to E_{CPE}(\mathbf{q}, \mathbf{d}, \mathbf{R})$; that is,

$$E_{CPE}(\mathbf{q}, \mathbf{d}, \mathbf{R}) = \mathbf{d}^T \cdot \mathbf{B} \cdot \mathbf{q} + \frac{1}{2} \mathbf{d}^T \cdot \mathbf{A} \cdot \mathbf{d}, \tag{20}$$

where

$$B_{kb} = D(R_{ab}) \int \int \frac{\widetilde{\varphi}_k^a(\mathbf{r}) \varphi^b(\mathbf{r}')}{|\mathbf{r} - \mathbf{r}'|} d^3 r d^3 r' \tag{21}$$

is a $3N \times N$ matrix of scaled Coulomb interactions between the CPE Gaussian-dipole and DFTB2 AO Slater monopole auxiliary basis sets;

$$A_{kk'} = \int \int \frac{\widetilde{\varphi}_k^a(\mathbf{r}) \widetilde{\varphi}_{k'}^b(\mathbf{r}')}{|\mathbf{r} - \mathbf{r}'|} d^3 r d^3 r' \tag{22}$$

is a $3N \times 3N$ matrix of Gaussian-dipole Coulomb interactions; and $D(R_{ab})$ is a switching function that removes the short-range Coulomb interactions. This switching function has the form: $D(R_{ab}) = 1$, if $R_{ab} > R_{hi}$; $D(R_{ab}) = 0$, if $R_{ab} < R_{lo}$; and

$$D(R_{ab}) = 1 - 10x^3 + 15x^4 - 6x^5, \tag{23}$$

otherwise; where $x = (R_{hi} - R_{ab}) / (R_{hi} - R_{lo})$; and $R_{hi} = R_{hi,a} + R_{hi,b}$ and $R_{lo} = R_{lo,a} + R_{lo,b}$ are a sum of atom parameters.

The CPE correction energy is minimized by solving $\partial E_{CPE}(\mathbf{q}, \mathbf{d}, \mathbf{R})/\partial d_k^a = 0$, which has an analytic result:

$$\mathbf{d} = -\mathbf{A}^{-1} \cdot \mathbf{B} \cdot \mathbf{q}. \tag{24}$$

In summary, given the short-range damping parameters $R_{lo,a}$ and $R_{hi,a}$, and Gaussian-dipole exponents ζ_a, the matrices $\mathbf{A}$ and $\mathbf{B}$ are computed (Eqs. 21–22), the response is then determined (Eq. 24), and the energy correction is then evaluated (Eq. 20).

The CPE energy correction is a modification of the W-functional, that is, $W[\delta\rho, \omega] \to J(\mathbf{q}, \mathbf{R}) + E_{CPE}(\mathbf{q}, \mathbf{d}, \mathbf{R})$, and depends on the density matrix through $\mathbf{q}$. Therefore, its derivatives with respect to $\mathbf{q}$ enter the Fock matrix;

$$\begin{aligned} F_{ij} = T_{ij} + V_{ij}^{(1)} \\ - \frac{1}{2} S_{ij} \left(\frac{\partial J(\mathbf{q}, \mathbf{R})}{\partial q^a} + \frac{\partial E_{CPE}(\mathbf{q}, \mathbf{d}, \mathbf{R})}{\partial q^a} \right. \\ \left. + \frac{\partial J(\mathbf{q}, \mathbf{R})}{\partial q^b} + \frac{\partial E_{CPE}(\mathbf{q}, \mathbf{d}, \mathbf{R})}{\partial q^b} \right). \end{aligned} \tag{25}$$

CPE0: The charge-independent model. If the Gaussian-dipole exponents did not depend on charge, then

$$\frac{\partial E_{CPE0}(\mathbf{q}, \mathbf{d}, \mathbf{R})}{\partial q^a} = \sum_b \sum_{k \in b} d_k^b B_{ka}. \tag{26}$$

CPEQ: The charge-dependent model. In this model, the CPE Gaussian-dipole exponent is made a function of the underlying DFTB2 partial charge

$$\zeta_a(q^a) = \zeta_a(0) e^{b_a q^a}, \tag{27}$$

where $\zeta_a(0)$ and b_a are atom parameters.

The purpose for making the model charge-dependent is to introduce some elementary, missing physics. In particular, anions are known to be more polarizable than neutral atoms, which are more polarizable than cations. The atomic polarizabilities of minimal valence basis models tend to have the opposite behavior because, as the AO degrees of freedom are extinguished, there is less opportunity to respond to external perturbations. Equation 27 reproduces the expected behavior when $b_a > 0$: as an atom becomes anionic $q^a < 0$, $\zeta_a(q^a)$ becomes smaller; the Coulomb self-energy A_{kk} decreases; and therefore, the polarizability increases; that is, the CPE polarizability correction of an isolated atom is inversely proportional to the dipole self-energy, $\Delta \alpha = 1/A_{kk}$.

Notice that the charge dependence takes an exponential-, as opposed to linear form. One can compute the polarizability of atoms and ions using standard ab initio methods to convince themselves of the exponential behavior, but it can also be gleaned intuitively from tables of experimental atomic electron affinities and ionization potentials, which can be used to plot the energy of an atom as a function of charge state. These plots are monotonic and exponential-like and are interpreted to suggest certain truisms: Excess electrons are less bound to the atom and are thus sensitive to external perturbations. The opposite is true when there are a deficient number of electrons. Moreover, the inner-shell electrons are successively bound with ever-increasing strength to the nucleus, making their response to external perturbations exponentially irrelevant.

Having now explained the exponential behavior and appreciating that, by doing so, this behavior now seems obvious; the reader may question why its explanation is even necessary. Our reason for explaining this in detail is because there have been recent works [56, 57] that may cause some readers to believe that the justification for including charge dependence is founded in the expansion of the Taylor series to third order; because, in this way, the second-order energy can be corrected for *linear* changes in the charge. This, of course, is a mathematically sound way for introducing charge dependence; however, it is unfortunate that a reader can easily misconstrue it into suggesting that the problem at hand is a symptom of a premature truncation of the Taylor series expansion. The symptoms, however, result from a poor auxiliary description of the response density and its restricted response in a limited AO basis. Indeed, a VE model with a sufficient AO and auxiliary basis, but in the absence of other severe approximations, reproduces closed-shell standard DFT results very well even when the VE Taylor expansion of the xc-functional is truncated at first order!

By making the ζ_a's a function of the DFTB2 partial charges, the $\mathbf{A}$ and $\mathbf{B}$ matrix elements now have an explicit

$\mathbf{q}$-dependence, and this complicates the Fock matrix correction

$$\frac{\partial E_{\mathrm{CPEQ}}(\mathbf{q}, \mathbf{d}, \mathbf{R})}{\partial q^a} = \sum_b \sum_{k \in b} d_k^b B_{ka}$$
$$+ \sum_{k \in a} \sum_b d_k^a \frac{\partial B_{kb}}{\partial q^a} q^b$$
$$+ \sum_{k \in a} \sum_b \sum_{k' \in b} d_k^a \frac{1}{2} \frac{\partial A_{kk'}}{\partial q^a} d_{k'}^b. \qquad (28)$$

If the only dependence of the matrix elements are through the CPE Gaussian-dipole exponent, as is the case for DFTB2, then their required derivatives can be expressed solely through a chain rule of those exponents; for example,

$$\frac{\partial B_{kb}}{\partial q^a} = \frac{\partial B_{kb}}{\partial \zeta_a}\bigg|_{\zeta_a = \zeta_a(q^a)} \frac{\partial \zeta_a(q^a)}{\partial q^a}, \quad \text{for } k \in a; \qquad (29)$$

however, if one were to apply the CPE correction to a third-order SCC-DFTB model (DFTB3) [56–58], whose Slater monopole exponents also have a $\mathbf{q}$-dependence, then an additional derivative of $\mathbf{B}$ would be necessary.

2.4 PBE/6-31G*-based VE models

As discussed above, DFTB2 is limited in both the completeness of its AO basis and the electrostatic description of its response density. In the previous sections, we described how one can try to mask these limitations by making a modest change or extension to the model. In this section, we describe a series of models that approach the problem from a different perspective: we start with a base model that reproduces standard PBE/6-31G* very well and then apply approximations that worsen the model. This allows us to better identify and quantify the limitations of the model, motivate and test new corrections, and convincingly demonstrate that those corrections reintroduce missing physics.

To describe these models, it is useful to discuss how they differ from a base model, which we call VE. As the name suggests, it is a direct application of the VE method (Eqs. 1–8), and it is our best attempt at avoiding all other approximations. The VE model is evaluated with the AOs and reference density resulting from standard PBE/6-31G* calculations, uses the PBE xc-functional within the zeroth-order reference energy and first-order potential, and uses the SPW92 xc-functional [59] within the second-order terms; that is, the second-order potentials, $v_{\mathrm{ref}}^{(2)}(\mathbf{r})$ and $w_{\mathrm{ref}}^{(2)}(\mathbf{r})$, are the second functional derivatives of SPW92. Had we used the PBE xc-functional for the second functional derivative and carried out the Taylor series expansion to infinite order, we would expect VE to exactly reproduce standard PBE/6-31G*; therefore, we will compare all of our VE models to standard PBE/6-31G*.

The other models that we describe employ various approximations; some of which are more severe than others. The benign approximations are relevant in our examinations insofar as they are commonly used in existing semiempirical models and yet provide certain computational advantages. Therefore, we will also introduce more severe approximations upon those, and in these cases, we will continue to compute their errors relative to standard PBE/6-31G*, but with the understanding that, even if we remedied the severe approximations with a new technique, we'd still expect the model to suffer from the errors attributed to the more benign approximations.

The remaining, closely related models are summarized below. Some of the approximations used in these models require more explanation than what is described here, and we defer those explanations to the ensuing sections that follow.

VE: The base model described by Eqs. 1–8, where the AO basis and reference densities result from 6-31G* atom calculations, and $E_{xc}[\rho, \omega] = E_{PBE}[\rho, \omega]$, except for the second functional derivatives, which use $E_{xc}[\rho, \omega] = E_{SPW92}[\rho, \omega]$. This model differs from standard PBE/6-31G* because of our truncation of the Taylor expansion to second order, and our uniform gas xc-approximation for the second functional derivatives.

VE0: The VE model, where the reference potential energy is approximated by a cluster expansion, that is,

$$V[\rho_{\text{ref}}, 0] \approx \sum_a V_a^{(0)} + \sum_{b>a} \Delta V_{ab}^{(0)}, \tag{30}$$

where

$$\Delta V_{ab}^{(0)} = V_{ab}^{(0)} - V_a^{(0)} - V_b^{(0)}, \tag{31}$$

$V_{ab}^{(0)} \equiv V[\rho_a + \rho_b, 0]$, and $V_a^{(0)} \equiv V[\rho_a, 0]$. If the cluster expansion was exact, then we would expect VE0 reproduce the VE results

VE1: The VE0 model, where the first-order potential energy is evaluated using a one- and two-body approximation (Eq. 9). If Eq. 9 was exact, we would expect VE1 to reproduce the VE0 results.

VE1S: The VE0 model, where the first-order potential (Eq. 8) is approximated by a sum of isolated atom potentials; that is, $v_{\text{ref}}^{(1)}(\mathbf{r}) \approx \sum_a v_a^{(1)}(\mathbf{r})$. If this approximation was exact, we would expect VE1S to reproduce the VE0 results.

VE1W: The VE0 model, where the first-order potential is approximated by a weighted cluster expansion (Eq. 48), described later in this text. If Eq. 48 was exact, we would expect VE1W to reproduce the VE0 results.

VEJ: The VE model, where the second-order xc-potentials are completely ignored; that is, $v_{\text{ref}}^{(2)}(\mathbf{r}) = w_{\text{ref}}^{(2)}(\mathbf{r}) = 0$. VEJ cannot describe spin polarization at all.

VEJ/1S(M): The VEJ model, where the AO products are represented in an auxiliary basis of atom-centered Slater monopole functions, whose exponents are chosen to reproduce the experimental hardness of the atoms. This auxiliary basis is analogous to that used in DFTB2; like DFTB2, the atomic charges are chosen by Mulliken partitioning. If the auxiliary basis exactly reproduced the AO products, then we'd expect VEJ/1S(M) to reproduce the VEJ results.

VEJ/2P,3D,4F: The VEJ model, where the AO products are represented with an auxiliary basis of Gaussian-multipole expansions (GME)

$$\varphi_{lm}^a(\mathbf{r}) = \frac{C_{lm}(\nabla_a)}{(2l-1)!!}$$
$$\times \sum_{\gamma=1}^{3} c_\gamma \left(\frac{\zeta_\gamma(\xi)}{\pi}\right)^{3/2} e^{-\zeta_\gamma(\xi)|\mathbf{r}-\mathbf{R}_a|^2}, \tag{32}$$

whose primitive contraction coefficients c_γ, and exponents $\zeta_\gamma(\xi)$, have been chosen to mimic a Slater multipole-like expansion with exponent ξ. The notation 2P,3D,4F indicates that period 1 elements use two GMEs of $l_{\max} = 1$, each corresponding to one of two Slater exponents; period 2 elements use three GMEs of $l_{\max} = 2$; and period 3 elements use four GMEs of $l_{\max} = 3$. The relationship between this generalized auxiliary basis and the AO products is discussed in the next section and described completely in Ref. [32]. If the auxiliary basis exactly reproduced the AO products, then we'd expect it to match VEJ's results.

2.5 Generalization of the auxiliary basis

This section reviews a method that has only recently been reported in Ref. [32]. Our purpose for reviewing that work here is to unify its motivation with the second grand challenge: The limited auxiliary basis description of the response density.

DFTB2 explicitly treats the electrostatic interactions of the response density and is therefore a major advance beyond the traditional first-order tight-binding models whence it is based. Not only does the reintroduction of this missing physics improve the description of interatomic interactions, but it makes DFTB2 a more transferable and accurate predictor of atomic charges. The DFTB2 response density is represented in a basis of Slater monopoles, whose exponents are chosen to reproduce the atom's chemical hardness [60, 61]. The response density is made to interact with itself electrostatically, and thus, the DFTB2 response density self-energy is a quadratic function of atomic partial charges.

The electron affinity and ionization potentials of an atom suggest that the energy of an atom is not a pure-quadratic function of charge, but is instead asymmetrical. As electrons are removed from the atom, the chemical hardness increases; because the next electron to be removed is closer to the nucleus, its interaction with the nucleus is stronger and thus requires even greater energy to eject. This asymmetry might motivate one to try and improve DFTB2 by making the Slater exponent a function of charge, so that the Slater self-energies mimic an exponential-like or asymmetrical behavior. Invoking charge dependence in this way is an ad hoc correction. By this, we mean that the charge dependence is being used to alleviate a particular symptom without regard to the underlying problem that gives rise to that symptom. As such, there is no limitation on the functional form of the charge dependence; that is, one could make the Slater exponents a linear function of charge, or an exponential function of charge, or a sixth-order spline polynomial of charge, etc.

DFTB3 improves upon DFTB2 by making the Slater exponents a linear function of charge and then justifies this mathematical form by extending the Taylor series expansion of the Kohn-Sham potential energy to third order [56–58]. When reading those works [56–58], a reader may misconstrue the use of a third-order expansion to mean that the underlying problem giving rise to the quadratic behavior of the hardness is a premature truncation of the Taylor series to second order. Our previous works suggest that this is not the case [26, 32]. The pure-quadratic response self-energy of an isolated atom, as modeled by DFTB2, results only from having limited the auxiliary basis so completely that both the diffuse and tightly bound electron densities are represented by the same spatial function and thus contribute equally to the self-energy. One would instead intuitively expect the excess electrons to be more diffuse and therefore contribute less to the atom's self-energy than the more tightly bound electrons.

We attribute the source of the error to the incomplete AO and auxiliary basis being used and *not* the truncation of the expansion. We've come to this conclusion by repeatedly testing the following elementary logic [26, 32]: (1) If the source of error was the lack of a third-order term, then the error should continue to manifest itself when all other approximations of the model are removed. (2) If a second-order model is used, and the error ceases to manifest itself when some approximation (other than the expansion order) is removed, then the source of the error is not the second-order truncation. As a simple demonstration, Fig. 1 plots the hardness of oxygen as a function of charge using a variety of approximations. The VE model lacks third- and higher-order expansion terms, but it agrees extremely well

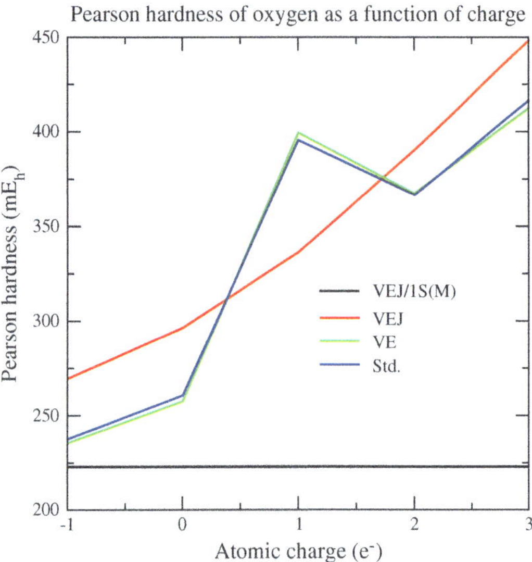

Fig. 1 The Pearson hardness of oxygen atom and ions for various models. Let $E(q)$ represent the energy of oxygen with an electric charge of q. We then compute the hardness as $[E(q-1) - E(q+1)]/2$. When the auxiliary basis is limited to a single, charge-independent Slater function [VEJ/1S(M)], the hardness is constant. The hardness becomes variable when more auxiliary functions are used (VEJ is equivalent to having used a complete auxiliary basis). Furthermore, upon including second-order LDA-xc (VE), the second-order model reproduces standard PBE/6-31G* (Std.) extremely well. By having removed the other approximations in the model and showing a strong agreement between the second-order Kohn-Sham expansion with its standard DFT counterpart, we conclude that the second-order expansion is not a premature truncation of the Kohn-Sham potential energy. If the lack of a third-order expansion term introduced errors of any consequence, then we'd expect those errors to continue to manifest themselves when the other approximations of the model are removed

with the standard PBE/6-31G* results; therefore, the lack of third- and higher-order expansion terms is not the source of error preventing an asymmetrical charge dependence.

From the above discussion, it should now be clear that an avenue for improving DFTB2, as an alternative to explicitly modeling a charge dependence or extending the order of the Taylor series expansion, would be to simply include more Slater functions: those which model the diffuse electrons, and those that model the inner-electrons. Furthermore, one could include higher-order multipoles, as this would: allow for an improved description of dipole polarizabilities, provided that the AO degrees of freedom were available; improve the strength and angular dependence of hydrogen bonds; and improve the nonbonded interactions, in general [62].

Of course, this is all rather easy to say, but how would you actually do it and still have it be practical? Although auxiliary basis sets have been used in traditional ab initio methods, the techniques they use typically require three-center integrals [63–69], whereas the efficiency of SCC-DFTB's technique is that it requires only two-center integrals and thus can be precomputed and splined as a function of atom separation. Our approach was chosen to preserve this property, and we call it a generalization of the auxiliary basis because, like SCC-DFTB, we will use the auxiliary basis to directly model the AO products.

Some readers might insist that SCC-DFTB is modeling the atomic response densities, *not the individual AO products!* So, to convince those readers, and to more clearly demonstrate how our method is a generalization, let us insert Eq. 11 into (10)

$$\delta\rho(\mathbf{r}) = \sum_a \sum_{ij\in a}(P_{ij} - P_{ij}^{(0)})\left[S_{ij}\varphi^a(\mathbf{r} - \mathbf{R}_a)\right]$$
$$+ \sum_{a\neq b}\sum_{i\in a}\sum_{j\in b}(P_{ij} - P_{ij}^{(0)})$$
$$\times \left[\frac{1}{2}S_{ij}\varphi^a(\mathbf{r} - \mathbf{R}_a) + \frac{1}{2}S_{ij}\varphi^b(\mathbf{r} - \mathbf{R}_b)\right]; \tag{33}$$

and now, upon comparing this to

$$\delta\rho(\mathbf{r}) = \sum_{ij}(P_{ij} - P_{ij}^{(0)})\chi_i(\mathbf{r})\chi_j(\mathbf{r}); \tag{34}$$

we see that $\chi_i(\mathbf{r})\chi_j(\mathbf{r}) \approx S_{ij}\varphi^a(\mathbf{r} - \mathbf{R}_a)$, for one-center AO products; and

$$\chi_i(\mathbf{r})\chi_j(\mathbf{r}) \approx \frac{1}{2}S_{ij}\varphi^a(\mathbf{r} - \mathbf{R}_a)$$
$$+ \frac{1}{2}S_{ij}\varphi^b(\mathbf{r} - \mathbf{R}_b), \tag{35}$$

for two-center AO products. The factors of S_{ij} and $S_{ij}/2$ are the coefficients of a simple partitioning that "map" the AO product onto the atoms' auxiliary basis.

Our generalization is to simply use more auxiliary functions; that is,

$$\chi_i(\mathbf{r})\chi_j(\mathbf{r}) \approx \sum_{k\in a}M_{ijk}^a\varphi_k^a(\mathbf{r} - \mathbf{R}_a), \tag{36}$$

for one-center AO products; and

$$\chi_i(\mathbf{r})\chi_j(\mathbf{r}) \approx \sum_{k\in a}M_{ijk}^a\varphi_k^a(\mathbf{r} - \mathbf{R}_a)$$
$$+ \sum_{k\in b}M_{ijk}^b\varphi_k^b(\mathbf{r} - \mathbf{R}_b), \tag{37}$$

for two-center AO products, where $\mathbf{M}^a$ and $\mathbf{M}^b$ are matrices of mapping coefficients, which depend on the orientation and separation of the two atoms.

When the two atoms are aligned along the z-axis, and the angular dependence of the AO and auxiliary functions are described by spherical harmonics, many of the mapping coefficients vanish due to symmetry [20]. Therefore, one can spline the nonzero values of these matrices as a function of separation along the z-axis, interpolate their values for a given separation, and then use Wigner-D matrices to

rotate those interpolated matrices into the molecular orientation [70].

The values of the nonzero matrix elements in the z-axis orientation are a *choice*. For example, we've described DFTB2 as using a Mulliken partitioning, but some have chosen to use CM3 charge mappings in their post-SCF analysis [71]. We have chosen to use is a constrained electrostatic fitting procedure [63, 67, 72–76]. The mathematical details of this procedure are fully presented in Ref. [32]; and one may choose differently; so we discuss this only briefly here. An electrostatic fitting procedure chooses the mapping coefficients in an attempt to reproduce the electrostatic potential of the AO product everywhere in space. It is a variational procedure that can be understood as minimizing the vector-norm squared difference between the AO product electric field with its auxiliary basis representation everywhere. By saying that we use a *constrained* electrostatic fitting procedure, we mean that Lagrange multipliers have been included within the variational solution to preserve the AO product multipole moments.

Just as DFTB2's Mulliken partitioning yielded partial atomic charges, the generalized auxiliary basis yields vectors of atomic charge moments, $\mathbf{m}^a : \delta\rho(\mathbf{r}) = -\sum_a\sum_{k\in a}m_k^a\varphi_k^a(\mathbf{r})$, where

$$m_k^a = -\sum_{ij\in a}(P_{ij} - P_{ij}^{(0)})M_{ijk}^a$$
$$- \sum_{i\in a}\sum_{j\notin a}(P_{ij} - P_{ij}^{(0)})M_{ijk}^a$$
$$- \sum_{i\notin a}\sum_{j\in a}(P_{ij} - P_{ij}^{(0)})M_{ijk}^a. \tag{38}$$

The spin density can be represented in an analogous way: $\omega(\mathbf{r}) = -\sum_a\sum_{k\in a}s_k^a\varphi_k^a(\mathbf{r})$, where the auxiliary spin-charge moment of atom a is

$$s_k^a = -\sum_{ij\in a}(P_{ij}^\alpha - P_{ij}^\beta)M_{ijk}^a$$
$$- \sum_{i\in a}\sum_{j\notin a}(P_{ij}^\alpha - P_{ij}^\beta)M_{ijk}^a$$
$$- \sum_{i\notin a}\sum_{j\in a}(P_{ij}^\alpha - P_{ij}^\beta)M_{ijk}^a. \tag{39}$$

These auxiliary representations of the densities transform the $W[\delta\rho, \omega]$ functional into a function of auxiliary charge moments, spin-charge moments, and atomic positions; $W[\delta\rho, \omega] \to W(\mathbf{m}, \mathbf{s}, \mathbf{R})$; and the σ-spin Fock matrix generalizes to

$$F_{ij}^\sigma = T_{ij} + V_{ij}^{(1)} - \sum_a\sum_{k\in a}M_{ijk}^a\phi_k^{a,\sigma}, \tag{40}$$

for one-center products; and

59

$$F_{ij}^{\sigma} = T_{ij} + V_{ij}^{(1)} - \sum_a \sum_{k \in a} M_{ijk}^a \phi_k^{a,\sigma}$$

$$- \sum_b \sum_{k \in b} M_{ijk}^b \phi_k^{b,\sigma}, \tag{41}$$

for two-center products; where

$$\phi_k^{a,\sigma} = \frac{\partial W(\mathbf{m}, \mathbf{s}, \mathbf{R})}{\partial m_k^a} + (-1)^{\delta_{\sigma,\beta}} \frac{\partial W(\mathbf{m}, \mathbf{s}, \mathbf{R})}{\partial s_k^a}, \tag{42}$$

is a spin-resolved auxiliary charge potential.

The VEJ/2P,3D,4F model computes the mapping coefficients as described above; its second-order energy is a Coulomb approximation; and the generalized auxiliary basis transforms Eq. 12 into

$$J(\mathbf{q}, \mathbf{R}) = \frac{1}{2} \sum_{ab} \sum_{k \in a} \sum_{k' \in b} m_k^a m_{k'}^b$$

$$\times \int \int \frac{\varphi_k^a(\mathbf{r}) \varphi_{k'}^b(\mathbf{r}')}{|\mathbf{r} - \mathbf{r}'|} d^3r \, d^3r'. \tag{43}$$

2.6 Decomposition of the nonadditive xc-potentials

This section introduces new, preliminary methods that we've been using to gain insight into the decomposition of the xc-potentials. Our purpose for describing them here is to motivate and inspire the development of new approximations that address the third grand challenge: Moving beyond the one- and two-body treatment of the first-order matrix elements.

A topic on which we have yet to publish, and which is a topic that few people have ever attempted to address, is new approximations that go beyond the one- and two-center treatment of $\mathbf{V}^{(1)}$ and approximations that consider the many-body character of the second-order VE energy [77]. We believe that this is because it is not immediately apparent how these nonadditive potentials decompose into one-centered functions and two-centered corrections, or whether they can be adequately expressed in that manner at all. This brings us to the purpose of this section: we will determine how the nonadditivity of the first- and second-order potentials decompose into atomic contributions. Our motivation is to gain the insight necessary to propose new VE integral approximations that explicitly consider all the atoms in the system and to develop the new mathematical techniques that those proposals would require.

What new mathematical techniques potentially need to be developed, and how would that be aided by the present exploration? (1) If the first-order potential was an additive function, or approximated as such, then the contribution of a third-atom requires the potential of only that atom. (2) If one needs to explicitly treat the three-body nonadditivity, then one ultimately requires the potential of atom triplets. (3) If the first-order potential can be expressed as a sum of one-body potentials, *or a sum of one-body potentials and two-body corrections*, then there is some possibility of reducing the calculation of $\mathbf{V}^{(1)}$ to one that requires no more than two-center integrals, even if it requires a loop over third atoms. There are various mathematical tricks that would could try to use for this purpose; for example, using auxiliary basis representations of the AOs and/or first-order potential, or using resolution-of-the-identity techniques [69]. But the problem we are trying to emphasize here is a general lack of insight as to whether it is even possible to avoid many-body corrections [as in case (2)] or not; and if so, how?

Previous descriptions of DFTB2 [30] have attempted to rationalize the use of Eq. 9 by explaining how it differs from a superposition of atom-centered potentials

$$v_{\text{ref}}^{(1)}(\mathbf{r}) \approx \sum_a v_a^{(1)}(\mathbf{r}). \tag{44}$$

Although Eq. 44 ignores the nonadditivity of the xc-potential, it at least recognizes that the system is composed of more than one or two atoms. Being that the only nonadditive component of $v_{\text{ref}}^{(1)}(\mathbf{r})$ is the xc-potential, it begs to question whether or not the error introduced by ignoring nonadditivity in Eq. 44 is acceptable. It is for this reason that we construct the VE1S model and include it in our comparisons within Table 1.

The potential resulting from Eq. 44 is too negative relative to the exact potential; that is, $v_{\text{ref}}^{(1)}(\mathbf{r}) - \sum_a v_a^{(1)}(\mathbf{r}) > 0$; however, a nonadditive multicenter function can be *cluster expanded* into additive single-center components and many-body corrections; for example,

$$v_{\text{ref}}^{(1)}(\mathbf{r}) = \sum_a v_a^{(1)}(\mathbf{r}) + \sum_{b > a} \Delta v_{ab}^{(1)}(\mathbf{r})$$

$$+ \sum_{c > b > a} \Delta v_{abc}^{(1)}(\mathbf{r}) + \cdots \tag{45}$$

where $v_{\text{ref}}^{(1)}(\mathbf{r}) - \sum_a v_a^{(1)}(\mathbf{r})$ is the nonadditive behavior of $v_{\text{ref}}^{(1)}(\mathbf{r})$, and

$$\Delta v_{ab}^{(1)}(\mathbf{r}) = v_{ab}^{(1)}(\mathbf{r}) - v_a^{(1)}(\mathbf{r}) - v_b^{(1)}(\mathbf{r}), \tag{46}$$

$$\Delta v_{abc}^{(1)}(\mathbf{r}) = v_{abc}^{(1)}(\mathbf{r}) - \Delta v_{ab}^{(1)}(\mathbf{r})$$

$$- \Delta v_{ac}^{(1)}(\mathbf{r}) - \Delta v_{bc}^{(1)}(\mathbf{r})$$

$$- v_a^{(1)}(\mathbf{r}) - v_b^{(1)}(\mathbf{r}) - v_c^{(1)}(\mathbf{r}), \tag{74}$$

are two- and three-body corrections, respectively. The two-body and larger-body corrections owe their existence solely to the nonadditivity of the xc-functional; therefore, in subsequent discussion, we refer to the nonadditivity of the first-order potential or the xc-potential interchangeably, depending on context.

Although the one-body sum (Eq. 44) is too negative (see e.g., Fig. 2), our experience is that the two-body

Table 1 Bond, angle, and dipole moment errors relative to standard PBE/6-31G* results for 52 molecules upon geometry optimization

Model	Bond μ_{SE}	Bond μ_{UE}	Angle μ_{SE}	Angle μ_{UE}	Dipole μ_{SE}	Dipole μ_{UE}	ΔE_{rxn} μ_{SE}	ΔE_{rxn} μ_{UE}
N	114		98		52		6	
Avg.	1.322 Å		112.546°		0.496 a.u.		158.637 kcal/mol	
VE	0.002	0.002	−0.006	0.177	−0.007	0.012	0.076	0.221
VE0	−0.013	0.014	−0.629	0.858	−0.005	0.015	−8.408	11.708
VE1	−0.473	0.473	0.785	10.666	0.120	0.373	369.505	511.706
VE1S	−0.128	>0.130	−1.497	>6.070	0.030	>0.185	−79.629	>130.563
VE1W	−0.008	0.011	−0.454	0.827	−0.006	0.013	−0.166	11.621
VEJ	0.012	0.013	−0.198	0.564	−0.045	0.052	−1.161	1.615
VEJ/1S(M)	−0.045	0.046	0.384	1.148	−0.207	0.257	−15.165	15.165
VEJ/2P,3D,4F	0.013	0.014	−0.258	0.577	−0.039	0.053	0.045	1.337

The statistics under the heading ΔE_{rxn} are adiabatic SCF energy differences corresponding to the homolytic bond dissociation reactions: $C_2H_6 + H_2 \rightarrow 2CH_4$, $C_2H_4 + 2H_2 \rightarrow 2CH_4$, $C_2H_2 + 3H_2 \rightarrow 2CH_4$, $N_2H_4 + H_2 \rightarrow 2NH_3$, $Si_2H_6 + H_2 \rightarrow 2SiH_4$, and $H_2O_2 + H_2 \rightarrow 2H_2O$. μ_{SE} and μ_{UE} are the signed and unsigned errors of the property, respectively, and whose units are shown in the row "Avg.", which displays the average reference value. "N" is the amount of data included in the statistics. The VE1S statistics exclude any contributions from Si_2H_6, which geometry optimizes into a hopelessly nonsensical configuration where all 6 hydrogens are clustered between the two silicons, at which point we can no longer coax SCF convergence. Thus, we expect the true VE1S errors to be larger than those reported here

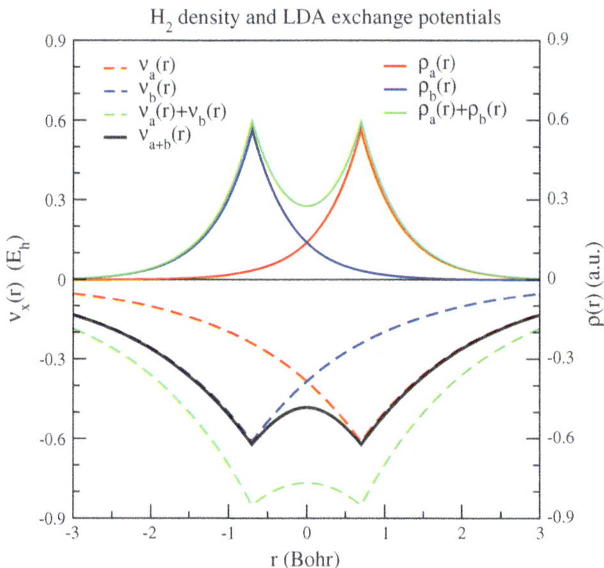

Fig. 2 Example of exchange-potential nonadditivity using the LDA exchange functional in molecular hydrogen. The hydrogen atom density is the quantum mechanically exact density, and the atoms are separated by 1.4011 Bohr. The x axis is the position along the internuclear axis. The sum of the isolated atom exchange potentials (*dashed green line*) is more negative than the molecular exchange potential (*solid black line*). At the bond midpoint ($r = 0$), the ratio of the summed potentials to the molecular potential is exactly $2:2^{1/3}$

corrections over-correct it (see e.g., Fig. 3), and the resulting potential is arguably worse than the simple sum; however, we have empirically found that the third- and higher-order many-body corrections act to cancel some of the two-body corrections. Specifically, the many-body corrections cancel those two-body corrections for the pairs

of atoms whose contribution to the reference density in the region of their overlap is dwarfed by a larger or intervening third atom. In other words, we have found that one can truncate the cluster expansion and weight the two-body corrections by some factor $0 < W_{ab} < 1$ that accounts for the many-body corrections in an effective way. We call this the *weighted cluster expansion*

$$v_{ref}^{(1)}(\mathbf{r}) \approx \sum_a v_a^{(1)}(\mathbf{r}) + \sum_{b>a} W_{ab} \Delta v_{ab}^{(1)}(\mathbf{r}), \qquad (48)$$

and this is the primary approximation in the VE1W model.

We have explored many forms of W_{ab}; the simplest of these is $W_{ab} = S_f(X_{ab} + X_{ba})$, where $S_f(x) = x$, if $x < x_0$; $S_f(x) = 1$, if $x > 1$;

$$S_f(x) = x_0 + (1 - x_0) \sum_{i=1}^{5} c_i \left(\frac{x - x_0}{1 - x_0} \right)^i, \qquad (49)$$

if $x_0 < x < 1$; $x_0 = 19/20$; $c_1 = 1$; $c_2 = 0$; $c_3 = 4$; $c_4 = -7$; $c_5 = 3$; $X_{ab} = S_{ab}^{ref}/\sum_c S_{cb}^{ref}$; $S_{aa}^{ref} = 0$; and $S_{ab}^{ref} = \int \rho_a(\mathbf{r})\rho_b(\mathbf{r})d^3r$. We don't want the reader to become overwhelmed by technical details that are presented for the sole purpose of ensuring the reproducibility of our results. Our discussion can be understood by interpreting the math as being a model that results in $W_{ab} \approx 1$ for those pairs of atoms that one would intuitively consider to be covalently bonded, and $W_{ab} \approx 0$ for those that are not.

The reader may question why the xc-potential is nonadditive, and why the simple sum of potentials (Eq. 44) leads to a net potential that is too negative, and why the two-body corrections over-correct the one-body sum. In brief, the xc-functional's nonadditivity results from a

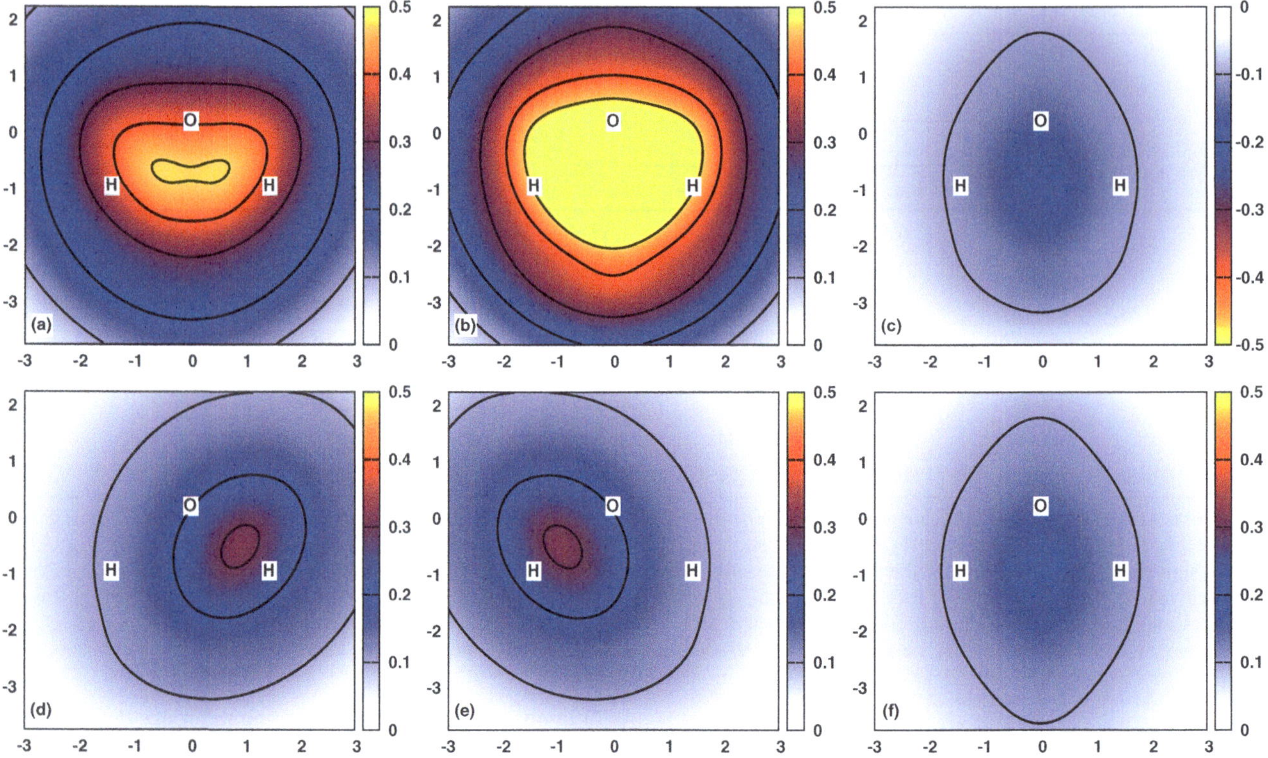

Fig. 3 Decomposition of water's PBE first-order potential $v_{\text{ref}}^{(1)}(\mathbf{r})$. The nonadditive potential is depicted in **a**, which is the sum of the two- and three-body corrections shown in **b** and **c**, respectively. The two-body correction (**b**) is the sum of individual pair-corrections $\Delta v_{ab}^{(2)}(\mathbf{r})$ depicted in **d**, **e**, and **f**. The colors represent the value of the potential along the plane formed by the atoms. The x and y axis are expressed in Bohr, and the *color scale* is in E_h. Note that potential in **c** is negative, whereas all other potentials are positive. The *color scheme* emphasizes that the magnitude of the three-body correction is so similar to the H–H pair-correction that these potentials cancel. As a result, the nonadditivity shown in **a** is well reproduced by the sum of **d** and **e**. The solid lines are isocontour values of 0.1, 0.2, 0.3, 0.4, and 0.47 E_h

dependence on the density to some *fractional order*, and the reason why the one-body sum is too negative is because that fractional order is *between one and two*. Let us clarify through a brief demonstration. Consider the uniform electron gas exchange functional $E_x[\rho] = -C_x \int \rho(\mathbf{r})^{4/3} d^3 r$, where $C_x = \frac{3}{4}\left(\frac{3}{\pi}\right)^{1/3}$. The first-order exchange potential is

$$v_x(\mathbf{r}) = -\frac{4C_x}{3}\rho(\mathbf{r})^{1/3}. \tag{50}$$

Let us evaluate $v_x(\mathbf{r})$ with a homonuclear dimer $\rho(\mathbf{r}) = \rho_a(\mathbf{r} - \mathbf{R}) + \rho_b(\mathbf{r} + \mathbf{R})$. At the bond midpoint $\mathbf{r} = 0$, it holds that $\rho(0) = \rho_a(-\mathbf{R}) + \rho_b(\mathbf{R}) = 2\rho_a(\mathbf{R})$ [since $\rho_a(\mathbf{r}) = \rho_b(\mathbf{r})$ and each are spherically symmetric], and the exchange potential is

$$v_{x,a+b}(0) = -2^{1/3}\left[\frac{4C_x}{3}\rho_a(\mathbf{R})^{1/3}\right], \tag{51}$$

whereas the sum of atom potentials is

$$v_{x,a}(0) + v_{x,b}(0) = -2\left[\frac{4C_x}{3}\rho_a(\mathbf{R})^{1/3}\right]. \tag{52}$$

We see that the sum of atom potentials is more negative by a ratio of $2:2^{1/3}$. With this insight, it is easy to explain why the two-body corrections (Eq. 46) over-correct the one-body sum (Eq. 44). Just as the sum of one-body potentials (Eq. 44) results in a net potential that is too negative, the subtraction of one-body potentials in Eq. 46 results in a two-body correction that is too positive.

At this point, we have completed our description of the approximations as they relate to the calculation of the core-Hamiltonian and to those models discussed in the Results section; however, we have a few comments regarding the second-order potentials. The observations that we describe here are of interest to the community, as they can be used to argue the validity of the approximations employed in spin-polarized SCC-DFTB models [38, 78], and also suggest ways that those approximations can be improved.

Unlike the first-order potential, the second-order potentials, $v_{\text{ref}}^{(2)}(\mathbf{r})$ and $w_{\text{ref}}^{(2)}(\mathbf{r})$, are not well behaved; they are strictly negative functions that approach $-\infty$ as $\rho_{\text{ref}}(\mathbf{r}) \to 0$. Thus, $w_{\text{ref}}^{(2)}(\mathbf{r})$ is poorly modeled by a simple sum of one-body $w_a^{(2)}(\mathbf{r})$'s. Furthermore, the many-body cluster corrections introduce as many large errors as they remove, and the expansion does not converge until it has been totally exhausted; however, we have found the following approximation to be almost exactly correct:

$$
\begin{aligned}
w_{\text{ref}}^{(2)}(\mathbf{r}) \approx & \sum_a L_a(\mathbf{r}) w_a^{(2)}(\mathbf{r}) \\
& + \sum_{b>a} W_{ab} \Big\{ [L_a(\mathbf{r}) + L_b(\mathbf{r})] w_{ab}^{(2)}(\mathbf{r}) \\
& - L_a(\mathbf{r}) w_a^{(2)}(\mathbf{r}) - L_b(\mathbf{r}) w_b^{(2)}(\mathbf{r}) \Big\},
\end{aligned}
\tag{53}
$$

and similarly for $v_{\text{ref}}^{(2)}(\mathbf{r})$, where $L_a(\mathbf{r}) = \rho_a(\mathbf{r})^2 / \sum_b \rho_b(\mathbf{r})^2$ is a spatial partition function, which one can interpret as being a "fuzzy" Voronoi-like cell. It holds the properties: $L_a(\mathbf{r}) \approx 1$ in the region of a; $L_a(\mathbf{r}) \approx 0$ near the region of any other atom; $\sum_a L_a(\mathbf{r}) = 1$ for all $\mathbf{r}$; and as a consequence of all the above, $L_a(\mathbf{r})L_{b \neq a}(\mathbf{r}) \approx 0$. The sum of partitioned one-body potentials in the *partitioned cluster expansion* is a much-improved representation of the second-order potentials (see Fig. 4), and the weighted two-body partitioned corrections are relatively small.

If one approximates $w_{\text{ref}}^{(2)}(\mathbf{r})$ by ignoring the two-body corrections, then the potential is described by the isolated atom potentials within their respective Voronoi cells, and the spin-polarization energy $E[\omega]$ becomes

$$
\begin{aligned}
E[\omega] &= \frac{1}{2} \int \omega(\mathbf{r})^2 w_{\text{ref}}^{(2)}(\mathbf{r}) d^3 r \\
&= \sum_a \frac{1}{2} \int \omega(\mathbf{r})^2 \Big[L_a(\mathbf{r}) w_a^{(2)}(\mathbf{r}) \Big] d^3 r \\
&= \sum_a \frac{1}{2} \int \omega(\mathbf{r}) \omega_a(\mathbf{r}) w_a^{(2)}(\mathbf{r}) d^3 r
\end{aligned}
\tag{54}
$$

where $\omega_a(\mathbf{r}) = L_a(\mathbf{r})\omega(\mathbf{r})$ is the atom-partitioned spin density about a. If $\omega(\mathbf{r})$ was represented by an atom-centered auxiliary basis and was chosen to partition the density in a manner analogous to $L_a(\mathbf{r})$, then the spin-polarization energy reduces to one- and two-center overlap-like integrals between auxiliary functions.

A different result is obtained if one employs the exact decomposition $\omega(\mathbf{r}) = \sum_b \omega_b(\mathbf{r}) = \sum_b L_b(\mathbf{r})\omega(\mathbf{r})$, and then approximates their overlap $L_a(\mathbf{r})L_{b \neq a}(\mathbf{r}) \approx 0$; that is,

$$
\begin{aligned}
E[\omega] &= \sum_{ab} \frac{1}{2} \int \omega(\mathbf{r})^2 L_b(\mathbf{r}) \Big[L_a(\mathbf{r}) w_a^{(2)}(\mathbf{r}) \Big] d^3 r \\
&\approx \sum_a \frac{1}{2} \int \omega_a(\mathbf{r})^2 w_a^{(2)}(\mathbf{r}) d^3 r \\
&\approx \sum_a E[\omega_a].
\end{aligned}
\tag{55}
$$

If one presumed that the $L_a(\mathbf{r})$'s are spherical, then Eq. 55 is no different than what is used in spin-polarized SCC-DFTB [38]; however, we feel that we've shown that the arguments needed to reach this result are more justified than what one might infer from reading Refs. [38] and [79]; and it also suggests ways that one can try to improve upon their work.

2.7 Computational details

2.7.1 DFTB2 with charge-dependent polarizability corrections

Figure 5 compares the polarizabilities of DFTB2, DFTB2+CPEQ, and DFTB2+CPE0 to standard B3LYP/6-31++G** results. The comparisons include 33 anionic, 142 neutral, and 35 cationic molecules taken from the QCRNA database [80]. The molecules were selected for having contained some combination of H, C, N, and O, but no other elements.

The H, C, N, and O DFTB2 parameters are described in Ref. [22], and these parameters remain unmodified in our CPE-corrected variations, which we henceforth refer to without the "DFTB2+" prefix. The CPEQ and CPE0 models have four additional atomic parameters: $\zeta_a(0), b_a, R_{\text{lo},a}$, and $R_{\text{hi},a}$. Although note that the CPE0 model's b_a parameter is zero for all atoms, these parameters were optimized for each model to B3LYP/6-311++G** dipole moments and B3LYP/6-31++G** isotropic polarizabilities, and the merit function used to judge the goodness of the fit is

$$
\chi^2 = \sum_{i=1}^{142} 80|\Delta\mathbf{q}^{(1)}|_i + \sum_{i=1}^{210} |\Delta\alpha|_i;
\tag{56}
$$

where $|\Delta\mathbf{q}^{(1)}|$ is the norm of the difference between the model and reference dipole moment vectors; $|\Delta\alpha|$ is the absolute difference between the model and reference isotropic polarizabilities; and the summation over dipole moment errors includes only neutral molecules. Both the reference and model calculations are evaluated at the B3LYP/6-31++G** geometry-optimized structures.

2.7.2 Generalization of the auxiliary basis

Table 1 compares the geometry-optimized bond, angle, dipole moment, and relative energy errors of various PBE/6-31G*-based models to standard PBE/6-31G*. The statistics include results from 52 molecules [81] taken from the G2/97 neutral small molecule test set [82], which were also used in our previous VE studies [26, 32]. For the purpose of discussing the generalization of the auxiliary basis, the reader should compare VEJ/1S(M) and VEJ/2P,3D,4F to VEJ, and note how well VEJ compares to VE. The parametrization of the VEJ/2P,3D,4F auxiliary basis Slater exponents is described elsewhere [32].

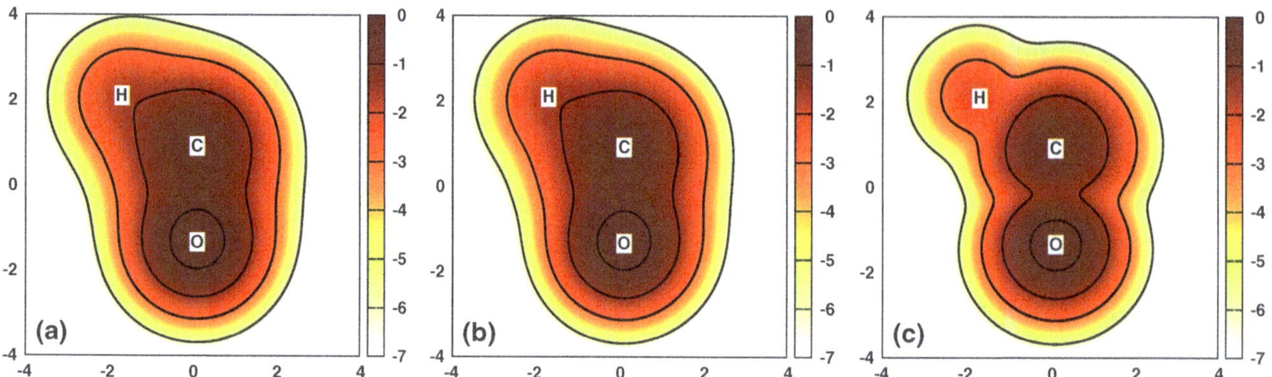

Fig. 4 Decomposition of the SPW92 second-order spin-polarization potential $w_{\text{ref}}^{(2)}(\mathbf{r})$ for the HCO molecule. The panels depict: **a** the potential, without approximation; **b** the one- and two-body partitioned cluster approximation (Eq. 53); and **c** the one-body partitioned cluster approximation; that is, $W_{ab} = 0$ for all pairs. The *colors* represent the value of these potentials in the plane formed by the atoms. The x and y axis are in units of Bohr, and the *color scale* is E_h. The *solid lines* are isocontour values of -0.7, -1.5, -3, and -6 E_h

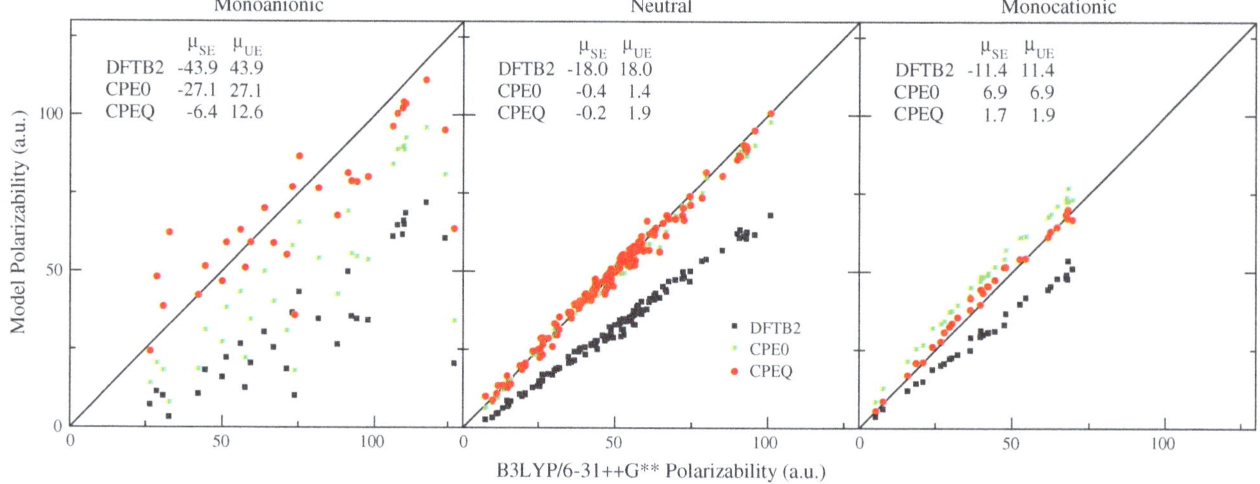

Fig. 5 Comparison of DFTB2, DFTB2+CPE0, and DFTB2+CPEQ polarizabilities to standard B3LYP/6-31++G** results. μ_{SE} and μ_{UE} are the mean signed and unsigned errors (a.u.), respectively

2.7.3 Decomposition and nonadditivity of the xc-potentials

Table 1 is used to compare the errors of the VE1, VE1S, and VE1W models, all of which are variations of the VE0 model.

Figure 3 displays the nonadditivity of the PBE first-order potential of water and its decomposition into two- and three-body corrections and the decomposition of the two-body correction into the individual pair-corrections.

3 Results

3.1 DFTB2 with charge-dependent polarizability corrections

Figure 5 shows that DFTB2 underpredicts the B3LYP/ 6-31++G** polarizabilities; the extent to which depends

on the ionic character of molecules. The mean unsigned relative percent errors of the anionic, neutral, and cationic molecules are 59, 38, and 26%, respectively. In other words, the anionic molecules need a larger polarizability correction than the neutrals, and neutrals need a larger correction than the cations.

To emphasize this further, we examine the results of a charge-independent model, CPE0. It greatly improves the neutral molecule polarizabilities, but the resulting cations are too polarizable, and the anions are still not polarizable enough.

However, the errors are largely corrected with the charge-dependent model, CPEQ. It reduces the neutral and cationic polarizability errors to 5%; and the anionic polarizability errors are two and three times smaller than the CPE0 and DFTB2 errors, respectively.

The DFTB2 mean unsigned dipole moment errors relative to the B3LYP reference is 0.175 a.u., and the CPE0 and CPEQ models reduce this to 0.167 and 0.166 a.u., respectively. As expected, the CPE0 and CPEQ models do not significantly improve the DFTB2 dipole moments, because the models have been chosen to screen the short-range interactions. The CPE method lacks the proper physics to describe covalent bonds and thus cannot be relied upon to improve molecular dipole moments. For this purpose, it would be more fruitful to improve the underlying DFTB2 model, for example, by increasing the number of AOs.

We note that there is very little difference between the DFTB2+CPEQ method of the present work and the MNDO/d+CPE method presented in Ref. [37]. The main difference is that we have included Fock matrix corrections (Eq. 28) in the present work. There are some more trivial differences as well, such as our use of Slater dipole CPE response functions in DFTB2+CPEQ, as opposed to the Gaussian-dipole functions used in MNDO/d+CPE. The only other difference is: the reference density about which the DFTB2+CPEQ correction is expanded is a collection of Slater charges, whereas the MNDO/d+CPE correction is expanded about a collection of point charges. The beauty of the CPE correction is that it is expanded about a reference density, and it doesn't matter what underlying method gave rise to that reference density. However, just as the CPE correction is an ad hoc correction to MNDO/d, so too is it an ad hoc correction to DFTB2.

3.2 Generalization of the auxiliary basis

Table 1 demonstrates that a Taylor series truncation of the xc-functional to first order (VEJ) has a small effect on geometries and a modest effect on dipole moments. Furthermore, it is capable of closely reproducing closed-shell molecule relative energies. Some new models have extended the Taylor series expansion to third order as a mechanism for including charge dependence [56–58], but the VEJ results presented here should clarify that one should take care to not misinterpret those models as having corrected a symptom originating from a premature truncation of the Taylor series expansion.

Approximating the second-order energy with a SCC-DFTB-like Slater monopole auxiliary basis [VEJ/1S(M)] significantly increases the errors, but these errors are alleviated when the auxiliary basis is extended to better reproduce the AO products (VEJ/2P,3D,4F).

Our previous work [32] showed that VEJ/1S(M) underpredicts the hydrogen bond strength of the water dimer by approximately 4 mE_h relative to both VEJ and standard PBE/6-31G*, and that this error was corrected when using a more sophisticated auxiliary basis.

3.3 Decomposition and nonadditivity of the xc-potentials

The introduction to this manuscript emphasized the severity of the VE1 approximation, and we show its results in Table 1. Its severity may seem intuitively obvious because it neglects all atoms in the system other than those coincident with the AO product, and so one may try to improve it by including the potentials caused by the other atoms. The VE1S model attempts to overcome this severe approximation by (incorrectly) assuming that the neutral-atom potentials are additive functions; however, this model is arguably just as bad as VE1. One can try to construe the numerical values in a manner to suggest that VE1S is a minor improvement, but both models optimize the molecular geometries into utterly nonphysical configurations.

The reader may notice that one of the molecules (Si_2H_6) fails to SCF converge once VE1S optimizes it into an unrealistic geometry, but it is best not to over-interpret this. The geometries of VE1 and VE1S are both extremely poor, and the fact that one has SCF converge difficulties when evaluated at a nonsensical configuration, whereas the other does not, does not indicate that its approximations are somehow better. Had we expanded the test set to include even larger molecules, we'd expect the VE1 model to eventually fail as well. We have tried a number of approximations, most of which we do not describe here; our experience suggests that larger molecules are more likely to optimize into strange configurations and then fail to SCF converge, because the errors introduced by the approximations increase with the system size. It is because of this that we feel our limited test set is sufficient to identify which approximations are severe without reaching that extreme limit of failure.

The VE1S first-order potential is too negative because it incorrectly assumed that the neutral-atom xc-potentials were additive, and as the size of the system increases, the magnitude of this error is compounded. To account for this nonadditivity, one might include two-body corrections. Alas, the resulting potential is then not negative enough, and the magnitude of its error increases with the size of the system. Like VE1S, it readily geometry optimizes to nonsensical configurations.

The prospect of explicitly including three-body or larger-body corrections is sufficiently daunting that it may be tempting to just give up and use VE1 or the VE1-like approximation that most semiempirical models use. To make an advance beyond that, we need to have some way to account for the many-body effects in some new way. To this end, Fig. 3 decomposes the xc-potential of water into its two- and three-body components, and we see that the three-body nonadditivity cancels the two-body correction between the H's. One can perform similar decompositions for other molecules, and from these, it seems that this is true for all pairs of atoms whose overlap is dwarfed by the densities of the other atoms. This general idea can inspire many different forms of the weighted cluster approximation, and Table 1 displays the statistics for a very simple model (VE1W). This model reduces the errors to those comparable to VE0, the model on which it is based.

4 Conclusion

Kohn-Sham expansion methods can be made to be as accurate and transferable as the density-functional methods on which they are based. The primary sources of errors in a second-order VE model are *not* a consequence of a premature truncation of the Kohn-Sham functional. Instead, the source of errors is much more obvious; in that, just as one would expect standard DFT models to suffer from approximations, so too would one expect the analogous VE models. The most severe VE approximations that we've encountered are: the limited size of the AO basis; the poor auxiliary basis representation of the AO products; and the one- and two-body treatment of the first-order interactions.

As a mechanism for discussing these approximations: we've introduced a new self-consistent *charge-dependent* dipole polarization correction model (DFTB2+CPEQ) that improves DFTB2 polarizabilities; we've reviewed our recently published work that generalizes the auxiliary basis; and we're examined the decomposition of the first-order potential into many-body components, and used these observations to gain the insight and motivation for the development of new approximations.

From these discussions, we hope to have communicated the following:

(1) The errors introduced from having truncated the Taylor series expansion to second order are small relative to the other errors that we've discussed.

(2) The beneficial effects of third-order models arise from introducing charge-dependent corrections, and these corrections mask the limited size of the AO basis.

(3) Like all minimal valence AO basis models, DFTB2 underpredicts dipole polarizabilities; however, this underprediction will not be alleviated by increasing the size of the AO basis alone, because it would still lack atomic dipole response. In lieu of generating an entirely new DFTB model, one can improve the polarizabilities with an auxiliary CPE response. If the CPE response is not made charge-dependent, then the anionic and cationic molecular polarizabilities will be under- and over-predicted, respectively; but the polarizability of various charge states can be improved

by making the CPE response functions charge-dependent.

(4) The VE response density can be modeled more accurately by increasing the radial and angular completeness of the auxiliary basis description of the AO products, and this can be done in a way that requires only two-center integrals, which can be precomputed and splined as a function of internuclear separation.

(5) The one- and two-body treatment of the first-order matrix elements is a severe approximation; however, it is not improved by representing the first-order potential as a sum of isolated atom potentials. Nor is it improved by modeling it with a sum of one-body potentials and two-body corrections. We have presented some initial evidence that the many-body nonadditivity of the xc-potential can be modeled by damping some of the two-body corrections.

(6) The second-order potentials are not well described by *any* truncated cluster approximation, but is well modeled by a partitioned cluster approximation. By inserting this approximate form into the expression for the spin-polarization energy, one can derive its model form used in spin-polarized DFTB2 methods by making some additional approximations.

It is the hope that the insights provided in this work will be instrumental in directing further research effort toward development of the next-generation of semiempirical quantum models for broad application.

Acknowledgments The authors are grateful for financial support provided by the National Institutes of Health (GM084149). Computational resources from the Minnesota Supercomputing Institute for Advanced Computational Research (MSI) were utilized in this work. This research was supported in part by the National Science Foundation through TeraGrid resources provided by the National Center for Supercomputing Applications and the Texas Advanced Computing Center under grant TG-CHE100072.

References

1. Hückel E (1931) Z Phys 70:204
2. Hückel E (1931) Z Phys 72:310
3. Hückel E (1932) Z Phys 76:628
4. Hückel E (1933) Z Phys 83:632
5. Pariser R, Parr RG (1953) J Chem Phys 21:767
6. Hoffmann R (1963) J Chem Phys 39:1497
7. Pople JA, Segal GA (1966) J Chem Phys 44:3289
8. Pople JA, Beveridge DL, Dobosh PA (1967) J Chem Phys 47:2026
9. Baird NC, Dewar MJS (1969) J Chem Phys 50:1262
10. Bingham RC, Dewar MJS, Lo DH (1975) J Am Chem Soc 97:1285
11. Dewar MJS, Thiel W (1977) Theor Chim Acta 46:89
12. Thiel W, Voityuk AA (1996) Theor Chim Acta 93:315
13. Dewar MJS, Zoebisch E, Healy EF, Stewart JJP (1985) J Am Chem Soc 107:3902

14. Stewart JJP (1989) J Comput Chem 10:221
15. Stewart JJP (2007) J Mol Model 13:1173
16. Clark T (2000) J Mol Struct (Theochem) 530:1
17. Winget P, Selçuki C, Horn A, Martin B, Clark T (2003) Theor Chem Acc 110:254
18. Winget P, Clark T (2005) J Mol Model 11:439
19. Rocha GB, Freire RO, Simas AM, P Stewart JJ (2006) J Comput Chem 27:1101
20. Slater JC, Koster GF (1954) Phys Rev 94:1498
21. Porezag D, Frauenheim T, Köhler T, Seifert G, Kaschner R (1995) Phys Rev B 51:12947
22. Elstner M, Porezag D, Jungnickel G, Elsner J, Haugk M, Frauenheim T, Suhai S, Seifert G (1998) Phys Rev B 58:7260
23. Tuttle T, Thiel W (2008) Phys Chem Chem Phys 10:2125
24. Kolb M, Thiel W (1993) J Comput Chem 14:775
25. Weber W, Thiel W (2000) Theor Chem Acc 103:495
26. Giese TJ, York DM (2010) J Chem Phys 133:244107
27. Perdew JP, Burke K, Ernzerhof M (1996) Phys Rev Lett 77:3865
28. Elstner M (2007) J Phys Chem A 111:5614
29. Seifert G (2007) J Phys Chem A 111:5609
30. Frauenheim T, Seifert G, Elstner M, Hajnal Z, Jungnickel G, Porezag D, Suhai S, Scholz R (2000) Phys Status Solidi B 217:41
31. Otte N, Scholten M, Thiel W (2007) J Phys Chem A 111:5751
32. Giese TJ, York DM (2011) J Chem Phys 134:194103
33. Giese TJ, York DM (2008) J Chem Phys 128:064104
34. Giese TJ, York DM (2008) J Chem Phys 129:016102
35. Giese TJ, York DM (2008) J Comput Chem 29:1895
36. Giese TJ, York DM (2007) J Chem Phys 127:194101
37. Giese TJ, York DM (2005) J Chem Phys 123:164108
38. Köhler C, Seifert G, Gerstmann U, Elstner M, Overhof H, Frauenheim T (2001) Phys Chem Chem Phys 3:5109
39. Kohn W, Sham L (1965) Phys Rev A 140:A1133
40. Frauenheim T, Seifert G, Elstner M, Niehaus T, Köhler C, Amkreutz M, Sternberg M, Hajnal Z, Di Carlo A, Suhai S (2002) J Phys Condens Matter 14:3015
41. Elstner M, Frauenheim T, Kaxiras E, Seifert G, Suhai S (2000) Phys Status Solidi B 217:357
42. Murdachaew G, Mundy CJ, Schenter GK (2010) J Chem Phys 132:164102
43. Maerzke KA, Murdachaew G, Mundy CJ, Schenter GK, Siepmann JI (2009) J Phys Chem A 113:2075
44. Chang DT, Schenter GK, Garrett BC (2008) J Chem Phys 128:164111
45. Matsuzawa N, Dixon DA (1992) J Phys Chem 96:6232
46. Fiedler L, Gao J, Truhlar DG (2011) J Chem Theory Comput 7:852
47. York DM, Yang W (1996) J Chem Phys 104:159
48. Nalewajski RF (1984) J Am Chem Soc 106:944
49. Mortier WJ, Van Genechten K, Gasteiger J (1985) J Am Chem Soc 107:829
50. Mortier WJ, Ghosh SK, Shankar S (1986) J Am Chem Soc 108:4315
51. Morales J, Martínez TJ (2001) J Phys Chem A 105:2842
52. Itskowitz P, Berkowitz ML (1997) J Phys Chem A 101:5687
53. York DM (1995) Int J Quantum Chem 56:385
54. Zhou B, Ligneres VL, Carter EA (2005) J Chem Phys 122:044103
55. Hobson EW (1892) Proc Lond Math Soc 24:55
56. Yang Y, Yu H, York DM, Cui Q, Elstner M (2007) J Phys Chem A 111:10861
57. Gaus M, Cui Q, Elstner M (2011) J Chem Theory Comput 7:931
58. Yang Y, Yu H, York D, Elstner M, Qiang C (2008) J Chem Theory Comput 4:2067
59. Perdew JP, Wang Y (1992) Phys Rev B 45:13244
60. Parr RG, Yang W (1984) J Am Chem Soc 106:4049
61. Pearson RG (1988) J Am Chem Soc 110:7684
62. Paxton AT, Kohanoff JJ (2011) J Chem Phys 134:044130
63. Dunlap BI (2000) J Mol Struct (Theochem) 529:37
64. Glaesemann KR, Gordon MS (2000) J Chem Phys 112:10728
65. Hamel S, Casida ME, Salahub DR (2001) J Chem Phys 114:7342
66. Ahlrichs R (2004) Phys Chem Chem Phys 6:5119
67. Sodt A, Subotnik JE, Head-Gordon M (2006) J Chem Phys 125:194109
68. Pedersen TB, Aquilante F, Lindh R (2009) Theor Chem Acc 124:1
69. Hohenstein EG, Sherrill CD (2010) J Chem Phys 132:184111. doi:10.1063/1.3426316
70. Choi CH, Ivanic J, Gordon MS, Ruedenberg K (1999) J Chem Phys 111:8825
71. Kalinowski JA, Lesyng B, Thompson JD, Cramer CJ, Truhlar DG (2004) J Phys Chem A 108:2545
72. Dunlap BI, Rösch N, Trickey SB (2010) Mol Phys 108:3167
73. Jung Y, Sodt A, Gill PW, Head-Gordon M (2005) Proc Natl Acad Sci 102:6692
74. Piquemal J, Cisneros G, Reinhardt P, Gresh N, Darden TA (2006) J Chem Phys 124:104101
75. Cisneros GA, Piquemal J, Darden TA (2006) J Chem Phys 125:184101
76. Elking DM, Cisneros GA, Piquemal J, Darden TA, Pedersen LG (2010) J Chem Theory Comput 6:190
77. Tu Y, Jacobsson SP, Laaksonen A (2006) Phys Rev B 74:205104
78. Zheng G, Witek HA, Bobadova-Parvanova P, Irle S, Musaev DG, Prabhakar R, Morokuma K (2007) J Chem Theory Comput 3:1349
79. Frauenheim T, Seifert G, Elstner M, Hajnal Z, Jungnickel G, Porezag D, Suhai S, Scholz R (2000) Phys Stat Sol 217:41
80. Giese TJ, Gregersen BA, Liu Y, Nam K, Mayaan E, Moser A, Range K, Nieto Faza O, Silva Lopez C, Rodriguezde Lera A, Schaftenaar G, Lopez X, Lee T, Karypis G, York DM (2006) J Mol Graph Model 25:423
81. The molecules in the test set are: BeH, C_2H_2, C_2H_4, C_2H_6, CH_2, CH_3, CH_4, CH_4O, CH_4S, CH_3Cl, CN, CS, HC, HCN, HCO, HF, HCl, LiH, NH, HO, H_2O, H_2O_2, OCl, NO, OS, O_2, CO, SiO, CO_2, SO_2, F_2, Cl_2, FCl, Li_2, LiF, Na_2, NaCl, N_2, NH_2, NH_3, N_2H_4, HOCl, H_2CO, P_2, PH_2, PH_3, S_2, H_2S, SiH_2, SiH_3, SiH_4, and Si_2H_6
82. Curtiss LA, Raghavachari K, Redfern PC, Pople JA (1997) J Chem Phys 106:1063

Theor Chem Acc (2012) 131:1197
DOI 10.1007/s00214-012-1197-8

REGULAR ARTICLE

The polarizing forces of water

Revati Kumar · Thomas Keyes

Received: 2 May 2011 / Accepted: 9 September 2011 / Published online: 11 March 2012
© Springer-Verlag 2012

Abstract A brief review of popular polarizable potentials for water, including both those parameterized to fit experimental properties, typically of the liquid, or electronic structure calculations on small clusters, is presented. The recently developed POLIR potential, which was parameterized to reproduce both ab initio calculations on clusters and the experimental liquid IR spectrum, is discussed, and some new results for both clusters and the liquid phase are shown, indicating its transferable nature.

Keywords Polarization · Vibrational spectra · Transferable potential

1 Introduction

Polarization can broadly be defined as the response of molecular electrostatic properties to an electric field, which may be applied externally or arise from the local environment, namely the surrounding molecules. The corresponding polarization energy and forces are nonadditive in nature, since the interaction between two molecules is influenced by the presence of other molecules. Most existing polarizable potentials [1–11] invoke dipole polarizable sites: In the presence of an electric field, a dipole is induced, which in turn changes the electric field elsewhere and interacts with the other charges, dipoles, induced dipoles, etc., of the system.

Published as part of the special collection of articles: From quantum mechanics to force fields: new methodologies for the classical simulation of complex systems.

R. Kumar · T. Keyes (✉)
Department of Chemistry, Boston Unviersity, Boston, MA, USA
e-mail: keyes@bu.edu

Note that even molecules without induced dipoles respond to an electric field by orienting their permanent dipoles, yielding a nonzero average system dipole. In this article, "polarization" denotes induced dipole polarization, which is our focus, not the "permanent dipole polarization" present in all models.

Polarization energies and forces are a class of many-body interactions. Consider three particles A, B, and C interacting with each other. The total "two-body" interaction is the sum of the interactions of A with B, B with C, and A with C. However, if the total interaction energy is not equal to the sum of these two-body energies, the remaining energy is referred to as the many-body energy (or in this case the three-body energy). Ab initio interaction energies are not pairwise additive, and the many-body interactions can be substantial. For example, ab initio calculations in water clusters have shown that while the two-body energy dominates and forms $\approx 80\%$ of the total energy, the many-body energies form the remaining $\approx 20\%$ [12].

Another indication of the importance of polarization in water is the change in dipole moment as one goes from the gas phase to the liquid. Gas-phase water molecules have a dipole moment of 1.85 D, while in the condensed phase, it goes up to an estimated range of 2.5–2.9 D [13, 14]. The increase is a result of polarization [15].

Polarization is not the only many-body interaction in nature, but it contributes significantly and, furthermore, provides a convenient framework to empirically include other many-body interactions in a force field [16, 17]. Furthermore, the short-range damping schemes [18] to be described below provide a treatment of classical charge penetration and an effective treatment of quantal exchange. Intramolecular charge transfer is treated via configuration-dependent charges, while intermolecular charge transfer is outside the scope of this work.

The earliest potentials developed for water were "effective" pair potentials like the SPC [19, 20] and TIPnP [21–26] families. These models have Coulomb interactions due to the partial charges, and Lennard–Jones type van der Waals interactions, both of which are two-body. However, by parameterizing with fits to experimental observables like the density and radial distribution functions of liquid water at around 300 K, they include, in an effective manner, the action of polarization. The resulting partial charges were found to give a higher dipole moment than the gas-phase monomer and closer to the value found in the liquid. While effective pair potentials do reasonably well for the liquid, they fail to describe water clusters [27, 28]. A number of them do poorly in reproducing the phases of ice and the melting temperature (unless they were parameterized to do so) [29, 30].

The ultimate goal is potentials that are transferable, that is, not limited to the conditions for which they were parameterized. Polarization, allowing many-body interactions and a dipole that varies with the environment, is a crucial element in this regard [31–33]. A nonpolarizable model that increases the monomer permanent dipole above the true monomer value to mimic the effect of polarization in the ambient liquid is obviously not transferable to small clusters [28], and probably not to high temperatures, since permanent dipole polarization is washed out by thermal agitation, while induced dipole polarization is not. Considerable effort has been directed to the development of polarizable force fields for water, because of its significance under diverse conditions and its evident signatures of polarization mentioned above, and because it acts as a test subject for general methodologies of force field development.

In this paper, we present a brief review of polarizable force fields for water, including the various techniques of implementing polarization. The manner of parameterization defines two main categories: They may be parameterized based either on experimental quantities associated with the condensed state, or on ab initio (electronic structure) calculations. The second half of the paper discusses some recent results using the flexible POLIR water potential recently developed in the Keyes group [10]. POLIR is a hybrid model in the sense that it was parameterized to both ab initio calculations on water clusters and the IR spectrum of the ambient liquid. POLIR does fairly well in predicting the properties of liquid water, despite having used no bulk information but the spectrum, and does remarkably well in reproducing additional spectroscopic properties that were not built in. In the original paper, Mankoo and Keyes showed that POLIR gets the HOH angle increase on going from the gas phase to the liquid. The calculated diffusion constant, radial distribution functions, and the dielectric constant of liquid water are reproduced reasonably well. In this paper, we further explore the properties of POLIR, both in clusters and in the liquid.

2 Theoretical methodology for polarization

In this section, we present a few of the methods used to describe polarization in force fields.

2.1 Point inducible dipoles

A brief description of the method of point inducible dipoles is provided by the following example. Consider a system of N water molecules with a dipole polarizability, α, on each oxygen atom. In addition, the H atoms and the O atoms have fixed charges, q_H and q_O, respectively, such that each molecule is neutral. The induced dipole moment, $\boldsymbol{\mu}_i$, on the O atom of molecule i is given by:

$$\boldsymbol{\mu}_i = \alpha \left[\mathbf{E}_i + \sum_{j \neq i} \mathbf{T}_{ij} \cdot \boldsymbol{\mu}_j \right], \tag{1}$$

where the two terms in the bracket are the field, $\mathbf{E}_i$, at the O atom of molecule i from the permanent charges excluding those on molecule i, and the field from other induced dipoles, where

$$\mathbf{E}_i = \sum_{j \neq i} \sum_{m=1,3} \frac{q_{j,m} \vec{r}_{i,1;j,m}}{r_{i,1;j,m}^3}. \tag{2}$$

The summation involves the atomic partial charges on molecule j at atomic site m, $q_{j,m}$, where $m = 1$ denotes the O-site on j, $m = 2, 3$ denotes the 2 H atoms on j, and $\vec{r}_{i,1;j,m}$ is the vector between the O-site on i and the atomic site m on j. Exclusion of intramolecular polarization by permanent charges is usual, but is not justified if one seeks to describe molecular vibrations. The dipole tensor $\mathbf{T}_{ij}$ is a 3×3 matrix whose elements are:

$$T_{ij}^{\beta\gamma} = \frac{3 r_{ij}^{\beta} r_{ij}^{\gamma}}{r_{ij}^5} - \frac{\delta_{\beta\gamma}}{r_{ij}^3}, \tag{3}$$

where β and γ denote the Cartesian components x, y, or z, $\delta_{\beta\gamma}$ is the Kronecker delta and $\vec{r}_{ij}$ refers to the vector between the O-sites of molecules i and j. The polarization energy is given by:

$$U_{\mathrm{pol}} = \sum_i \left[-\boldsymbol{\mu}_i \cdot \mathbf{E}_i - \sum_{j \neq i} [\boldsymbol{\mu}_i \cdot \mathbf{T}_{ij} \boldsymbol{\mu}_j] + \frac{1}{2\alpha} \boldsymbol{\mu}_i \cdot \boldsymbol{\mu}_i \right] \tag{4}$$

The induced dipoles are obtained from Eq. 1, and the straightforward solution involves matrix inversion. Unfortunately, this proves to be too expensive for most problems, and hence, iterative schemes are often used to obtain the induced dipoles and evaluate Eq. 4. One can also use an extended Lagrangian formalism wherein the induced dipoles are treated as dynamical variables with an associated mass and are propagated in time [34].

The above equations can be generalized to include higher permanent moments (p) than charges alone. Eq 1 then takes the more general form,

$$\boldsymbol{\mu}_i = \alpha \cdot [\mathbf{E}^p + \mathbf{E}^{ind}]. \tag{5}$$

A problem with point dipoles is that, at short distances, the polarization energy diverges [18, 35]. For a diatomic parallel to the field, the "polarization catastrophe" occurs at an interatomic distance of $(4\alpha_1 \alpha_2)^{1/6}$, where α_1 and α_2 are the polarizabilities of the two atoms/sites. In models with distributed atomic polarizabilities in which intramolecular induced dipole interactions are allowed, the distances can be short enough to cause this catastrophe. A number of methods have been proposed to overcome the problem, the most widely used of which is Thole's damping [18]. The electric field at site i, $\mathbf{E}_i$, and the elements of the dipole tensor, $\mathbf{T}_{ij}$, between sites i and j are given by

$$\mathbf{E}_i = \sum_{j \neq i} f_3(r_{ij}) \frac{q_j(r_{ij})\vec{r}_{ij}}{r_{ij}^3}, \tag{6}$$

where we note that the charge may be position dependent, and

$$T_{ij}^{\beta\gamma} = f_5(r_{ij}) \frac{3r_{ij}^\beta r_{ij}^\gamma}{r_{ij}^5} - f_3(r_{ij}) \frac{\delta_{\beta\gamma}}{r_{ij}^3}. \tag{7}$$

The damping functions, $f_3(r_{ij})$ and $f_5(r_{ij})$, are

$$f_3(r_{ij}) = 1 - \exp\left[-a\left[\frac{r_{ij}}{(\alpha_i\alpha_j)^{\frac{1}{6}}}\right]^b\right] \tag{8}$$

and

$$f_5(r_{ij}) = 1 - \left[1 + a\left[\frac{r_{ij}}{(\alpha_i\alpha_j)^{\frac{1}{6}}}\right]^b\right]\exp\left[-a\left(\frac{r_{ij}}{(\alpha_i\alpha_j)^{\frac{1}{6}}}\right)^b\right]. \tag{9}$$

The value of the exponent, b, is typically set to 3, although 4 and noninteger values have also been used [10, 36, 37]. The damping constant, a, was originally set to 0.5 by Thole, and a number of groups since then have used a as a parameter to be optimized [6, 7, 10, 11, 37, 38]. Different damping constants have been invoked for the dipole tensor and the electric field, and for intramolecular interactions and intermolecular interactions [7, 10, 37]. Alternative damping functions have been proposed, including a linear Thole type and a Gaussian [39, 40].

2.2 Drude oscillator/charge-on-spring models

The underlying principle behind the Drude oscillator [41] and charge-on-spring [42, 43] models is that two charges of

equal and opposite sign represent a dipole, which changes as the separation of the charges changes [44]. In the case of water, one of the charges $(-q_{pol})$ is fixed at the O-site, and the other charge (also called Drude particle, with $+q_{pol}$) is attached to the first charge via a harmonic spring (see Fig. 1), with an equilibrium separation of zero and a spring constant of k_{pol}. The position of the second charge is

$$\mathbf{r}_O + q_{pol}\mathbf{E}/k_{pol}. \tag{10}$$

The induced dipole is given by the product of the field-induced separation of the charges, q_{pol}, $\mathbf{E}/k_{pol}$, and the magnitude of the charge, q_{pol},

$$\boldsymbol{\mu} = q_{pol}^2\mathbf{E}/k_{pol}, \tag{11}$$

so the polarizability obeys $\alpha = q_{pol}^2/k_{pol}$. One difference between the charge-on-spring model of Yu et al. [42] and the Drude model (SWM4-DP) [41] is that the former has three sites, with the O atom carrying a charge q_O along with the induced dipole charge of $-q_{pol}$, whereas the latter has four sites, with the O atom carrying just the charge $-q_{pol}$, and the charge q_O of the oxygen shifted to an "M-site".

There are a number of ways that Eq. 11 can be solved to get the induced dipoles, for example, iterative methods as in the charge-on-spring implementation by the Yu et al. [42], extended Lagrangian type methods wherein the "Drude particle" is treated as a dynamical variable and propagated in time similar to the position vectors, etc [41].

2.3 Fluctuating charge models

In these models, the environment, through the electrostatic potential, is assumed to change the electronegativity of the atomic sites, and charge flow takes place in order to equalize the electronegativity [45]. The potential energy of a system with N charges is given by

$$U = \sum_{i=1}^{N} u_i + \sum_{i<j} \frac{q_i q_j}{r_{ij}} + U_{vdW}, \tag{12}$$

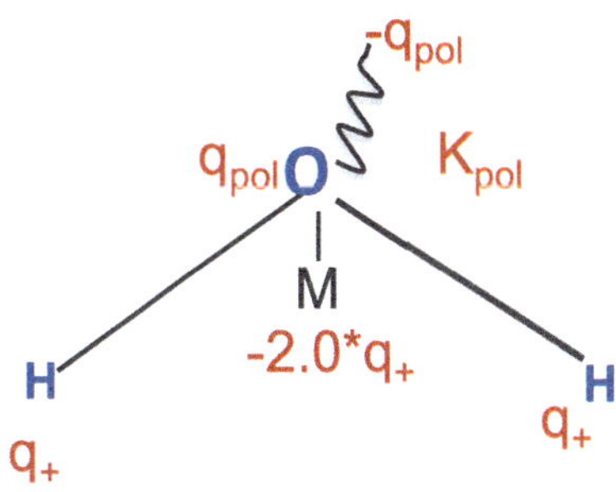

Fig. 1 A schematic diagram of the Drude oscillator model

where

$$u_i = u_0 + \chi_i q_i + \frac{1}{2} J_i q_i \tag{13}$$

is the energy of the creation of a partial charge, q_i. The derivative of U with respect to q_i gives the electronegativity per unit charge of atom i, and hence minimizing the energy with respect to the atomic charges, that is,

$$\frac{\delta U}{\delta q_i} = 0, \tag{14}$$

is equivalent to equalizing the electronegativities. Equation 14 gives rise to a set of coupled equations that can be solved either iteratively or, as in the Drude oscillator case, one can use an extended Lagrangian scheme wherein the charges are also considered to be dynamical variables. During the electronegativity equalization process, one can impose charge neutrality either on the molecule or simply on the entire system. In the latter case, charge transfer between molecules takes place, which in some case can be so high as to be physically unrealistic [44].

3 Types of polarizable models

One can divide polarizable models into two types, depending on whether they are primarily parameterized on experimental properties of the condensed phase, or on electronic structure calculations on water clusters. A brief description of some potentials belonging to these two categories is presented in the following two subsections.

3.1 Models parameterized to bulk experimental properties

Some of the earliest polarizable potentials, POL1 [1], RPOL [3], etc., were based on the work of Applequist et al. [46], wherein atomic polarizabilities were derived by fitting molecular polarizabilities to the experimental values. The Applequist-type models had both intra- and intermolecular polarization, but comparatively, small values of the atomic polarizabilities prevented a polarization catastrophe.

The POL1 model of Caldwell et al. used the Applequist values for the atomic polarizabilities and fit the remaining terms in the potential to get the liquid density and internal energy. The gas-phase molecular polarizability in POL1 is lower than the experimental value, and a modified version with the correct value, RPOL, was developed by Dang. The atomic polarizabilities were removed in the Dang-Chang [5] model and replaced by a single polarizability (with the gas-phase value of 1.44 Å) at a massless "M" site close to the O atom along the O–H bisector, which also carries the

partial charge corresponding to the O atom. The partial charges and the position of the M-site were parameterized to reproduce the gas-phase dipole and quadrupole moments. Although the remaining parameters were essentially determined by fits to experimental properties of the liquid phase like the internal energy and diffusion constant, care was taken to obtain a reasonable description of cluster energies as well. However, this model consistently predicts lower three-body energies compared to the ab initio results for isomers of the water hexamer cluster [11].

Fluctuating charge as a mechanism of polarization was first introduced by Berne et al. in their SPC-FQ and TIP4P-FQ potentials [45], which are rigid with either the TIP4P or SPC geometry. An advantage of these models is that the force calculation is less time-consuming than the usual iteration to obtain the induced dipole, and the intermolecular interactions (apart from the van der Waals term) retained the Coulombic form with simple charge–charge interactions. Using the extended Lagrangian formalism, they were only 1.1 times slower than the corresponding fixed point charge models. Unfortunately, SPC-FQ and TIP4P-FQ have no out-of-plane polarizability [28].

The charge-on-spring models were parameterized to fit experimental quantities like the heat of vaporization and the density [42]. The SWM4-DP and SWM4-NDP (Drude) potentials were also parameterized to bulk experimental properties [41, 47]. The result is a good description of ambient liquid water, but not of clusters [28]. The SWM4-DP monomer has a much lower polarizability of 1.04 Å^3 as opposed to the experimental gas-phase value of 1.44 Å^3. However, it should be noted that the goal has been to efficiently introduce polarization into large biological simulations, for example, proteins, which are typically carried out at ambient conditions. The models now encompass not simply water, but other molecules like alkanes and aromatic compounds [48–50].

3.2 Ab initio-based models

Recent advances in electronic structure theory have made accurate high-level calculations on water clusters easily accessible, and force fields have been accordingly parameterized based upon two broadly different decomposition schemes of the ab initio energy:

1. Consider a system of n water molecules. For the sake of simplicity, let them be rigid, with the monomer energy denoted U_1. The energy of a dimer formed by molecules i and j is $U(ij)$, the energy of a trimer of molecules i, j, and k is $U(ijk)$, and so on. The ab initio interaction energy $\Delta U_n(1, 2, 3, \ldots, n)$ can be decomposed with a cumulant expansion [12, 51–53],

72

$$\Delta U_n = U(1,2,3,\ldots,n) - nU_1$$
$$= \sum_{i=1}^{n-1} \Delta U(ij) + \sum_{i=1}^{n-2}\sum_{j>i}^{n-1}\sum_{k>j}^{n} \Delta U(ijk) + \cdots \tag{15}$$

where $\Delta U(ij)$ and $\Delta U(ijk)$ are the two- and three-body energies,

$$\Delta U(ij) = U(ij) - 2U_1, \tag{16}$$
$$\Delta U(ijk) = U(ijk) - 3U_1 - (\Delta U(ij) + \Delta U(ik)$$
$$+ \Delta U(jk)). \tag{17}$$

This many-body decomposition scheme has been used to develop a number of water models. For example, in DPP [7] and DPP2 [11], the Thole damping of the polarizability was parameterized, so that the model recovers the three-body energy of water hexamers and the remaining terms were fit to the energies of a selection of water dimers (two-body energies).

2. One may also decompose the interaction energy into distinct physical contributions like the electrostatic component, the induction term, and the dispersion term. One of the earliest schemes is the Kitaura–Morokuma analysis [54]. Recently, the ALMO [55] and the SAPT [56] decompositions have been used to develop model potentials.

The two classes of decomposition method are not independent. For example, with SAPT, one can determine the two- and three-body contributions to the polarization energy, the exchange energy, and so on. These kinds of decompositions were used to parameterize, in part, the CC-pol [9] model, while DPP2 also used them to parameterize terms like the two-body dispersion energy.

One of the more widely used ab initio-based models is the AMOEBA potential [6]. It is flexible, with simple expressions for the zero-field monomer energy, including anharmonic contributions, for the bond stretching and bending terms, along with a term that couples the stretch and the bend. AMOEBA has permanent multipoles up to the quadrupole on each atom, obtained from distributed multipole analysis [57]. The atomic polarizabilities were taken from the work of Thole and, since intramolecular induced dipole–induced dipole interactions are allowed, Thole damping is included, with exponent $b = 3$. The damping constant, a, was determined by a fit to water cluster energies, and the van der Waal's interaction terms were fit so as to reproduce both the water dimer energies and some bulk properties like the liquid density.

Many of the models discussed so far like DPP, DPP2, CC-Pol, and TTM [58] are rigid. As one goes from the gas phase to the liquid, the HOH angle goes from $104.5°$ to around $107°$ [59]. However, most flexible potentials exhibit a decrease in the angle. In AMOEBA, the equilibrium angle in the monomer has been set to $108.5°$ to get agreement with liquid properties [6].

Accurate fits of the water monomer potential energy and dipole surface by Partridge and Schwenke [60], based on high-level ab initio calculations, proved to be a key ingredient in the development of accurate flexible members of the previously rigid TTM family by Burnham and Xantheas (TTM2-F [61]), Fanourgakis and Xantheas (revised TTM2-F [62] and TTM3-F [38]), and Burnham et al. (TTM4-F [37]). An important aspect of these models is their ability to reproduce the increase in the HOH angle upon condensation, as well as the vibrational spectra of water clusters, liquid water, and ice. The earlier flexible TTM potentials were not as successful as the later versions, namely TTM3-F and TTM4-F.

Interestingly, water in the gas phase dissociates to neutral radicals, which is the limiting case of the Partridge–Schwenke dipole surface [38]. However, in the liquid, the dissociation products are ions, and therefore, TTM3-F uses an appropriately modified dipole surface [38]. On the other hand, TTM4-F attempts to reproduce both the monomer dipole surface and the polarizability surface [37]. One should note that these models were parameterized using the results of electronic structure calculations and thus lack the effect of nuclear quantization, which could be included by the use of path integral methods like centroid molecular dynamics [63–65].

While most ab initio-based polarizable models include only dipole polarizabilities, the ASP potential of Millot and Stone includes both dipole and quadrupole polarizabilities [66]. It was parametrized to fit electronic structure calculations on both the monomer and the dimer and was later reparameterized (ASP-VRT) to fit spectroscopic data from vibrational-rotational-tunneling experiments [67].

The list of models mentioned in this section is by no means exhaustive. The broadly applicable algorithms for polarizable force field development, SIBFA (Sum of Interactions Between Fragments Ab initio computed) [68] and EFP (Effective Fragment Potential) [69], based on ab initio energy decomposition schemes, have associated water potentials. The GEM model of Piquemal et al. [70] is an extension of the SIBFA method wherein the atomic charge densities were fit to Gaussian s-type functions. As mentioned earlier, water acts as a "test subject" for development, and in that spirit, the SIBFA and the AMOEBA approaches have been extended to systems of biological interest [71, 72].

Although the polarizable potentials mentioned above can be around twenty times slower than simple point charge models in MD simulations [73], that price must be paid for a realistic description of systems with significant many-body effects, such as charged systems like excess electrons in water and heterogeneous systems like water in cavities [74–76].

3.3 POLIR potential

As mentioned in the introduction, POLIR is a hybrid water potential. It has permanent charges, permanent dipoles, and dipole polarizabilities on the O and H atoms. The charges depend on the internal geometry. If one labels one of the hydrogen atoms as H_1 and the other as H_2, then the charge on H_1 is given by

$$q_{H_1} = q^0 + c1 \cdot (r_{OH_1} + r_{OH_2} - 2 \cdot r_{OH}^{eq}) + c3 \cdot (r_{OH_1} - r_{OH_2}) \tag{18}$$

The charge on H_2 can be similarly written, and the charge on the O atom is given by $-(q_{H_1} + q_{H_2})$. The interaction terms are intermolecular charge–charge, inter- and intra-molecular charge–permanent dipole and charge-induced dipole, dipole–dipole (which includes both permanent and induced dipoles), and intermolecular van der Waals, a function of the O–O distance. The charge–charge, charge–dipole, and dipole–dipole interactions are all damped using the Thole method. The zero-field monomer energy is given by the Partridge–Schwenke function, and therefore, in order to avoid double counting, the zero-field intramolecular charge–dipole and dipole–dipole energies are subtracted from the total energy.

POLIR takes nothing but the monomer potential energy function from Partridge–Schwenke. The charge, q^0, and all other elements of the dipole surface are determined in the fitting process and have no relation to PS. Note also that these parameters have a single value, they are not redetermined for different states of water.

The Thole damping constants, the constants $c1$ and $c3$ in the geometry-dependent charges, the polarizabilities, and the van der Waals terms are determined by fits to properties of the monomer (dipole surface, polarizability surface, etc.), dimer PES scan, and the pentamer minimum energy.

POLIR is constructed with the idea that polarization is so important for spectroscopy that the quality of the spectroscopic predictions is a good indicator of the quality of the overall description of polarization. Correspondingly, optimizing a potential for vibrational and intermolecular spectra is a good way to optimize the treatment of polarization. The equilibrium hydrogen charge, q^0, is determined primarily by fitting the O–H stretch region of the IR spectrum in the ambient liquid. Every time q^0 was varied, the potential parameters were refit to ensure that the potential reproduced the dimer PES, the monomer properties (e.g. the dipole moment) etc.

In the fit process, the correct frequency and absolute intensity occur simultaneously. Since the intensity is dominated by induced dipoles [77], the importance of polarization for the frequency shift is strongly suggested. Perhaps interestingly, the result is $q^0 = 0.42$, the value obtained [20] in the SPC/E potential, which attempts an effective treatment of polarization within a pairwise additive approach.

Note that POLIR is the only potential that includes the charge–dipole interaction on bonded OH atoms. It is difficult to see why this term should be negligible for vibrational spectroscopy. When an induced dipole changes, physics requires that its interaction with the charge on a bonded neighbor changes, which changes the vibrational frequency.

POLIR is designed to be used with classical simulations, in conjunction with the most simple, essential quantum corrections. In calculating the IR spectra, we use an harmonic intensity correction and the Berens–Wilson anharmonic frequency correction [78], which is a redshift of 184 cm^{-1} in the monomer, and, most encouragingly, not much different in the liquid and ionic solvation shell. Further details of the parameterization and the calculation of the IR spectra are given in Refs. [10, 33].

Because POLIR has been parameterized partially to the liquid-state experimental IR spectrum, and because of the anharmonic frequency correction, some effects of nuclear quantization have been implicitly taken into account. Therefore, an explicit treatment, for example, with centroid MD, would lead to some amount of double counting and does not fit with our approach. The internal energy of the ambient liquid using POLIR is -12.8 kcal/mol, and the experimental value is -9.9 kcal/mol [21]. The discrepancy could be due to nuclear quantization, but the agreement seems satisfactory for a quantity that was not incorporated in the parameterization. Although we have already demonstrated transferability for the potential energy of minimum energy configurations of clusters, this could be an issue for the average total energy.

3.3.1 Cluster properties

The original POLIR paper examined the IR harmonic spectra of a few water clusters [10]. Now, we explore in greater detail the energetic properties of an important water cluster—the hexamer; significant low-lying isomers are shown in Fig. 2. We compare the energies calculated with POLIR and other polarizable potentials with the ab initio energies (CCSD(T) [79] energies). Since some of these models are rigid, with the bond angles and bond lengths set to the gas-phase monomer, we took the ab initio clusters (from Ref. [11]) optimized with rigid monomer constraints. From the root mean square deviation in Table 1, it is clear that POLIR does very well in predicting the cluster energies (Fig. 3). We performed a many-body decomposition of the energies, as in Sect. 3.2, and the three-body energies (Fig. 4, Table 2) are once again quite impressive, all the more so given that the model was not parameterized to any hexamer properties.

Fig. 2 The four isomers of the
hexamer water cluster

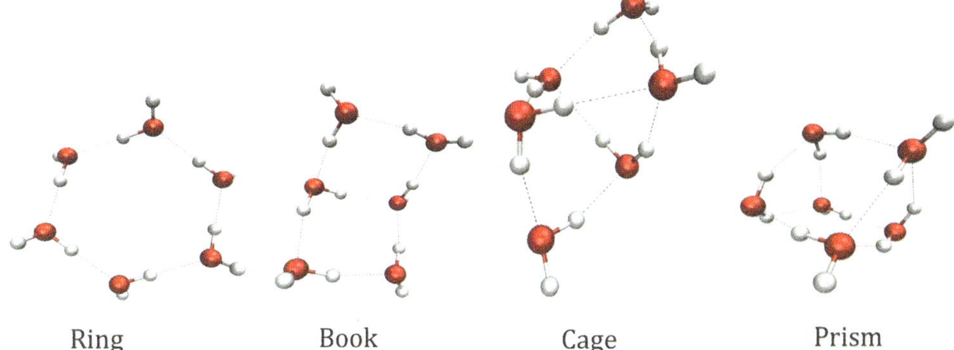

Table 1 The root mean square deviation of the energies of the four isomers of the water hexamer cluster for the various models with respect to the ab initio CCSD(T) values

Model	RMSD with respect to CCSD(T)
DC	5.50
AMOEBA	0.47
TTM3-F	3.12
POLIR	0.39

The DC, AMOEBA, TTM3-F, and CCSD(T) results were taken from Ref. [11]

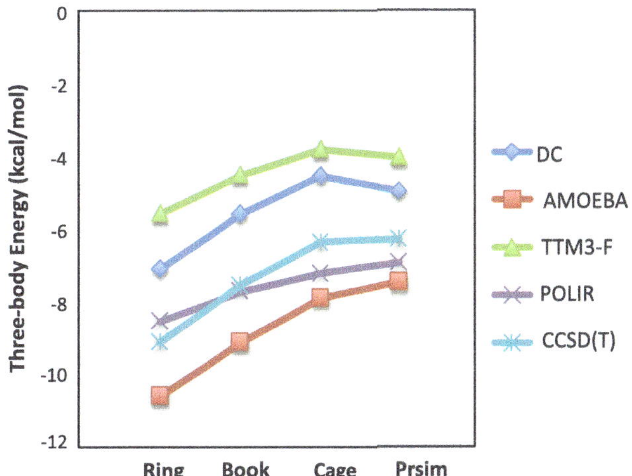

Fig. 4 The three-body energies of the four isomers of the water hexamer cluster for the various models along with the ab initio CCSD(T) results. The DC, AMOEBA, TTM3-F, and CCSD(T) results were taken from Ref. [11]

Table 2 The root mean square deviation of the three-body energies of the four isomers of the water hexamer cluster for the various models with respect to the ab initio CCSD(T) values

Model	RMSD with respect to CCSD(T)
DC	1.79
AMOEBA	1.43
TTM3-F	2.87
POLIR	0.61

The DC, AMOEBA, TTM3-F, and CCSD(T) results were taken from Ref. [11]

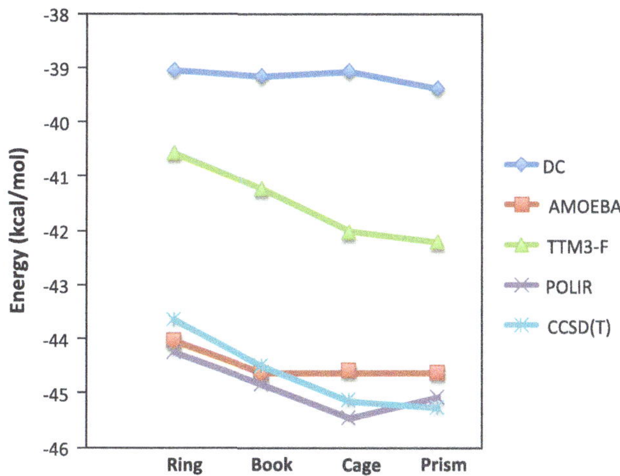

Fig. 3 The energies of the four isomers of the water hexamer cluster for the various models along with the ab initio CCSD(T) results. The DC, AMOEBA, TTM3-F, and CCSD(T) results were taken from Ref. [11]

Raman spectra of POLIR with experimental results. We also examined the HOH angle distribution.

3.3.2 Liquid-state properties

In order to calculate liquid-state properties, we performed MD simulations in the NVE ensemble at 300 K with periodic boundary conditions, using the Ewald summation [80] technique. The simulation box contained 128 water molecules at a density of 0.997 g/cc. The simulations were used to compare the rotational dynamics as well as the

3.3.2.1 Rotational dynamics Mankoo and Keyes [10] obtained a diffusion constant of 2.4 cm^2/s, close to the experimental value of 2.3 cm^2/s [81]. Here, we continue the test of the dynamical aspects of POLIR. The rotational dynamics of water have been studied in great detail using both IR spectroscopy and NMR techniques [82–84]. These

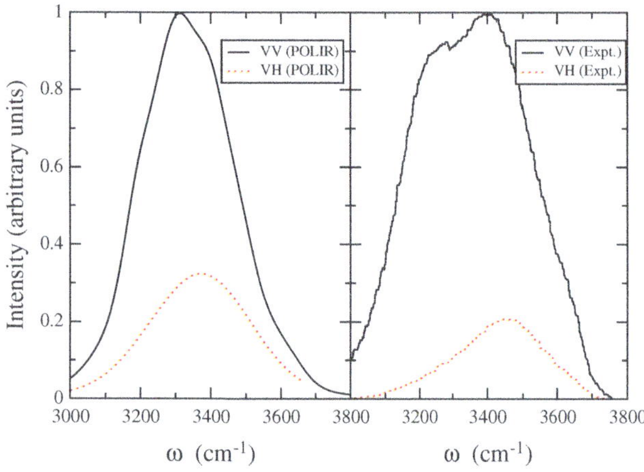

Fig. 5 The Raman spectra from POLIR and the experimental Raman spectra from Ref. [82]

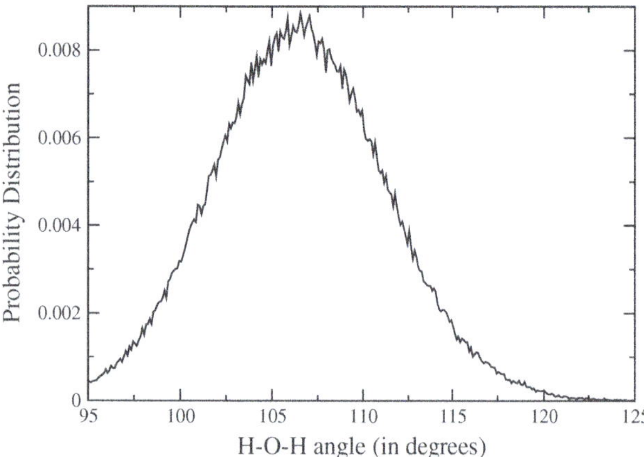

Fig. 6 The distribution of HOH angles in liquid water calculated with POLIR

experiments provide the characteristic rotational times for various vectors like the H–H vector, the O–H vector, and the vector normal to the plane of the molecule. The NMR rotational time for any unit vector, $\hat{u}$, is given by the integral (from 0 to ∞) of the correlation function, $C_{uu}(t)$, where

$$C_{uu}(t) = \langle P_2(\hat{u}(t) \cdot \hat{u}(0)) \rangle, \tag{19}$$

and $P_2(x)$ is the second Legendre function. We obtain a decay constant of 2.9 ps for both the H–H and O–H vectors, which compares well to the experimental values of 2.36 ps [82] and 2.4 ps [84], respectively. For the unit vector normal to the plane of the molecule, we find a characteristic time of 1.7 ps, while the experimental value is 1.8 ps [84].

3.3.2.2 Raman spectroscopy Raman spectra (Fig. 5) were calculated from the correlation functions of the appropriate elements of the polarizability tensor of the entire system [85],

$$I_{VV}(\omega) \propto Q(\omega) \int\limits_{-\infty}^{\infty} e^{-i\omega t} \langle \bar{\alpha}(0) \cdot \bar{\alpha}(t) + \frac{2}{15} Tr(\boldsymbol{\beta}(0) \cdot \boldsymbol{\beta}(t))) \rangle \mathrm{d}t \tag{20}$$

and

$$I_{VH}(\omega) \propto Q(\omega) \int\limits_{-\infty}^{\infty} e^{-i\omega t} \langle \frac{1}{10} Tr(\boldsymbol{\beta}(0) \cdot \boldsymbol{\beta}(t)) \rangle \mathrm{d}t \tag{21}$$

Here, $\bar{\alpha}$ is one-third the trace of the polarizability tensor, $\boldsymbol{\alpha}$, and $\boldsymbol{\beta}$ is the traceless anisotropic part.

$$\boldsymbol{\alpha} = \bar{\alpha}\boldsymbol{I} + \boldsymbol{\beta}. \tag{22}$$

In Fig. 5, we present the VV and VH spectra for POLIR in the O–H region, as well as the corresponding experimental spectra [86]. The calculated spectra are in reasonable agreement with experiment, although our VV line is somewhat too narrow. The Raman spectra were not used in the parameterization, and hence, this qualitative agreement is a true test of POLIR.

3.3.2.3 Bond angle POLIR gives an average bond angle of around 107°, in good agreement with experiment. We calculated the distribution of the bond angles (Fig. 6) and found it to be surprisingly broad, with significant populations ranging from as low as 95° to as high as 112°! This matches results from ab initio simulations [87] to the TTM2-F potential [59]. It is puzzling that rigid models do as well as they do in describing the liquid, given that they lack the high angular flexibility shown by "real" water. Examining the compensating effects in these rigid models might prove interesting.

4 Conclusion

For many years, theoretical and computational chemistry and physics focused on pair potentials, due to their tractability. It is now clear, however, that many-body energies and forces are not small corrections in water; they are extremely important. The need to go beyond pairwise additivity has been masked somewhat by the use of pair potentials optimized for a particular thermodynamic state, providing an effective treatment of many-body effects. However, such potentials are not transferable.

Polarization is both a real many-body effect and allows parameterization of other many-body effects. Consequently, considerable recent effort has been devoted to the development of polarizable potentials for water, with application to environments ranging from small clusters to the condensed phases. These models have been parameterized to fit either experimental properties or ab initio calculations on clusters, or in some cases both ("hybrid").

The importance of polarization is perhaps best seen by expanding the list of "test" properties beyond the usual pair distribution function, diffusion constant, etc., to spectroscopic properties. The vibrational IR intensity/molecule increases by a factor of $18\times$ upon condensation [88], and the increase is due to polarization. The IR spectrum is seeing induced dipoles; that is, if the dipole is decomposed into its permanent and induced parts, the dipole correlation that determines the spectrum is dominated by the induced-induced term [77].

The POLIR potential was developed based on the idea that spectroscopic properties must enter the parameterization, as optimizing the spectra optimizes the treatment of polarization, which is important for everything else. POLIR has shown promising results for clusters, liquid water, solvated ions [33], and ice. While flexible, polarizable, models like POLIR are computationally challenging, the accurate treatment of the many-body polarization forces results in a transferable force field capable of reproducing spectroscopic properties.

Acknowledgments Acknowledgment is made to the donors of the American Chemical Society Petroleum Research Fund and to the National Science Foundation (grant CHE 0848427) for support of this research. The authors thank Prof. Brian Space for the digitized data of the experimental Raman spectrum and Prof. Kenneth Jordan for the structures of the isomers of the water hexamer cluster.

References

1. Caldwell J, Dang LX, Kollman PA (1990) J Am Chem Soc 112(25):9144–9147
2. Cieplak P, Kollman P, Lybrand T (1990) J Chem Phys 92(11):6755–6760
3. Dang LX (1992) J Chem Phys 97(4):2659–2660
4. Bernardo DN, Ding Y, Krogh-Jespersen K, Levy RM (1994) J Phys Chem 98(15):4180–4187
5. Dang LX, Chang T-M (1997) J Chem Phys 106:8149
6. Ren P, Ponder J (2003) J Phys Chem B 107:5933
7. Defusco A, Schofield DP, Jordan KD (2007) Mol Phys 105:2681
8. Li J, Zhou Z, Sadus RJ (2007) J Chem Phys 127(15):154509
9. Bukowski R, Szalewicz K, Groenenboom GC, van der Avoird A (2007) Science 315:1249
10. Mankoo PK, Keyes T (2008) J Chem Phys 129(3):034504
11. Kumar R, Wang F-F, Jenness GR, Jordan KD (2010) J Chem Phys 132(1):014309
12. Xantheas SS (2000) Chem Phys 258(2–3):225–231
13. Dyke TR, Muenter JS (1973) J Chem Phys 59(6):3125–3127
14. Gubskaya AV, Kusalik PG (2002) J Chem Phys 117(11):5290–5302
15. Adams D (1981) J Nat 293:447–449
16. Astrand PO, Linse P, Karlstroem G (1995) Chem Phys 191(1–3):195–202
17. Liu Y-P, Kim K, Berne BJ, Friesner RA, Rick SW (1998) J Chem Phys 108(12):4739–4755
18. Thole BT (1981) Chem Phys 59(3):341–350
19. Berendsen HJC, Postma WF, van Gunsteren WF, J, H (1981) Intermolecular forces. Reidel, Dordrecht
20. Berendsen HJC, Grigera JR, Straatsma TP (1987) J Phys Chem 91:6269
21. Jorgensen WL, Chandrasekhar J, Madura JD, Impey RW, Klein ML (1983) J Chem Phys 79:926
22. Mahoney MW, Jorgensen WL (2001) J Chem Phys 114:363
23. Horn HW, Swope WC, Pitera JW, Madura JD, Dick TJ, Hura GL, Head-Gordon T (2004) J Chem Phys 120(20):9665–9678
24. Rick SW (2004) J Chem Phys 120(13):6085–6093
25. Abascal JLF, Sanz E, Fernández RG, Vega C (2005) J Chem Phys 122(23):234511
26. Abascal JLF, Vega C (2005) J Chem Phys 123(23):234505
27. Kumar R, Christie RA, Jordan KD (2009) J Phys Chem B 113(13):4111–4118
28. Kiss PT, Baranyai A (2009) J Chem Phys 131(20):204310
29. Vega C, McBride C, Sanz E, Abascal JLF (2005) Phys Chem Chem Phys 7:1450–1456
30. Fernández RG, Abascal JLF, Vega C (2006) J Chem Phys 124(14):144506
31. Gregory JK, Clary DC, Liu K, Brown MG, Saykally RJ (1997) Science 275(5301):814–817
32. Wu JC, Piquemal J-P, Chaudret R, Reinhardt P, Ren P (2010) J Chem Theory Comput 6(7):2059–2070
33. Kumar R, Keyes T (2011) J Am Chem Soc 133:9441–9450
34. Sala J, Guàrdia E, Masia M (2010) J Chem Phys 133(23):234101
35. Stone AJ (1996) The theory of intermolecular forces. Clarendon Press, Oxford
36. Keyes T, Napoleon RL (2011) J Phys Chem B 115(3):522–531
37. Burnham CJ, Anick DJ, Mankoo PK, Reiter GF (2008) J Chem Phys 128(15):154519
38. Fanourgakis GS, Xantheas SS (2008) J Chem Phys 128(7):074506
39. van Duijnen PT, Swart M (1998) J Phys Chem A 102(14):2399–2407
40. Masia M, Probst M, Rey R (2005) J Chem Phys 123(16):164505
41. Lamoureux G, Alexander D, MacKerell J, Roux B (2003) J Chem Phys 119(10):5185–5197
42. Yu H, Hansson T, van Gunsteren WF (2003) J Chem Phys 118(1):221–234
43. Yu H, van Gunsteren WF (2004) J Chem Phys 121(19):9549–9564
44. Rick SW, Stuart S (2002) J rev Comput Chem 18:89–146
45. Rick SW, Stuart SJ, Berne BJ (1994) J Chem Phys 101(7):6141–6156
46. Applequist J, Carl JR, Fung K-K (1972) J Am Chem Soc 94(9):2952–2960
47. Lamoureux G, Harder E, Vorobyov IV, Roux B Jr, ADM (2006) Chem Phys Lett 418(1–3):245–249
48. Vorobyov IV, Anisimov VM, MacKerell AD (2005) J Phys Chem B 109(40):18988–18999
49. Lopes PEM, Lamoureux G, Roux B, MacKerell AD (2007) J Phys Chem B 111(11):2873–2885
50. Baker CM, Lopes PEM, Zhu X, Roux B, MacKerell AD (2010) J Chem Theor Comput 6(4):1181–1198
51. Kumar R, Skinner JL (2008) J Phys Chem B 112(28):8311–8318
52. Harkins D, Moskowitz J, Stillinger FH (1970) J Chem Phys 53:4544
53. Xantheas S (1994) J Chem Phys 100:7523
54. Kitaura K, Morokuma K (1976) Int J Quantum Chem 10(2):325–340
55. Khaliullin RZ, Cobar EA, Lochan RC, Bell AT, Head-Gordon M (2007) J Phys Chem A 111(36):8753–8765
56. Jeziorski B, Moszyński R, Szalewicz K (1994) Chem Rev 94:1887–1930
57. Stone AJ, Alderton M (1985) Mol Phys 56:1047
58. Burnham CJ, Xantheas SS (2002) J Chem Phys 116:1500
59. Fanourgakis GS, Xantheas SS (2006) J Chem Phys 124(17):174504

60. Partridge H, Schwenke DW (1997) J Chem Phys 106(11):4618–4639
61. Burnham CJ, Xantheas SS (2002) J Chem Phys 116(12):5115–5124
62. Fanourgakis G, Xantheas S (2006) J Phys Chem A 110(11):4100–4106
63. Cao J, Voth GA (1993) J Chem Phys 99(12):10070–10073
64. Cao J, Voth GA (1994) J Chem Phys 100(7):5093–5105
65. Cao J, Voth GA (1994) J Chem Phys 100(7):5106–5117
66. Millot C, Stone A (1992) Mol Phys 77:439–462(24)
67. Goldman N, Leforestier C, Saykally R (2005) J Philos Trans A Math Phys Eng Sci 363(1827):493–508
68. Piquemal J-P, Chelli R, Procacci P, Gresh N (2007) J Phys Chem A 111(33):8170–8176
69. Day PN, Jensen JH, Gordon MS, Webb SP, Stevens WJ, Krauss M, Garmer D, Basch H, Cohen D (1996) J Chem Phys 105(5):1968–1986
70. Piquemal J-P, Cisneros GA, Reinhardt P, Gresh N, Darden TA (2006) J Chem Phys 124(10):104101
71. Ponder JW, Wu C, Ren P, Pande VS, Chodera JD, Schnieders MJ, Haque I, Mobley DL, Lambrecht DS, DiStasio RA, Head-Gordon M, Clark GNI, Johnson ME, Head-Gordon T (2010) J Phys Chem B 114(8):2549–2564
72. Roux C, Bhatt F, Foret J, de Courcy B, Gresh N, Piquemal J-P, Jeffery CJ, Salmon L (2011) Proteins Struct Funct Bioinf 79(1):203–220
73. Fanourgakis G, Tipparaju V, Nieplocha J, Xantheas S (2007) Theor Chem Acc 117:73–84
74. Jorgensen WL, McDonald NA, Selmi M, Rablen PR (1995) J Am Chem Soc 117(47):11809–11810
75. Sommerfeld T, DeFusco A, Jordan KD (2008) J Phys Chem A 112(44):11021–11035
76. Bucher D, Rothlisberger U (2010) J Gen Physiol 135(6):549–554
77. Ahlborn H, Xingdong J, Space B, Moore P (1999) J Chem Phys 111:10622
78. Berens P, Wilson KR (1981) J Chem Phys 74:4872
79. Raghavachari K, Trucks GW, Pople JA, Head-Gordon M (1989) Chem Phys Lett 157(6):479–483
80. Smith W (1998) CCP5 Newslett 46:18
81. Krynicki K, Green CD, Sawyer DW (1978) Faraday Discuss Chem Soc 66:199–208
82. Jonas J, DeFries T, Wilbur DJ (1976) J Chem Phys 65(2):582–588
83. Tan H-S, Piletic IR, Fayer MD (2005) J Chem Phys 122(17):174501
84. Ropp J, Lawrence C, Farrar TC, Skinner JL (2001) J Am Chem Soc 123(33):8047–8052
85. McQuarrie DA (2000) Statistical mechanics. University Science Books, Mill Valley
86. Hare D, Sorensen C (1990) J Chem Phys 93:25–33
87. Cristobal A, Kyoungrim B, Jiali G (1998) In: Combined quantum mechanical and molecular mechanical methods, chap 4. American Chemical Society, Washington, DC, pp 35–49
88. Bertie JE, Ahmed MK, Eysel HH (1989) J Phys Chem 93(6):2210–2218

Theor Chem Acc (2012) 131:1159
DOI 10.1007/s00214-012-1159-1

Polarization effects in protein–ligand calculations extend farther than the actual induction energy

Pär Söderhjelm

Received: 11 May 2011 / Accepted: 30 September 2011 / Published online: 2 March 2012
© Springer-Verlag 2012

Abstract The various roles that polarizabilities play in the calculation of protein–ligand interaction energies with a polarizable force field are investigated, and the importance and distance dependence of some common approximations is determined for each of these roles separately, using quantum-mechanical calculations as the reference. It is found that the pure induction energy, if defined as the energetic gain from the charge redistribution upon interaction between the protein and ligand, is a rather short-ranged effect that becomes independent of the exact implementation at distances above ~ 4 Å. On the other hand, the polarization between the protein residues in the assembly of a protein from separately computed fragments (as is routinely done in force field development) has a significant effect on the computed interaction energies, even for residues as far as 15 Å from the ligand. Finally, polarization improves the transferability of partial charges, but the simple polarization model used in, for example, the Amber force field explains only 14–19% of the conformational variation of the charges. In all cases, more advanced polarization models, especially involving anisotropic polarizabilities, seem to give significantly better descriptions of these effects. The study suggests that an accurate treatment of polarization can be important even in systems where the actual induction energy is small in magnitude.

Keywords Polarizable force field · Induction · Distributed polarizabilities · Amber · Intermolecular interactions · Ligand binding

1 Introduction

The reliability of a molecular simulation (e.g., a molecular dynamics or Monte Carlo simulation) is limited in at least three ways. First, the system that you simulate is typically much smaller than the real system. Second, the amount of sampling you can afford may not be enough to cover all relevant regions of phase space sufficiently well to give statistically converged results. Third, the potential-energy surface that you use in the simulation is only an approximation to the real (normally unknown) potential-energy surface.

For complex systems, such as macromolecules in water, the requirements in the two first aspects normally force you to use an empirical molecular mechanics (MM) force field as your potential-energy function. It has been noted that as computer resources are growing, the accuracy of the force field may become the limiting factor in many applications [1]. Force fields used in this area usually divide the potential energy into bonded and nonbonded terms. The bonded terms typically consist of bond, angle, and dihedral terms and describe the short-range part of the intramolecular interactions. The nonbonded terms include at least an electrostatic term and a Van der Waals term (containing dispersion and repulsion). It describes intermolecular interactions, as well as intramolecular interactions between

Published as part of the special collection of articles: From quantum mechanics to force fields: new methodologies for the classical simulation of complex systems.

Electronic supplementary material The online version of this article (doi:10.1007/s00214-012-1159-1) contains supplementary material, which is available to authorized users.

P. Söderhjelm (✉)
Department of Chemistry and Applied Biosciences,
Computational Science, ETH Zürich, USI-Campus,
via Giuseppe Buffi 13, 6900 Lugano, Switzerland
e-mail: par.soderhjelm@phys.chem.ethz.ch

parts of a big molecule that are not directly bound, but may be close in space.

Although it is widely recognized that electrostatics is a key to many interesting problems (e.g., protein–ligand interaction), most force fields still use a simple Coulomb interaction between atom-centered fixed partial charges. A promising route to increasing the accuracy is to explicitly include electronic polarization. The most common methods use either distributed point polarizabilities, fluctuating charges, or shell models [2–4]. Other ways to improve the electrostatics are to include higher-order multipoles (dipoles, quadrupoles, etc.) [5] or to use smeared charges that describe the charge distribution better than point charges [6].

Polarizable force fields date back to the 1960s [7] and were early applied to biological systems [8, 9], but only during the last decade, they have been incorporated into the most widely used simulation packages for macromolecules, either as polarized variants of established force fields, for example, Amber-02 and PFF [10, 11], or de novo developments such as Amoeba [12]. All of these are based on atomic isotropic dipole polarizabilities. The models differ mainly in the specific values of the polarizabilities and in the treatment of intramolecular polarization, and these choices are related.

Two early models have particularly influenced the development. Applequist showed that by allowing full coupling between all induced dipoles, it was possible to find atomic polarizabilities that reproduce the total polarizabilities of many molecules [13], but the anisotropy was often overestimated. Thole modified this model by introducing damping functions that corrected some of the error from using point dipoles at short range [14, 15]. In this way, the atomic polarizabilities could have larger and more realistic values, and the excessive anisotropy could be avoided. Roughly speaking, the Amber-02 model is based on the Applequist polarizabilities, albeit with modified coupling rules, whereas the Amoeba force field is based on the Thole model.

Other polarizable force fields, such as SIBFA, EFP, and NEMO, aim specifically at reproducing quantum-mechanical (QM) interaction energies and have a more complex functional form [5, 16, 17]. They normally derive their polarizabilities directly from QM calculations on the interacting monomers. Typically, they do not include coupling between the polarizabilities and thus have to use anisotropic atomic polarizabilities to reproduce the anisotropy of the molecular polarizabilities. Damping functions can nevertheless be used to account for the lack of Pauli effects in the intermolecular interactions [18–20].

Despite that the atomic polarizabilities obtained from QM calculations are not very similar for atoms of a certain element or atom type [21], most empirical force fields use only 8–15 atom types for polarizabilities, with 1–4 different polarizabilities for each element for the normal amino acids; some even use the same value for all non-hydrogen atoms [8]. It has been shown that improved accuracy is obtained using specific atomic polarizabilities [22]; thus, polarizabilities may well be treated in the same way as partial charges. Very recently, we showed that conformationally averaged QM-derived polarizabilities are a good choice for such atomic polarizabilities, as they give results that are fairly close to those obtained with conformation-specific polarizabilities [21].

It is difficult to directly assess the accuracy of a certain polarization model, mainly because of two reasons. First, the potential energy surface is extremely difficult to measure experimentally. One is normally limited to using quantities like solvation free energies or binding free energies and indirectly test the force field's ability to reproduce these quantities [23], which of course is made more difficult by the other simulation issues mentioned above. Even if studies have started to appear that compare the performance of nonpolarizable and polarizable force fields for biological problems [24], the precision of these studies is usually too poor to allow any detailed assessment of the polarization model itself. Alternatively, one can compare the results with QM interaction energies of high quality [25]. Very recently, an article pair has been published that systematically tries several variants of intramolecular polarization and tests the accuracy of the corresponding polarization models [26, 27]. As a reference, the authors use QM interaction energies at the MP2/aug-cc-pVTZ level. They find that all variants are better than additive force fields in reproducing the total interaction energy, but that a damping function [28] combined with exclusion of polarization through 1 or 2 bonds gives the best results.

Second, only the total potential energy is an observable; its decomposition is ambiguous. Thus, any improvements in the accuracy of the polarization term may be obscured by inaccuracies in the other terms and limits in the parametrization procedure. For example, in the above-mentioned assessment [27], the charges were adequately refitted for each polarization model, but it is still possible that the Van der Waals parameters, which are normally fitted for a particular treatment of electrostatics, unintentionally favor any of the models. One way to solve this issue is to use a specific energy decomposition scheme, for example, the restricted variational space method [29, 30], to match each contribution individually [16]. Another way to deal with the problem is to use the change in electrostatic potential as a reference, either at the monomer [31] level or the dimer level [32]. A third way is to only consider the total energy, but reassure that all the remaining terms are treated in a consistent way, for example, by taking them from supermolecular calculations of smaller subsystems [33]. All these studies conclusively show that the inclusion of polarization can improve the accuracy of the total potential energy, but that a careful treatment is necessary for this to happen.

On the other hand, there are many issues that are common to all polarizable force fields and therefore can be studied in a more general sense. One example is whether the transferability of the partial charges between various systems and conformations is improved by the inclusion of polarization. Another relevant question (for reasons of computational efficiency) is how far out from the active site we have to keep a good description of the polarization, that is, where we can shift to a simpler model. In studies of protein–ligand interaction and protein shift of absorption spectra, we tried to answer the latter question [34, 35]. To our surprise, there was a long-range (10–20 Å) sensitivity to changes in the treatment of polarization; for example, the effect of switching from anisotropic to isotropic polarizabilities in all residues separated by more than 10 Å from the ligand was 13 kJ/mol for a charged ligand [34].

There are two contributions to this energy difference. First, there is the actual induction energy, caused by charge redistribution in the protein due to the electric field from the ligand and vice versa. Second, there is a portion of the electrostatic interaction energy between the protein and the ligand, caused by polarization of each protein residue in the electric field from other residues. The aim of the current study is to study these effects separately and characterize the error convergence of each of them. For completeness, we also consider a third effect that polarizabilities may have in force fields: we investigate whether the intramolecular polarization improves the transferability of the electrostatic model among various conformations, as earlier studies have indicated [36–39]. For all tests, we use the same test system as in ref. [34], the avidin–biotin interaction. From these tests, a more complete picture of the role of polarization in protein–ligand systems will be obtained.

2 Methods

2.1 The distributed point polarizability model

When a molecule is subjected to an electric field F, the induced dipole moment within the linear-response approximation is given by

$$\mu_{ind} = \alpha \cdot F \quad (\text{e.g. } \mu_x = \alpha_{xx} F_x + \alpha_{xy} F_y + \alpha_{xz} F_z) \tag{1}$$

where α is called the *polarizability tensor* of the molecule and is sometimes replaced by a scalar quantity (the *isotropic polarizability*), defined by $(\alpha_{xx} + \alpha_{yy} + \alpha_{zz})/3$.

To accurately describe the polarization occurring when two molecules interact, knowledge of the molecular polarizability tensors is normally not sufficient. First, the electric field is not homogeneous, that is, it varies in different parts of the molecule. Second, the induced dipole moment is seldom a useful quantity, because it does not specify the local charge redistribution. A solution to these problems is the *distributed point polarizability model*, which is the most common model in polarizable force fields, used in QM-mimicking methods such as SIBFA, EFP, and NEMO [5, 16, 17], as well as in simpler methods [8, 10, 12]. In this model, the response of each molecule is described as a set of induced dipoles, located at certain positions (most commonly the atomic nuclei). At the position of the polarizability α_i, the induced dipole μ_i^{ind} is given by:

$$\mu_i^{ind} = \alpha_i \cdot F_i = \alpha_i \cdot \left[F_i^{stat} - \sum_{j \neq i} g_{ij} \mu_j^{ind} \nabla\nabla\left(\frac{1}{r_{ij}}\right) \right] \tag{2}$$

where the electric field F_i in that position has been divided into contributions from the static charge distribution and from other induced dipoles, and a (possibly distance-dependent) scale factor g_{ij} has been introduced to allow for damping or exclusion of close-lying interactions. Equation 2 defines a linear system of equations, which can be solved through matrix inversion or by iteration.

The induction energy is given by

$$E^{ind} = -\frac{1}{2} \sum_{i=1}^{N} \mu_i^{ind} \cdot F_i^{stat} \tag{3}$$

where the factor 1/2 comes from the fact that the (positive) work of polarization cancels half of the interaction energy of the dipoles with the field. The energy contribution from a pair of induced dipoles is completely canceled by the work of polarization, so Eq. 3 contains only the static field; the coupling is included only through the values of μ_i^{ind}.

The static field F_i^{stat} is normally computed from the multipole expansion:

$$F_i^{stat} = -\sum_{j \neq i} h_{ij} \left[q_j \nabla\left(\frac{1}{r_{ij}}\right) + \mu_j \cdot \nabla\nabla\left(\frac{1}{r_{ij}}\right) \right.$$
$$\left. + \Theta_j \cdot \nabla\nabla\nabla\left(\frac{1}{r_{ij}}\right) \cdots \right] \tag{4}$$

where h_{ij} is another scale factor possibly modifying the field from close-lying multipoles. In the simplest versions of polarizable force fields, only the first term (i.e. partial charges) is used.

The reasons for letting g_{ij} or h_{ij} deviate from unity vary in the two cases. For the coupling between polarizabilities in Eq. 2, the induced dipoles may become infinite at small distances (the "polarization catastrophe"), and the polarization can become unphysical even before that happens. Although it is possible to reproduce the molecular polarizability by a set of fully coupled atomic polarizabilities, significantly smaller values must be used than with an uncoupled set [13]. A better solution is to damp the dipole interaction tensor for overlapping charge distributions [14, 28]. In the Amber polarizable force field (ff02) [10], no damping is used, but contributions from atoms that are directly bound (1–2 interactions) or separated by two bonds (1–3 interactions) are omitted, and the 1–4 interactions are scaled by 5/6. In contrast, for most methods using polarizabilities from QM calculations, the intramolecular coupling of polarizabilities is completely omitted, because it is already implicitly included through the QM calculation. The explicit and implicit coupling schemes have been compared [40]. For medium-sized molecules such as amino acids, the implicit scheme has been found adequate (except possibly for the aromatic amino acids) [32]. A mixed approach have been proposed for fragmentation approaches [33]; this scheme is illustrated in Fig. 1 and will be denoted the *LoProp exclusion rules*.

On the other hand, for the response to the static field, which we will refer to as *static polarization*, the exclusion rules are dictated by the parametrization procedure. Clearly, when the multipoles of a molecule are obtained from a single QM calculation, h_{ij} should be zero within the molecule, as all intramolecular effects are already included in the multipoles. A corresponding rule has been derived for fragmentation approaches [33], as illustrated in Fig. 1. However, the choice of h_{ij} is arbitrary, as long as the

permanent multipoles are adjusted accordingly, as in, for example, the Amber ff02 model, in which $h_{ij} = g_{ij}$ [10]. For a rigid molecule, this choice does not affect the total interaction energies, whereas for a flexible molecule, the static polarization can be used to approximate how the charge distribution varies with changes in the molecular geometry [28, 36, 38, 39, 41].

The individual electrostatic and induction contributions are not comparable for models with different exclusion rules [33]. Therefore, we define the *pure induction energy* between two molecules as the energy caused by the charge redistribution of each molecule in response to the field from the *other* molecule, that is, when making the particular choice $h_{ij} = 0$ for all intramolecular interactions. The pure induction energy is always negative, in contrast to the direct difference of the energies in Eq. 3 which can have any sign [33].

Several methods to obtain distributed polarizabilities from QM calculations have been suggested [3]. A common approach is to partition the molecular polarizabilities, either in real space (e.g., the atoms-in-molecules approach) [42] or in terms of the basis set [43–46]. Alternatively, the polarizabilities can be fitted to molecular polarizabilities or to large sets of induction energies [22, 47–50]. For the most accurate descriptions in this study, we use LoProp multipoles (up to quadrupoles, $L = 2$) and anisotropic polarizabilities, located at the atomic nuclei and the midpoints of covalent bonds [46]. The multipoles are obtained by distributing the molecular charge density using an orthogonal basis of "nearly local" functions; the polarizabilities are obtained by distributing the molecular response to homogeneous electric fields using the same basis. The accuracy of the LoProp method has been tested before [31].

2.2 System

As an example of a protein–ligand interaction, we use avidin interacting with biotin. This system has previously been subject to several studies [51–54]. In particular, it has been used for testing the conformational dependence of charges [55] and polarizabilities [21] and for benchmarking the *Polarizable Multipole Interaction with Supermolecular Pairs* (PMISP) method [33, 34, 56].

For all calculations, we use a single snapshot of the avidin–biotin system, obtained from a molecular dynamics simulation using the polarizable Amber-02 force field [10]. The setup of the molecular dynamics simulation has been described before (the 02ohp simulation in ref. [54]). The geometry corresponds to one of the snapshots used to compute binding affinities with the PMISP method [56], but we removed the three extra biotin molecules that were considered part of the protein. The coordinate file used in all tests is provided as supplementary information.

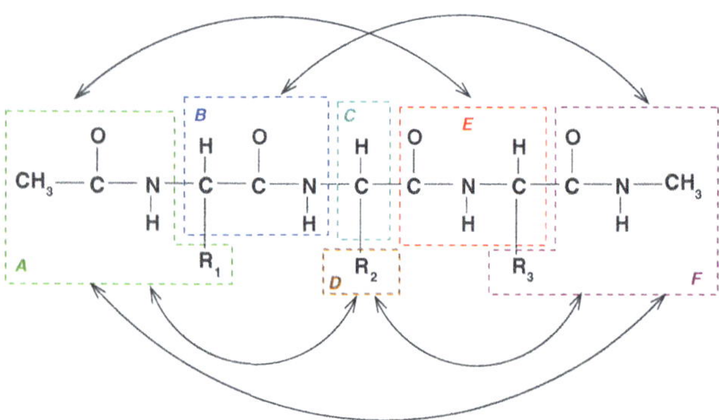

Fig. 1 Illustration of the MFCC procedure and the LoProp exclusion rules in the computation of properties (multipoles and polarizabilities) for a capped peptide. The atoms are divided into six groups. Individual QM calculations are performed on the subsystems *ABC*, *BCDE*, *CEF* and the concaps *BC* and *CE*, respectively. In all cases, the dangling bonds are capped with hydrogen atoms. The properties of the whole system are then computed as $TOT = ABC + BCDE + CEF − BC − CE$. As shown in ref. [33], intramolecular static polarization should only occur between atoms that have not been in the same QM calculation, that is, in this example only between the pairs of groups connected by *arrows*. The same rule is used for polarizability coupling

The protein is cut into 494 fragments: the residues of the protein (all being standard amino acids). The cuts are done through the peptide bonds and each fragment is capped with $-COCH_3$ and $-NHCH_3$ groups at the N and C termini, respectively. The four cystine linkages are cut through the disulfide bonds and each fragment is capped with a $-SCH_3$ group.

The protein fragments are mainly taken to interact separately with the ligand, but in some calculations fragment pairs are used, some of which are covalently linked. To assemble the properties (distributed multipoles and polarizabilities) of such fragment pair, the molecular fractionation with conjugate caps (MFCC) approach [57] is used (see Fig. 1), in which a third fragment (*concap*) is constructed, consisting of the capping groups of the two fragments merged together, for example, forming a $CH_3CONHCH_3$ molecule for a peptide link. The properties of the concap fragment are subtracted from the sum of properties of the two capped fragments, and the same procedure is used for energies [33].

The distance between the ligand and a fragment is defined as the minimal distance between any ligand atom and any fragment atom. Quantities like the distance-dependent mean absolute error are computed by dividing the systems into bins of width 1 Å according to their distance; the error at a distance r then includes systems with distance between r and $r + 1$ Å.

2.3 Computational details

In the test of induction energy, the supermolecular energies are first computed at the MP2/cc-pVTZ level, whereas the properties are computed at the B3LYP/6-31G* level, as this has been found to be a good approximation [34]. For the further analysis of the error, Hartree–Fock (HF) theory with the 6-31G* basis set is used for both supermolecular energies and properties. In the test of protein assembly, the HF/6-31G* level is used throughout. The derivation of charges in the test of conformational dependence is done at the B3LYP/cc-pVTZ level to be consistent with both the procedure used in Amber ff02 [10] and with the supermolecular MP2/cc-pVTZ energies. All supermolecular energies are corrected for basis set superposition errors by the *counterpoise* procedure [58].

All QM calculations are done with the MOLCAS software [59] except the derivation of charges in the test of conformational dependence, which is done using the Gaussian software [60] followed by the RESP procedure [61], possibly employed in an iterative procedure where the AMBER 10 program [62] is used to calculate the induced dipoles and the corresponding contribution to the electrostatic potential is subtracted from the QM potential [10]. Except for this procedure, all classical interaction energies are computed by local programs that can handle all sorts of exclusion rules. The conversion from the AMBER topology file is exact; for a given system, the local software gives the same polarization energy as AMBER.

3 Results

We investigate the effect of polarizabilities on the induction energy, protein assembly, and conformational dependence, respectively. The main tests performed are schematically summarized in Fig. 2.

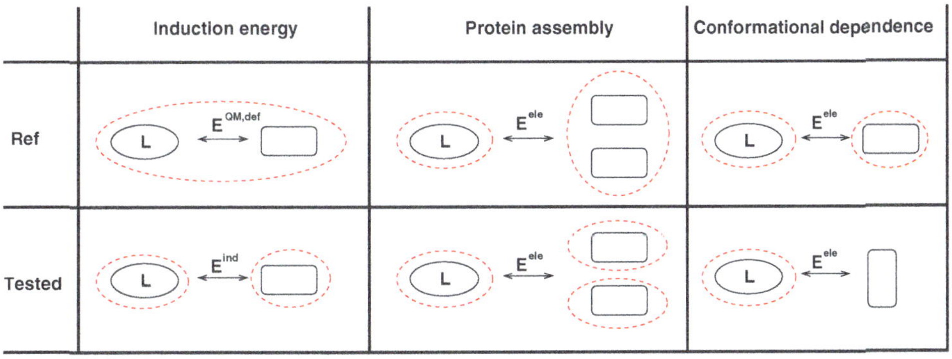

Fig. 2 Schematic overview of the three main tests performed in this article. Each column shows one test, labeled by the corresponding section heading and with the *top square* representing the reference quantity and the *bottom square* representing the tested quantity. The L-labeled *ellipse* is the ligand and the *rounded rectangle* any one of the (capped) protein residues. For the induction energy, the reference is the QM deformation energy, whereas the tested quantities are classical induction energies. For the protein assembly, the reference is the electrostatic interaction energy using multipoles computed from the residue pair (covalent or noncovalent neighbors) treated as a single molecule, whereas the tested quantity is the corresponding energy using multipoles assembled from separate residue calculations. For the conformational dependence, the reference is the electrostatic energy with the residue in its right conformation, whereas the tested quantities use charges not for the right conformation (e.g., averaged)

3.1 Induction energy

Long-range effects of approximations in the polarization model have been observed before [34, 35]. We now investigate how much of these effects are related to the pure induction energy, as opposed to the effects of the assembly of the protein from multiple fragments. The induction energy itself is inherently nonadditive. However, as the nonadditive effects have been analyzed before [33, 34], we focus instead on the interactions between the ligand and one fragment at a time.

The most natural question is how well the supermolecular energy is modeled by a given force field. In our case, we start with a very good description (a LoProp model with multipoles up to quadrupoles and anisotropic polarizabilities) and introduce approximations, one at a time. A similar test with a wider range of interactions has recently been performed for other polarization models [27].

To avoid careful parametrization of the short-range terms, we define the error as

$$R^{\text{tot}} = E^{\text{sup}} - E^{\text{ele}} - E^{\text{ind}} - E^{\text{vdw}} \qquad (5)$$

where E^{sup} is the BSSE-corrected supermolecular interaction energy, E^{ele} is the electrostatic energy, E^{ind} is the induction energy, and E^{vdw} is the Van der Waals term from the Amber force field [63]. It has been shown that, for distances larger than ~ 6 Å, the repulsion and dispersion contributions to the QM energy are well approximated by the Van der Waals term [34].

We calculated this error for all 494 fragment–ligand dimers. Unfortunately, even using a high multipole level, the errors in the electrostatic energy dominate in the whole distance range. Therefore, it was impossible to accurately assess the induction energy. For example, the isotropic approximation has a negligible effect compared with the total error, as shown in Fig. 3. Note that the fragments interacting directly with the ligand show substantial errors because of the crude short-range potential; for example, 10 fragments give an error larger than 2 kJ/mol. To proceed, we need to compare the induction energy with the particular part of the supermolecular interaction energy that depends on the deformation of the charge density, often called the deformation energy. However, this quantity is only rigorously defined at the HF level. Therefore, we first verified that the distribution of the errors (for distances larger than 6 Å) was roughly independent of the quantum-chemical method, with mean values for MP2 and HF of -0.03 and 0.06 kJ/mol, respectively, and standard deviations of 0.11 and 0.15 kJ/mol, respectively (the Van der Waals energy was only subtracted in the MP2 case as it contains mainly dispersion at these distances). At the HF level, we then define the error in the induction energy as

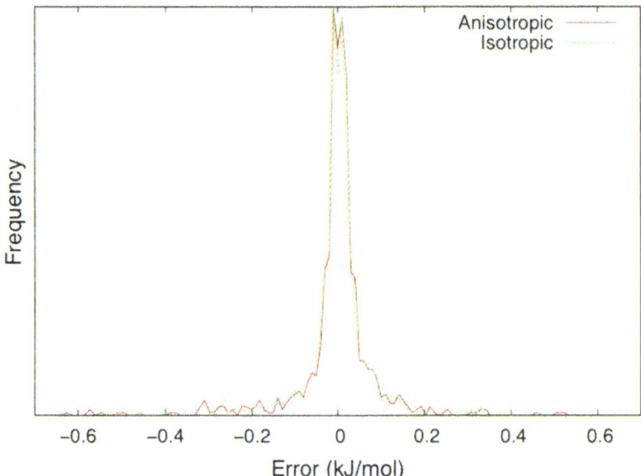

Fig. 3 Statistical distribution of the error in the total potential energy for each ligand–residue dimer computed with anisotropic and isotropic polarizabilities, respectively. The following outliers resulting from close interactions were removed from the figure: residue N12 (-10 kJ/mol error for anisotropic model), D13 (-1), Y33 (-4), T35 (-6), V37 (-1), T38 (-14), A39 (-19), T40 (-2), Q70 (3), F72 (3), S73 (-15), S75 (1), F79 (1), W97 (1), S101 (-1), N118 (-15), concap T38-A39 (-18), concap A39-T40 (-7), and concap F72-S73 (1 kJ/mol). The residue numbering refers to PDB structure 1AVD [64]

$$R^{\text{ind}} = E^{\text{sup}} - E^{\text{HL}} - E^{\text{ind}} \qquad (6)$$

where E^{HL} is the Heitler–London energy, also known as the first-order energy E_1, that is, the energy obtained in a supermolecular HF calculation if the unperturbed monomer orbitals are used without any subsequent SCF iteration. Because E^{HL} describes the electrostatic and exchange-repulsion energies exactly, the remaining error R^{ind} is a good indicator of the accuracy of the induction energy. The first two terms on the right-hand side of Eq. 6 constitute the deformation energy, but it should be noted that in this study, unlike many intermolecular decomposition schemes, this quantity is corrected for BSSE.

The average error as a function of the distance from the ligand is shown in Fig. 4 for three multipole levels ($L = 0$, $L = 1$, and $L = 2$, i.e. up to charges, dipoles, and quadrupoles, respectively) and two polarizability levels (anisotropic and isotropic). Higher multipoles ($L = 3$) were also tested but showed no significant difference from $L = 2$. Detailed results for representative complexes at different distances, as well as structures of these complexes, are given in Table S1 and Figure S1 in the supplementary information.

The large errors at short distances (below 4 Å) are mainly due to the short-range effects of charge penetration, charge transfer, and other neglected effects in the point-polarizability model. Note that the errors in the final force field would possibly be smaller, because you would try to model these effects (and the corresponding electrostatic

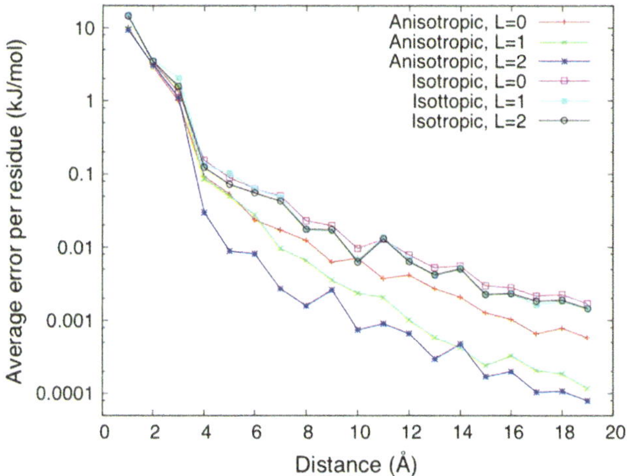

Fig. 4 Mean absolute error in the induction energy per residue as a function of the distance from the ligand

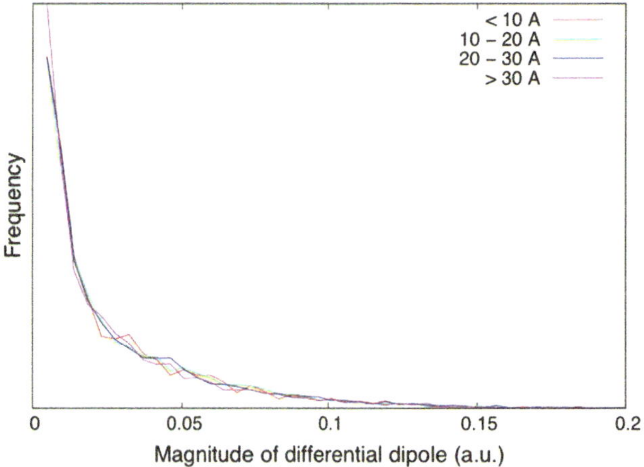

Fig. 5 Distribution of the magnitude of the difference between the induced dipole using anisotropic and isotropic polarizabilities. Each curve includes only the residues in a particular distance range from the ligand

effects) in an effective way through the repulsion term. At larger distances, on the other hand, the possibility to reduce the error by fitting disappears, and the error can be seen as a lower bound on the error of any force field, although it should be noted that the errors in the other terms are usually larger. In this far range, clear trends are obtained. As expected, anisotropic polarizabilities give lower errors than isotropic ones. With anisotropic polarizabilities, the accuracy also increases with increasing multipole level, but this effect is almost absent with isotropic polarizabilities, showing that once the isotropic approximation has been introduced, there is no need of having higher multipoles than point charges to compute the electrostatic field (although it may of course be important for the electrostatic term).

However, the main result from Fig. 4 is that, even for the worst model using point charges and isotropic polarizabilities, the errors quickly become negligible as the distance from the ligand increases. Thus, we can conclude that the pure induction is not responsible for the observed long-range effects of polarizabilities.

3.2 Protein assembly

A force field for a protein is usually assembled using parameters for smaller fragments, taken either directly from QM calculations or from a library. If the force field is polarizable, each fragment is polarized by the field from the surrounding fragments so that its charge distribution within the protein differs from that of the isolated fragment. This change, in turn, affects the electrostatic interaction with the ligand. Because this is an indirect effect (many-body effect), one would expect it to be smaller than the pure induction. However, electric fields in the protein

are often strong due to the proximity of charged residues. Thus, the statically induced dipoles may become significant and interact strongly with the charged ligand, as these interactions has a formal r^{-2} dependence (compared to r^{-4} for the ion–induced dipole interaction).

Of course, this effect depends on the polarization model used. A comparison of the individual induced dipoles in the assembled avidin protein (without the ligand) modeled by anisotropic and isotropic polarizabilities, respectively, shows that the deviation is randomly distributed with average magnitude of 0.026 a.u. (corresponding to $\sim 80\%$ of the average magnitude of the induced dipoles themselves) and independent of the distance from the ligand, as displayed in Fig. 5. Although the energetic effect of each of these individual differences is randomly distributed and rather small, in average 0.3 kJ/mol at 5 Å and 0.1 kJ/mol at 15 Å, the large number of contributions add up to a total energy contribution that is typically 3–6 kJ/mol for the residues outside of 15 Å and 4–11 kJ/mol outside 5 Å, depending on the particular geometry and QM method (results not shown). This is in agreement with Fig. 2 of ref. [34].

To investigate this effect more systematically, we use smaller subsystems consisting of only two fragments (each fragment being one capped amino acid) and the ligand (see Fig. 2). Two sets of fragment pairs were created. The first set consists of the 1,008 fragment pairs that have at least one atom–atom distance that is <2.5 Å, but are not covalently linked. The second set consists of the 494 fragment pairs that are directly covalently linked (eight of which share a cystine link, the rest a peptide bond). We test how well various polarization models describe the charge redistribution within each fragment pair upon association. As a measure of the charge redistribution for a given polarization model M, we calculate the change in electrostatic interaction energy between the fragment pair (F_i, F_j) and the ligand:

$$\Delta E_{ij}^{M} = E_{\mathrm{ele}}(L \leftrightarrow F_{ij}^{M}) - E_{\mathrm{ele}}(L \leftrightarrow F_{i}^{M}) - E_{\mathrm{ele}}(L \leftrightarrow F_{j}^{M}) \tag{7}$$

where $E_{\mathrm{ele}}(L \leftrightarrow F_{i}^{M})$ denotes the classical electrostatic interaction energy between the ligand and F_i. Whereas the ligand is always treated in the same way, with distributed multipoles up to quadrupoles, obtained from a LoProp calculation, we vary the representation M of the fragments. This variation includes the source of the polarizabilities (from a quantum-chemical LoProp calculation or from the Amber library), the form of the polarizabilities (anisotropic or isotropic; in the bonds or only in the atomic nuclei), the origin of the multipoles (LoProp multipoles up to quadrupoles or Amber charges), and the choice of exclusion rules for the polarizability coupling (LoProp or Amber). A *null model* without polarization is also included for comparison. All the tested models are listed in Table 1.

The fragment dimer F_{ij} is constructed by taking the two separate fragments described by M and letting them polarize each other. For the covalently bound pairs, the MFCC procedure is used, so that a concap term is added to Eq. 7 (see Sect. 2).

As a reference for these calculations, Eq. 7 is evaluated using multipoles for the fragment pair computed in a *single* QM calculation (again by the LoProp approach and up to quadrupoles). The mean unsigned error

$$R^{M} = \frac{1}{N} \sum_{i<j}^{N} |\Delta E_{ij}^{M} - \Delta E_{ij}^{\mathrm{ref}}| \tag{8}$$

for each method over each set of dimers is shown in Table 2. To verify that the trends are not influenced by short-ranged effects, the average over only the distant pairs (minimal distance from the ligand larger than 5 Å) is also shown. The full distance dependence for some of the methods is shown in Fig. 6 for the nonbonded set.

The setup of these calculations ensures that we specifically test the error in the polarization part of the assembly. As expected, the most accurate treatment (*a*), using LoProp multipoles and anisotropic polarizabilities in both atoms and bonds, gives the lowest error, from ~ 0.04 kJ/mol for distant pairs up to 0.5 kJ/mol at short range. Replacing the anisotropic polarizabilities with their isotropic counterparts has a significant effect, increasing the error by 50–65%. In agreement with the results for the pure induction, the further approximation to use point charges instead of multipoles has a smaller effect, and these two effects are almost additive. The removal of the polarizabilities in the bonds (by dispersing them onto the atoms) has a negligible effect.

The change from LoProp (*ix0*) to Amber (*ambf*) polarization increases the error by 66–74 %. This change can be divided into two steps: the change of the values of the polarizabilities and the change of exclusion rules (the g_{ij} in Eq. 2). As shown in Table 2, both steps give significant (and almost additive) contributions to the error, but the change of values has the largest effect. The Amber polarizabilities were not devised to reproduce QM calculations, as has been pointed out before [21]. Therefore, the results for the Amber model are not alarming. Although there is much room for improvement—the error is three times larger than for the best model—the Amber model still gives significantly better results than the null model, which assembles the pairs without considering polarization at all. It should be noted that, although the mean absolute error per pair is <1 kJ/mol for all models, the many pairs may add up to a substantial total error, for example, 70 kJ/mol for the null model.

Table 1 Summary of the polarization variants used to test the protein assembly

Name	Polarizabilities		Multipoles	Exclusion rule
	Source	Anisotropic		
a	LoProp	Yes	LoProp	LoProp
i	LoProp	No	LoProp	LoProp
a0	LoProp	Yes	Amber	LoProp
i0	LoProp	No	Amber	LoProp
ix0	LoProp (no bonds)	No	Amber	LoProp
ix0f	LoProp (no bonds)	No	Amber	Amber
amb	Amber	No	Amber	LoProp
ambf	Amber	No	Amber	Amber
null	No polarization			

The polarizabilities can come from LoProp or Amber ff02 and can be anisotropic or isotropic. The multipoles used in the polarization can be either LoProp multipoles (up to quadrupoles) or Amber ff02 charges. The exclusion rule can be either LoProp (no intramolecular polarization) or Amber (only 1–2 and 1–3 interactions omitted). For comparison, a null model without polarization is also included

Table 2 Mean absolute error in ΔE (Eq. 8; thousandths of kJ/mol) for the two sets of fragment pairs (the nonbonded pairs and the covalent pairs) using various polarization methods for treating the charge redistribution within the pair

Averages are taken over all pairs (all) or those outside of 5 Å (far). Numbers within brackets are the corresponding errors including also the induction energy

Method	Nonbonded pairs		Covalent pairs	
	All ($N = 1{,}008$)	Far ($N = 922$)	All ($N = 494$)	Far ($N = 452$)
a	73 (86)	48 (48)	72 (79)	47 (47)
i	112 (213)	79 (82)	89 (164)	56 (58)
a0	117	75	76	50
i0	137	97	97	62
ix0	129	93	99	62
ix0f	162	117	110	65
amb	198 (343)	134 (137)	97 (243)	62 (67)
ambf	225 (374)	154 (158)	114 (288)	73 (78)
null	503 (1,325)	347 (384)	189 (1,049)	129 (183)

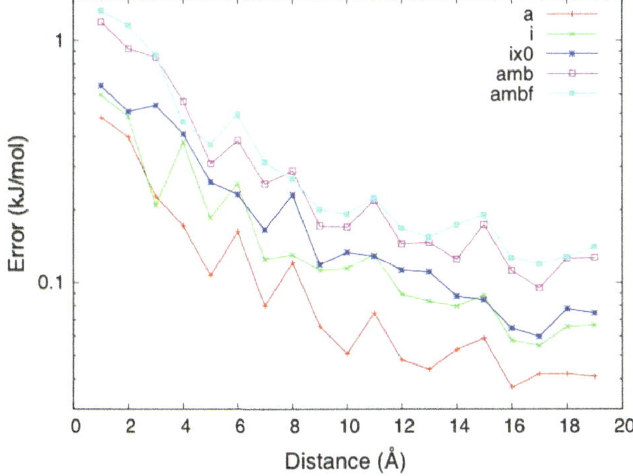

Fig. 6 Mean absolute error (Eq. 8) in the assembly of the nonbonded pairs as a function of the minimal distance between any of the pair fragments and the ligand

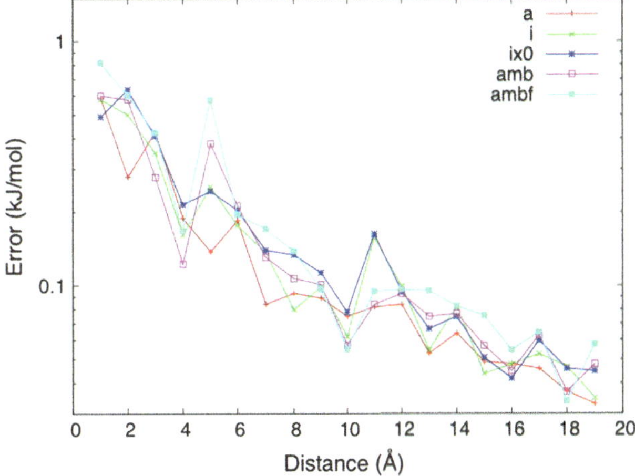

Fig. 7 Mean absolute error (Eq. 8) in the assembly of the covalent pairs as a function of the minimal distance between any of the pair fragments and the ligand

The fact that the LoProp exclusion rules give better results than the Amber exclusion rules, independently of the values of the polarizabilities, offers some physical insight. As described in Fig. 1, the coupling between polarizabilities within each fragment is ignored with the LoProp exclusion rules, because the coupling is implicitly included under the assumption that the electric field is uniform over the fragment. The Amber rules, on the other hand, try to model the coupling explicitly by only excluding coupling between atoms separated by one or two bonds. The results indicate that, for fragments that are as small as amino acids, it is better to ignore the coupling between polarizabilities.

The corresponding distance dependence for the covalently bound set is shown in Fig. 7. For this set, the results are less clear, but most of the trends remain. As expected, the treatment of intramolecular polarization is more difficult: all models give larger errors relative to the null model (e.g., 36% over the distant covalent pairs for the best model, compared to 14% for the distant nonbonded pairs;

see Table 2). On the other hand, the polarization effects for the assembly of covalently linked residues are smaller in absolute value and thus less important than those in, for example, hydrogen bonds.

For some of the methods, the errors including the actual induction energy between the ligand and the fragment pairs are also shown in Table 2. As can be seen, the additional error for the far set introduced by the induction energy is negligible for all models (except the null model, which does not include the induction energy). This confirms the result from the previous section that the approximations affect the induction energy only at short range.

3.3 Conformational dependence

In the final test, we examine the possible advantage of using a polarizable model to enhance the transferability of partial charges between various conformations of the same amino acid. To this end, we derived RESP charges for all capped residues (in their particular conformation) of

Table 3 Mean absolute error per residue (kJ/mol) for various charge sets over all 494 residues (all) or the 453 residues with distance >6 Å (far), using either the *qm0* results or supermolecular (super) energies as the reference

Reference	qm0 (all)	qm0 (far)	Super (far)
qm0	–	–	0.10
qm1 + pol	0.05	0.03	0.10
cons0	0.31	0.18	0.22
cons1 + pol	0.25	0.15	0.18
ff02 + pol	1.28	1.00	1.03

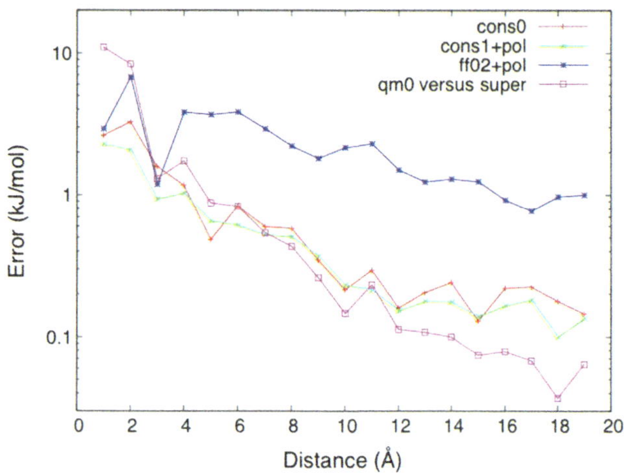

Fig. 8 Mean absolute error per residue (kJ/mol) introduced by the averaging of charges over conformations, with (*cons1*) and without (*cons0*) an additional polarization model, as well as with the standard *ff02* charges. For comparison, the *qm0 versus super* curve shows the deviation between the supermolecular MP2/cc-pVTZ interaction energy and the electrostatic interaction energy with the *qm0* charges

avidin, using an iterative procedure [10] that makes them consistent with the Amber ff02 polarizabilities. For comparison, we also computed RESP charges at the same QM level but without polarization. Thus, for each capped residue, we have two descriptions that nearly reproduce the same QM electrostatic potential: one with only partial charges (*qm0*) and one with partial charges and isotropic polarizabilities (*qm1*). Two other charge sets (*cons0* and *cons1*) were constructed by averaging the *qm0* and *qm1* charges, respectively, over all occurrences of a given amino acid in the protein. Finally, we include the standard Amber-02 charges (*ff02*). There is no standard charge set for a nonpolarizable model at this exact QM level. In all cases involving polarization, we employ the Amber-02 polarizabilities [10]. For the ligand, we always use the same charges (*qm0*) and no polarization.

We use the electrostatic interaction energy between the ligand and each residue described by *qm0* as the reference and report the mean absolute errors over all residues, or over residues with a distance >6 Å, in Table 3. First, we verify that the *qm0* and *qm1* descriptions give the same interaction energies. As shown in Table 3, the mean absolute difference between the interaction energies with these models is only 0.05 kJ/mol. Thus, before averaging, the nonpolarizable (*qm0*) and polarizable (*qm1*) descriptions are roughly equivalent.

Using the averaged charges instead of those derived for exactly the right conformation gives an average error of 0.25–0.31 kJ/mol over all residues and 0.15–0.18 kJ/mol over the distant residues. Interestingly, the error is consistently lower for the polarizable model (by 14–19%), showing that the polarization accounts for some of the conformational dependence of the charges. However, earlier investigations have reported much larger improvements for more advanced polarization models [36, 38, 39], so the Amber model is not optimal. Clearly, the conformational dependence of charges (and higher multipoles) is an unsolved problem.

To see how the accuracy relates to the overall force field accuracy, we also compare these electrostatic interaction energies with the supermolecular MP2/cc-pVTZ interaction energies, with the Amber Van der Waals energy subtracted as in Sect. 3.1. The results are shown in Table 3 for the far set (the comparison is not relevant for shorter distances because several terms are replaced by the standard Van der Waals energy). As expected, the errors are larger, but not by much.

The errors with the standard charges from the *ff02* protein library are much larger: the average error is 1.3 kJ/mol over all residues and 1.0 kJ/mol over the distant residues. This is somewhat unexpected, as one would expect the *ff02* charges to be similar to the *cons1* charges, as they are derived in similar ways. However, a detailed analysis shows that most of the error comes from the capping $-CH_3$ groups, which are significantly more negative (in both the reference calculations and the QM calculations) than the corresponding protein atoms that *ff02* tries to model.

The full distance dependence of the errors is shown in Fig. 8. For short distances, all the charge sets give large errors, although the use of polarization (*cons1*) seems to avoid the largest problems. At a distance of ∼4 Å, the errors with *cons0* and *cons1* charges quickly drop below 1 kJ/mol per residue, whereas the *ff02* error remains large. The comparison with supermolecular results shows that for distances larger than ∼7 Å, the conformational averaging is a larger source of inaccuracy than the point charge representation itself.

4 Conclusions

Based on the results of these tests, one may look at the computation of a protein–ligand interaction energy with a polarizable force field as a three-step process, with the polarization playing different roles in each step.

First, each protein fragment is internally polarized. Force fields often assume that different conformations of the same amino acid have the same partial charges. Polarization enables a physically motivated variation of the electrostatic properties of different conformations. We tested this effect by constructing artificial Amber-like polarizable and non-polarizable force fields with conformationally averaged partial charges and comparing the resulting electrostatic interaction energies with those using charges for the right conformation. Indeed, with the Amber polarization model, the error caused by conformational averaging is 14–19% smaller with polarization than without.

Second, each protein fragment is polarized by the surrounding fragments, so that the electrostatic properties are altered compared with what they would be in isolated amino acids. We found that approximations in the polarization model have a large effect on the induced dipoles in the whole protein and thus on the electrostatic interaction energy with the ligand, even for fragments as far as 15 Å from the ligand (e.g., $\sim$0.1 kJ/mol per residue for the isotropic approximation). By investigating the assembly of residue pairs using a series of polarization models, it was found that polarization between residues interacting by, for example, hydrogen bonds is more important than between covalently linked residues, but on the other hand it is also easier to model. Upon switching from a very accurate model to the Amber model, the steps particularly found to increase the error are the removal of anisotropy and the change of the values of the polarizabilities. However, the introduction of intramolecular coupling of the polarizabilities also has a consistently negative impact on the accuracy. Encouragingly, all tested polarization models were found to give significantly better results than a nonpolarizable model.

Finally, the protein is polarized by the ligand and vice versa, giving the pure induction energy. By comparing the induction energy from various polarization models with its quantum-mechanically computed counterpart, we found that the approximation to make the polarizabilities isotropic has a larger effect than, for example, the reduction of the multipole level, but that approximations done in residues separated by more than $\sim$4 Å from the ligand have a negligible effect on the energy compared with the total error of the force field.

These results indicate that a careful treatment of polarization may be important even in cases where the actual induction energy is small. They also provide some guidance how to improve current polarization models. Apparently, anisotropic polarizabilities and rigorous exclusion rules are essential to achieve quantitative agreement with QM calculations.

Acknowledgments Financial support from the Swedish research council (623-2009-821) is greatly acknowledged.

References

1. Ponder JW, Case DA (2003) Adv Protein Chem 66:27
2. Rick SW, Stuart SJ (2002) Potentials and algorithms for incorporating polarizability in computer simulations. In: Lipowitz KB, Boyd DB (eds) Reviews in computational chemistry, Wiley, New York, p 89
3. Cieplak P, Dupradeau FY, Duan Y, Wang JM (2009) J Phys Condens Matter 21(33):ARTN 333102
4. Warshel A, Kato M, Pisliakov AV (2007) J Chem Theory Comput 3:2034
5. Engkvist O, Åstrand P-O, Karlström G (2000) Chem Rev 100:4087
6. Piquemal J-P, Cisneros GA, Reinhardt P, Gresh N, Darden TA (2006) J Chem Phys 124:104101
7. Pullman B, Claverie P, Caillet J (1967) Proc Natl Acad Sci 57(6):1663
8. Warshel A, Levitt M (1976) J Mol Biol 103:227
9. Gresh N, Pullman B (1980) Biochim Biophys Acta 608(1):47–53
10. Cieplak P, Caldwell J, Kollman P (2001) J Comput Chem 22:1048
11. Maple JR, Cao Y, Damm W, Halgren TA, Kaminski GA, Zhang LY, Friesner RA (2005) J Chem Theory Comput 1:694
12. Ponder J, Wu C, Ren P, Pande VS, Chodera JD, Schnieders MJ, Haque I, Mobley DL, Lambrecht DS, DiStasio RA Jr (2010) J Phys Chem B 114(8):2549–2564
13. Applequist J, Carl JR, Fung K-K (1972) J Am Chem Soc 94:2952
14. Thole BT (1981) Chem Phys 59:341
15. Van Duijnen PT, Swart M (1998) J Phys Chem A 102(14):2399–2407
16. Gresh N, Cisneros GA, Darden TA, Piquemal J-P (2007) J Chem Theory Comput 3:1960
17. Gordon MS, Freitag MA, Bandyopadhyay P, Jensen JH, Kairys V, Stevens WJ (2001) J Phys Chem A 105:293
18. Gresh N (1995) J Comput Chem 16(7):856–882
19. Day PN, Jensen JH, Gordon MS, Webb SP, Stevens WJ, Krauss M, Garmer D, Basch H, Cohen D (1996) J Chem Phys 105:1968
20. Masia M, Probst M, Rey R (2005) J Chem Phys 123:164505
21. Söderhjelm P, Kongsted J, Ryde U (2011) J Chem Theory Comput 7:1404–1414
22. Elking D, Darden T, Woods RJ (2007) J Comput Chem 28(7):1261–1274
23. Shirts MR, Pitera JW, Swope WC, Pande VS (2003) J Chem Phys 119:5740
24. Jiao D, Golubkov PA, Darden TA, Ren P (2008) Proc Natl Acad Sci 105:6290–6295
25. Xantheas SS, Burnham CJ, Harrison RJ (2002) J Chem Phys 116:1493
26. Wang JM, Cieplak P, Li J, Hou TJ, Luo R, Duan Y (2011) J Phys Chem B 115(12):3091–3099
27. Wang JM, Cieplak P, Li J, Wang J, Cai Q, Hsieh MJ, Lei HX, Luo R, Duan Y (2011) J Phys Chem B 115(12):3100–3111
28. Ren P, Ponder JW (2002) J Comput Chem 23:1497
29. Stevens WJ, Fink WH (1987) Chem Phys Lett 139:15

30. Bagus PS, Hermann K, Bauschlicher CWJ (1984) J Chem Phys 80:4378
31. Söderhjelm P, Krogh JW, Karlström G, Ryde U, Lindh R (2007) J Comput Chem 28:1083
32. Söderhjelm P, Öhrn A, Ryde U, Karlström G (2008) J Chem Phys 128:014102
33. Söderhjelm P, Ryde U (2009) J Phys Chem A 113:617
34. Söderhjelm P, Aquilante F, Ryde U (2009) J Phys Chem B 113:11085
35. Söderhjelm P, Husberg C, Strambi A, Olivucci M, Ryde U (2009) J Chem Theory Comput 5:649
36. Engkvist O, Åstrand P-O, Karlström G (1996) J Phys Chem 100:6950
37. Tiraboschi G, Fournié-Zaluski MC, Roques BP, Gresh N (2001) J Comput Chem 22(10):1038–1047
38. Holt A, Karlström G (2008) J Comput Chem 29:1084
39. Holt A, Karlström G (2008) J Comput Chem 29:1905
40. Nakagawa S, Mark P, Ågren H (2007) J Chem Theory Comput 3:1947
41. Rasmussen TD, Ren P, Ponder JW, Jensen F (2007) Int J Quantum Chem 107:1390
42. Ángyán JG, Jansen G, Loos M, Hättig C, Heß BA (1994) Chem Phys Lett 219:267
43. Garmer DR, Stevens WJ (1989) J Phys Chem 93:8263
44. Stone AJ (1985) Mol Phys 56:1065
45. Le Sueur CR, Stone AJ (1993) Mol Phys 78:1267
46. Gagliardi L, Lindh R, Karlström G (2004) J Chem Phys 121:4494
47. Stout JM, Dykstra CE (1995) J Am Chem Soc 117(18):5127–5132
48. Celebi N, Ángyán JG, Dehez F, Millot C, Chipot C (2000) J Chem Phys 112:2709
49. Williams GJ, Stone AJ (2003) J Chem Phys 119:4620
50. Soteras I, Curutchet C, Bidon-Chanal A, Dehez F, Ángyán JG, Orozco M, Chipot C, Luque FJ (2007) J Chem Theory Comput 3(6):1901–1913
51. Miyamoto S, Kollman PA (1993) Proteins Struct Funct Genet 16:226
52. Wang J, Dixon R, Kollman PA (1999) Proteins Struct Funct Genet 34:69
53. Kuhn B, Kollman PA (2000) J Med Chem 43:3786
54. Weis A, Katebzadeh K, Söderhjelm P, Nilsson I, Ryde U (2006) J Med Chem 49:6596
55. Söderhjelm P, Ryde U (2009) J Comput Chem 30:750
56. Söderhjelm P, Kongsted J, Ryde U (2010) J Chem Theory Comput 6:1726
57. Zhang DW, Zhang JZH (2003) J Chem Phys 119:3599
58. Boys SF, Bernardi F (1970) Mol Phys 19:553
59. Molcas 7, University of Lund, Sweden (2007) See http://www.teokem.lu.se/molcas
60. Frisch MJ, Trucks GW, Schlegel HB, Scuseria GE, Robb MA, Cheeseman JR, Scalmani G, Barone V, Mennucci B, Petersson GA, Nakatsuji H, Caricato M, Li X, Hratchian HP, Izmaylov AF, Bloino J, Zheng G, Sonnenberg JL, Hada M, Ehara M, Toyota K, Fukuda R, Hasegawa J, Ishida M, Nakajima T, Honda Y, Kitao O, Nakai H, Vreven T, Montgomery JA Jr, Peralta JE, Ogliaro F, Bearpark M, Heyd JJ, Brothers E, Kudin KN, Staroverov VN, Kobayashi R, Normand J, Raghavachari K, Rendell A, Burant JC, Iyengar SS, Tomasi J, Cossi M, Rega N, Millam JM, Klene M, Knox JE, Cross JB, Bakken V, Adamo C, Jaramillo J, Gomperts R, Stratmann RE, Yazyev O, Austin AJ, Cammi R, Pomelli C, Ochterski JW, Martin RL, Morokuma K, Zakrzewski VG, Voth GA, Salvador P, Dannenberg JJ, Dapprich S, Daniels AD, Farkas Ö, Foresman JB, Ortiz JV, Cioslowski J, Fox DJ (2009) Gaussian 09 revision A.1. Gaussian Inc., Wallingford, CT
61. Bayly CI, Cieplak P, Cornell WD, Kollman PA (1993) J Phys Chem 97:10269
62. Case DA, Darden TA, Cheatham TE III, Simmerling CL, Wang J, Duke RE, Luo R, Crowley M, Walker RC, Zhang W, Merz KM, Wang B, Hayik S, Roitberg A, Seabra G, Kolossvary I, Wong KF, Paesani F, Vanicek J, Wu X, Brozell SR, Steinbrecher T, Gohlke H, Yang L, Tan C, Mongan J, Hornak V, Cui G, Mathews DH, Seetin MG, Sagui C, Babin V, Kollman PA (2008) Amber 10. University of California, San Francisco
63. Cornell W, Cieplak P, Bayly C, Gould I, Merz KM, Ferguson D, Spellmeyer D, Fox T, Caldwell J, Kollman P (1995) J Am Chem Soc 117:5179
64. Pugliese L, Coda A, Malcovati M, Bolognesi M (1993) J Mol Biol 231:698

Theor Chem Acc (2012) 131:1153
DOI 10.1007/s00214-012-1153-7

REGULAR ARTICLE

Recent applications and developments of charge equilibration force fields for modeling dynamical charges in classical molecular dynamics simulations

Brad A. Bauer · Sandeep Patel

Received: 12 May 2011 / Accepted: 27 September 2011 / Published online: 2 March 2012
© Springer-Verlag 2012

Abstract With the continuing advances in computational hardware and novel force fields constructed using quantum mechanics, the outlook for non-additive force fields is promising. Our work in the past several years has demonstrated the utility of polarizable force fields, in our hands those based on the charge equilibration formalism, for a broad range of physical and biophysical systems. We have constructed and applied polarizable force fields for small molecules, proteins, lipids, and lipid bilayers and recently have begun work on carbohydrate force fields. The latter area has been relatively untouched by force field developers with particular focus on polarizable, non-additive interaction potential models. In this review of our recent work, we discuss the formalism we have adopted for implementing the charge equilibration method for phase-dependent polarizable force fields, lipid molecules, and small-molecule carbohydrates. We discuss the methodology, related issues, and briefly discuss results from recent applications of such force fields.

Keywords Force field · Molecular dynamics simulations · Quantum mechanics · Polarizable · Charge equilibration

Published as part of the special collection of articles: From quantum mechanics to force fields: new methodologies for the classical simulation of complex systems.

B. A. Bauer · S. Patel (✉)
Department of Chemistry and Biochemistry,
University of Delaware, Newark, DE 19716, USA
e-mail: sapatel@udel.edu

1 Introduction

Computational chemistry is an indispensable tool in the arsenal of today's chemist, biochemist, and biologist. It is thus not surprising that the computational modeling community continues to push for advanced methods and models at a feverish pace. The advances have become ever faster in the recent decades due to the increased computational resources emerging as high-performance computer hardware becomes available to the masses at commodity prices. Along with the advances in hardware and software, algorithmic and force field developments have impacted the state-of-the-art profoundly. Though the issues of sampling on accurate (free) energy surfaces are intimately coupled, work continues along both fronts fairly along independent lines, largely due to the fact that both problems are profoundly difficult to tackle simultaneously with current resources. This review article deals with the aspect of molecular interaction models, and in particular, models based on non-additive electrostatic interactions. For the purposes of this review, we consider non-additive electrostatic models as those that allow for the variation of the local molecular electrostatic environment (i.e., charges, dipole moments, and higher-order electrostatic moments) based on some pre-defined, systematic theoretical formalism. Work in the recent decades has realized the need to revisit the implications of electrostatic polarizability in defining the quality of molecular simulations. As a result, the development of models that explicitly model polarizability has been steadily progressing. Considering the formalisms for modeling molecular polarizability in a classical treatment, point-dipole (and higher-order multipole) [1–11], Drude oscillator [12–21], ab initio-inspired methods [10, 22–27], and charge equilibration/fluctuating charge [28–44] models are currently being developed.

A detailed comparison among these approaches is beyond the scope of the current work, although recent reviews have addressed such comparisons [45]. The charge equilibration approach, which is used extensively in this work, is discussed in detail in the next section. In general, polarization methods consider the induction of a dipole moment ($\boldsymbol{\mu}_{\text{ind}}$) in the presence of an electric field ($\mathbf{E}$):

$$\boldsymbol{\mu}_{\text{ind}} = \alpha \mathbf{E}. \tag{1}$$

These formalisms differ in the treatment of the polarizability, α. For the purposes of model development and parameterization, α is often determined empirically. There is not a consensus on the magnitude of α in the condensed phase for numerous molecular species. Consequently, the value of polarizability in some force fields is determined empirically. The nature of the condensed-phase polarizability is either treated in an ad hoc manner via scaling to reproduce certain target properties of condensed-phase and gas-phase cluster models. In other cases, the gas-phase polarizability is applied directly for models of the condensed phase; the parameters of the model then implicitly account for the variation of polarizability with environment (condensed phase versus gas phase).

2 Force fields

2.1 Charge equilibration force fields

We next consider details of the CHEQ method and considerations in our specific implementation of the method. An additive (or *non-polarizable*) formalism for force fields is based on the construct that all atomic partial charges are fixed throughout the course of the simulations. Alternatively, we can consider the variation of atomic partial charge using the charge equilibration (CHEQ) formalism [28–37, 39]. The CHEQ formalism is based on Sanderson's idea of electronegativity equilibration [28, 29] in which the chemical potential is equilibrated via the redistribution of charge density. In a classical sense, charge density is reduced to partial charges, Q_i on each atomic site i. The charge-dependent energy for a system of M molecules containing N_i atoms per molecule is then expressed as

$$E_{\text{CHEQ}}(\vec{R}, \vec{Q}) = \sum_{i=1}^{M} \sum_{\alpha=1}^{N} \chi_{i\alpha} Q_{i\alpha} + \frac{1}{2} \sum_{i=1}^{M} \sum_{j=1}^{M} \sum_{\alpha=1}^{N_i} \sum_{\beta=1}^{N_j} J_{i\alpha j\beta} Q_{i\alpha} Q_{j\beta}$$
$$+ \frac{1}{2} \sum_{i=1}^{MN'} \sum_{j=1}^{MN'} \frac{Q_i Q_j}{4\pi\epsilon_0 r_{ij}} + \sum_{j=1}^{M} \lambda_j \left(\sum_{i=1}^{N} Q_{ji} - Q_j^{\text{Total}} \right) \tag{2}$$

where the χ terms represent the atomic electronegativities that control the directionality of electron flow and J terms

represent the atomic hardnesses that control the resistance to electron flow to or from the atom. Although these parameters are derived from the definitions of electron affinity and ionization potential, they are treated as empirical parameters for individual atom types. Heterogeneous hardness elements that describe the interaction between two different atom types are calculated using the combining rule [46] on the parameterized homogeneous hardness elements $\left(J_{ii}^{\circ} \right)$:

$$J_{ij}(R_{ij}, J_{ii}^{\circ}, J_{jj}^{\circ}) = \frac{\frac{1}{2}\left(J_{ii}^{\circ} + J_{jj}^{\circ} \right)}{\sqrt{1.0 + \frac{1}{4}\left(J_{ii}^{\circ} + J_{jj}^{\circ} \right)^2 R_{ij}^2}} \tag{3}$$

where R_{ij} is the distance between atoms i and j. This combination locally screens Coulombic interactions but provides the correct limiting behavior for atomic separations greater than approximately 2.5 Å. The standard Coulomb interaction between sites not involved in the dihedral, angle, or bonded interactions (indicated by primed summation) with each other is included as the third term in Eq. 2. The second term in Eq. 2 represents the local charge transfer interaction, which is usually restricted to within a molecule or an appropriate charge normalization unit, that is, no intermolecular charge transfer. Charge is constrained via a Lagrange multiplier, λ, which is included for each molecule as indicated in the last term of Eq. 2. We remark that use of multiple charge normalization units can modulate molecular polarizability by limiting intramolecular charge transfer to physically realistic distances. Such an approach controls previously observed superlinear polarizability scaling [41, 43, 47], which also manifests as the polarization catastrophe (as observed in point polarizable force fields) [8, 41], while developing a construct for piecing together small molecular entities into macromolecules.

Charge degrees of freedom are propagated via an extended Lagrangian formulation imposing a molecular charge neutrality constraint, thus providing for electronegativity equilibration at each dynamics step. The system Lagrangian is:

$$L = \sum_{i=1}^{M} \sum_{\alpha=1}^{N} \frac{1}{2} m_{i\alpha} \left(\frac{\mathrm{d}r_{i\alpha}}{\mathrm{d}t} \right)^2 + \sum_{i=1}^{M} \sum_{\alpha=1}^{N} \frac{1}{2} m_{Q,i\alpha} \left(\frac{\mathrm{d}Q_{i\alpha}}{\mathrm{d}t} \right)^2$$
$$- E(Q, r) - \sum_{i=1}^{M} \lambda_i \sum_{\alpha=1}^{N} Q_{i\alpha} \tag{4}$$

where the first two terms represent the nuclear and charge kinetic energies, the third term is the potential energy, and the fourth term is the molecular charge neutrality constraint enforced on each molecule i via a Lagrange multiplier λ_i. The fictitious charge dynamics are determined using a charge "mass" with units of (energy time^2/charge^2). This is analogous to the use of an adiabaticity parameter in fictitious wavefunction dynamics in Car Parinello (CP) type methods [31, 48]. Charges are thus propagated based on the forces arising from differences between the average electronegativity of a molecule and the instantaneous electronegativity at an atomic site.

2.2 Polarizability in charge equilibration models

When the electrostatic energy expression for a single molecule comprised of N atoms is differentiated with respect to charge and set equal to zero,

$$\frac{\partial E}{\partial Q_i} = \chi_i^0 + J_{ii}^{\circ} Q_i + \sum_{j \neq i}^{N} J_{ij} Q_j = 0, \tag{5}$$

a set of N equations can be solved to determine the set of charges minimizing the energy. In matrix form, this set of equations can be recast as

$$\mathbf{JQ} = -\boldsymbol{\chi} \tag{6}$$

where $\mathbf{J}$ is the atomic hardness matrix, $\mathbf{Q}$ is the atomic charge vector, and $\boldsymbol{\chi}$ is the atomic electronegativity vector. Written explicitly, the matrix representation is:

$$\begin{bmatrix} J_{11} & J_{12} & \cdots & \cdots & J_{1N} \\ J_{21} & \ddots & & & \vdots \\ \vdots & & \ddots & & \vdots \\ \vdots & & & \ddots & \vdots \\ J_{N1} & \cdots & \cdots & \cdots & J_{NN} \end{bmatrix} \begin{bmatrix} Q_1 \\ Q_2 \\ \vdots \\ \vdots \\ Q_N \end{bmatrix} = - \begin{bmatrix} \chi_1 \\ \chi_2 \\ \vdots \\ \vdots \\ \chi_N \end{bmatrix}. \tag{7}$$

These equations can be augmented to include the charge conservation constraint,

$$\sum_{i=1}^{M} Q_i = Q_{\text{net}}, \tag{8}$$

to produce a modified set of equations:

$$\mathbf{J'Q'} = -\boldsymbol{\chi'}; \tag{9}$$

this is written explicitly as:

$$\begin{bmatrix} J_{11} & J_{12} & \cdots & \cdots & J_{1N} & 1 \\ J_{21} & \ddots & & & \vdots & 1 \\ \vdots & & \ddots & & \vdots & \vdots \\ \vdots & & & \ddots & \vdots & \vdots \\ J_{N1} & \cdots & \cdots & \cdots & J_{NN} & 1 \\ 1 & \cdots & \cdots & \cdots & 1 & 0 \end{bmatrix} \begin{bmatrix} Q_1 \\ Q_2 \\ \vdots \\ \vdots \\ Q_N \\ \lambda \end{bmatrix} = - \begin{bmatrix} \chi_1 \\ \chi_2 \\ \vdots \\ \vdots \\ \chi_N \\ Q_{\text{net}} \end{bmatrix} \tag{10}$$

In Eq. 10, we have enforced the constraint that charge is conserved within the molecule, or equivalently, that the sum of atomic partial charges equals the desired net charge. As alluded to in the previous section, it is appropriate to

limit the extent of intramolecular charge transfer by introducing additional charge constraints. For instance, the total charge of the molecule is constrained to Q_{net} while additionally requiring the sum of charges for a subset of atoms within the molecule to equal a specified quantity, $Q_{\text{net}, k}$. We refer to this subset of atoms as a charge conservation unit. Since the net charge over the entire molecule must be maintained, the following condition must hold:

$$Q_{\text{net}} = \sum_{k=1}^{M} Q_{\text{net},k}, \tag{11}$$

in which M denotes the number of charge conservation units. We can expand our previous illustration of the system of equations generated with a single charge constraint (Eq. 10) to show the set of equations when a molecule is broken into two charge conservation units:

$$\begin{bmatrix} J_{11} & \cdots & J_{1h} & J_{1(h+1)} & \cdots & J_{1N} & 1 & 0 \\ \vdots & \ddots & & & & \vdots & \vdots & \vdots \\ J_{h1} & & J_{hh} & & & \vdots & 1 & 0 \\ J_{(h+1)1} & & & J_{(h+1)(h+1)} & & \vdots & 0 & 1 \\ \vdots & & & & \ddots & \vdots & \vdots & \vdots \\ J_{N1} & \cdots & J_{Nh} & J_{N(h+1)} & \cdots & J_{NN} & 0 & 1 \\ 1 & \cdots & 1 & 0 & \cdots & 0 & 0 & 0 \\ 0 & \cdots & 0 & 1 & \cdots & 1 & 0 & 0 \end{bmatrix}$$

$$\begin{bmatrix} Q_1 \\ \vdots \\ Q_h \\ Q_{h+1} \\ \vdots \\ Q_N \\ \lambda_1 \\ \lambda_2 \end{bmatrix} = - \begin{bmatrix} \chi_1 \\ \vdots \\ \chi_h \\ \chi_{h+1} \\ \vdots \\ \chi_N \\ Q_{\text{net},1} \\ Q_{\text{net},2} \end{bmatrix}. \tag{12}$$

For this example, atomic sites 1 through h are grouped into one charge conservation unit, while atomic sites $h + 1$ through N are grouped into a second charge normalization unit. In the augmented atomic hardness matrix, values of 1 in element $(i, N + 1)$ or $(N + 1, i)$ denote that atom i is assigned to the first charge conservation unit (which has a net charge of Q_{net}, 1). Similarly, values of 1 in elements $(i, N + 2)$ or $(N + 2, i)$ denote that atom i is assigned to the second charge conservation unit (with net charge Q_{net}, 2).

The molecular polarizability in the CHEQ formalism can be calculated as [41]:

$$\alpha_{\gamma\beta} = \mathbf{R}_\beta^T \mathbf{J}'^{-1} \mathbf{R}_\gamma \tag{13}$$

where $\mathbf{J}'$ is the atomic hardness matrix augmented with the appropriate rows and columns to treat the charge conservation constraints. $\mathbf{R}_\gamma$ and $\mathbf{R}_\beta$ are the γ and β Cartesian coordinates of the atomic position vectors (which are also augmented to appropriately match the dimensions of the atomic hardness matrix).

3 Phase-dependence of molecular polarizability

A variety of recent theoretical investigations involving ab initio calculations with polarizable continuum solvent, the partitioning of cluster polarizabilities, and the temperature/density dependence of dielectric constants of fluids reasonably establish that the surrounding condensed-phase environment can significantly affect the polarizability of a solvated molecule. Krishtal et al. have previously reported that the average intrinsic polarizability of water molecules decreases as the size of a cluster increases and also as the number and type of hydrogen bonds on a molecule increases [49]. The notion of decreasing polarizability in condensed regions is further supported by the ab initio calculations of Morita involving water clusters [50] that suggest that the condensed-phase polarizability of water should be 7–9% lower than that of the gas-phase value. The spatial constraints imposed by condensed-phase environments limit the number of accessible excited states and diffuse character of the electron density distribution as dictated by Pauli's exclusion principle [20, 50]. A recent study by Schropp and Tavan [51] further suggests that the average effect of the inhomogeneous electric fields *within* the molecular volume of a single water molecule is consistent with classical parameterizations of polarizable water force fields in which the molecular polarizability is assigned a value around 68% of the gas-phase value. Our lab has also explored qualitatively the nature of intrinsic molecular polarizability and its reduction for water and monovalent halide anions using a cluster-based approach coupled with Hirshfeld partitioning [52]. Acknowledging that our approach does not consider explicitly effects of intermolecular charge transfer, we observe 40–50% reduction of the gas-phase intrinsic molecular polarizability of anions; the results of this study, along with the studies mentioned above, suggest that the exact absolute value of polarizability reduction in the condensed phase is still an unresolved issue. While these results suggest a reduction of polarizability within the condensed phase, the implications for the rate and nature of the decrease remain unclear. Similarly, a self-consistent analytic formalism capable of correlating changes in molecular polarizability to atomic or molecular properties remains undetermined.

While it has been observed that metrics such as aggregation number, hydrogen bonding, or local density are associated with a phase-dependent decrease in molecular polarizability, such metrics are impractical from the perspective of a molecular dynamics simulation. Consequently, a relationship between the polarizability and an atomic property that smoothly and monotonically transitions from one phase to another is desirable in establishing a simple functional form for polarizability change between phases.

One potentially useful parameter for modeling phase-dependent changes in polarizability is the dipole moment of the molecule. Both experiment and theoretical calculations such as ab initio molecular dynamics simulations demonstrate water's molecular dipole moment increases moving from gas phase to condensed-phase environments [53–55]; currently, there is no consensus on an exact value of the average condensed-phase dipole moment of water. Molecular dipole moment can be readily calculated from classical molecular dynamics simulations from the atomic positions and partial charges. If a rigid water geometry is chosen, the calculation is even further simplified in that the dipole moment depends solely on the magnitude of the associated atomic partial charges. Furthermore, atomic hardnesses determine molecular polarizability within the charge equilibration formalism. Thus, a plausible approach to modeling a phase-dependent polarizability in water lies in coupling the atomic charges to the atomic hardness parameters. A similar approach has been previously implemented by Rappé and Goddard [30] for the hydrogen atom in which a linear charge dependence is introduced into the corresponding atomic hardness value. Most generally, each atomic hardness function will depend simultaneously on all partial charges within the molecule; however, a simpler approach (as adopted in our work) entails modulating or scaling all of the atomic hardness values within a molecule using a single parameter based on the polarization state of the molecule. Since the average molecular dipole moment for water appears to be correlated with the magnitude of the negative partial charge on the oxygen atom, we simplify our model by coupling the atomic hardnesses directly to the oxygen partial charge (carried by a lonepair M-site).

As can be inferred from Eq. 13, polarizability for a rigid water molecule within the CHEQ formalism (such as TIP4P-FQ [31]) is influenced by the magnitude of the atomic hardnesses. We introduce charge-dependent polarizability by establishing a charge dependence in the atomic hardness elements:

$$\mathbf{J}(Q_M) = g(Q_M)\mathbf{J}. \tag{14}$$

where $g(Q_M)$ is an imposed scaling function dependent on the M-site charge. In order to maintain proper gas and condensed-phase charge distributions, it is also necessary to scale the electronegativity values in addition to the hardnesses. While reasonable results may be obtained by employing the same scaling factor for both the electronegativities and the hardnesses, finer control of the condensed-phase dipole moment distribution is afforded by a tunable factor for the electronegativity scaling. Such flexibility also permits better control of the bulk dielectric constant that may be obtained from the average condensed-phase dipole moment [56]. We introduce an empirical χ-scaling

$$\chi(Q_M) = \frac{g(Q_M)}{h(Q_M)}\chi = [(1-p) + pg(Q_M)]\chi \tag{15}$$

where p is an empirical parameter that controls the extent to which χ is scaled relative to the hardness scaling function. For $p = 1$, the scaling on electronegativity values is equivalent to that in the hardnesses; similarly, a value of $p = 0$ would correspond to no scaling (i.e., a constant electronegativity with no charge-dependent scaling). The resulting charge-dependent, intramolecular CHEQ energy expression is then:

$$E(Q) = \sum_{i=1}^{N} \chi_i(Q_M)Q_i + \frac{1}{2}\sum_{i=1}^{N} J_{ii}^{\circ}(Q_M)Q_i^2$$
$$+ \sum_{i<j}^{N} J_{ij}(Q_M)Q_iQ_j + \lambda\left(\sum_{i=1}^{N} Q_i - Q_{total}\right). \tag{16}$$

Unfortunately, the explicit introduction of a charge dependence into the molecular hardness matrix results in the additional complication that the polarizability expression in Eq. 13 is no longer exact. Consequently, the corresponding system of equations for the equilibrium charges (and polarizabilities) is now a non-linear system and must be solved by an iterative approach. In situations where the equilibrium charge distribution is not strongly perturbed by the implicit charge dependence, a slightly modified version of Eq. 13

$$\alpha_{\beta\gamma}(Q_M) \approx \frac{\alpha_{\beta\gamma}}{\xi(Q_M)} - \left(\frac{\nabla_M g(Q_M)\langle R_\beta|\mathbf{J}^{-1}(\mathbf{r})|\hat{\mathbf{M}}\rangle}{|g(Q_M)|^2 h(Q_M)}\right)$$
$$\left[p\chi_M\langle R_\beta|\mathbf{J}^{-1}(r)|\hat{\mathbf{M}}\rangle + \frac{1}{2}\mu_\gamma\right] \tag{17}$$

is convenient for obtaining a leading-order approximation of the polarizability in the absence of a fully non-linear treatment [57]. In the above expression, $\alpha_{\beta\gamma}$ is the $\beta\gamma$-element of the gas-phase molecular polarizability tensor, $\nabla_M g(Q_M)$ is the derivative of the scaling function with respect to Q_M, $\mathbf{R}_\beta$ is the β-position vector, $\hat{\mathbf{M}}$ is a matrix that selects elements associated with the M-site (since we have chosen our hardness elements to only depend on the oxygen charge), and μ_γ is the γ-component of the dipole moment. We see the charge-dependent

polarizability differs from the unscaled (gas-phase) value by a multiplicative factor $\xi(Q_M) = 1/[g(Q_M)h(Q_M)]$ and additive terms, which are related to the M-site hardness and dipole moment, respectively. These additive terms of equal magnitude and opposite sign are small compared to the first term and do not greatly influence $\alpha(Q_M)$. If we neglect the additive terms and consider the limit in which $p = 1$, Eq. 17 reduces to

$$\alpha_{\beta\gamma}(Q_M) \approx \frac{\alpha_{\beta\gamma}}{g(Q_M)} \tag{18}$$

from which it is clearly seen that $\alpha(Q_M)$ modulates the gas-phase polarizability via an inverse relationship with the scaling function, $g(Q_M)$. While Eq. 18 is effective for illustrative purposes, we have employed Eq. 17 for calculations of the condensed-phase polarizability.

With an explicit charge-dependent polarizability using a simple scaling function $g(Q_M)$, it is relevant to discuss the nature and form of this scaling function. While there is no formal theory connecting charge and polarizability, general trends provide guiding insight. In prior work, Rappé and Goddard [30] have employed atomic hardnesses that depend linearly on charge. In the context of molecular dynamics simulations, such an approach would necessitate some degree of charge bounding to prevent unfavorable over-polarization or under-polarization and to establish consistent polarizabilities in the gaseous and condensed phases. In light of this, we have chosen to employ an error function that applies constant scaling in the purely condensed-phase and gaseous regions and approximately linear scaling in the intermediate region. The use of the error function is also preferred as it allows smooth transitions between each region, which is necessary to avoid discontinuities in the forces. Thus, we choose a scaling function of the form

$$g(Q_M) = a - b\,\mathrm{erf}(c(d - Q_M)) \tag{19}$$

since it incorporates additional empirical parameters that can be utilized to model the desired relationship between polarizability and charge. Parameters a and b collectively define the polarizability at the gaseous and condensed-phase limits. The rate of polarizability change with charge for non-isolated molecules is controlled by c. Collectively, c and d describe the onset of scaling and the range of charges over which polarizability changes.

We apply the above methodology to design a water model capable of capturing the variation of molecular polarizability in different electrochemical environments. The resulting TIP4P-QDP model (Transferable Intermolecular Potential 4-Site with Charge-Depenedent Polarizability) is based on the application of the charge-dependent scaling function to the original TIP4P-FQ model. We retain the TIP4P-FQ geometry in the TIP4P-QDP model;

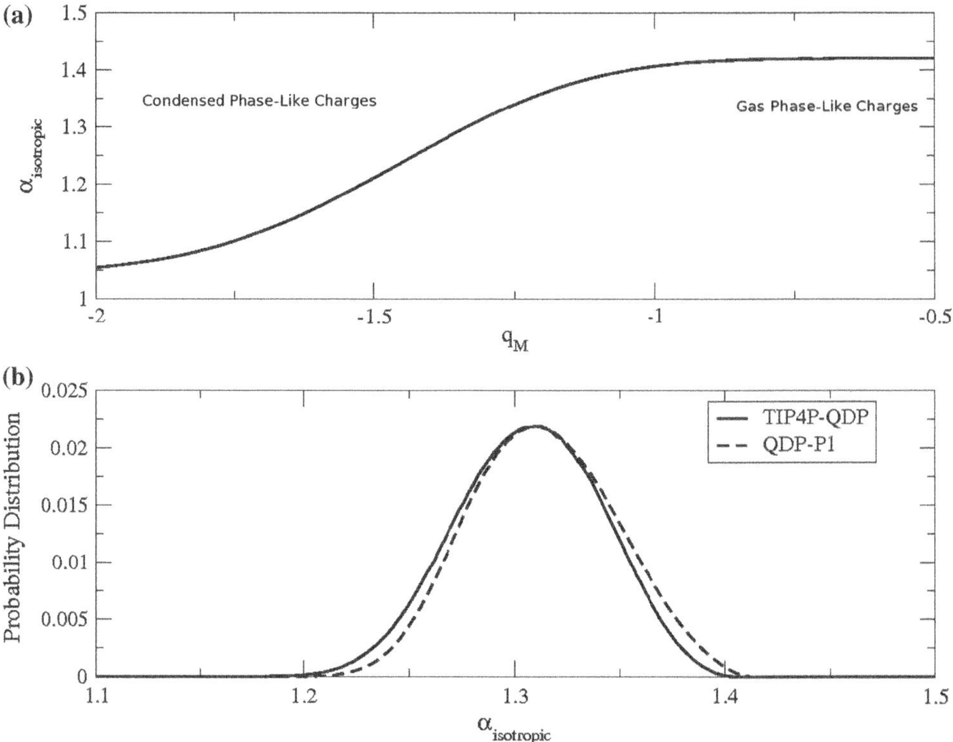

Fig. 1 Molecular polarizability scaling function and distribution for TIP4P-QDP and QDP-P1 water models. **a** Molecular polarizability (in Å^3) as a function of Q_M. **b** Distribution of molecular polarizabilities within the condensed phase. Polarizabilities were calculated using Eq. 17 and the charges from simulation. Data from Bauer et al. [57]

however, atomic hardnesses, electronegativities, and non-bonded parameters were adjusted to capture the desired physics associated with phase-dependent polarizability while reproducing target gas-phase and condensed-phase properties. Of primary interest, we reparameterize the hardness values to reproduce a reasonable gas-phase polarizability of 1.4 Å^3, which is notably higher than the empirical value offered by TIP4P-FQ ($\alpha \approx 1.12$ Å^3). We note that the resulting TIP4P-QDP gas-phase hardnesses maintain approximately the same relative magnitudes as the original TIP4P-FQ hardnesses. The error function form (Eq. 19) is then parameterized with the caveat that the gas-phase polarizability remains unchanged. In the condensed phase, the hardnesses are scaled by a value of $g(Q_M) > 1$ for charges greater than the equilibrium gas-phase charges. The parameters of the scaling function which influence the height, slope, and inflection are determined empirically such that the resulting polarizability distribution is centered about an appropriate condensed-phase polarizability value. For TIP4P-QDP, the average polarizability in the condensed phase is calculated to be 1.309 ($\pm$ 0.001) Å^3, approximately 17% higher than the static condensed-phase polarizability of TIP4P-FQ and about 11% less than the experimental gas-phase value. The average condensed-phase value approximately reflects a 6.5% reduction in the molecular polarizability relative to the TIP4P-QDP gas-phase value, which agrees well with the estimated range of 7–9% reduction calculated by Morita [50] from first principles. The scaling function allows for TIP4P-QDP

molecules to have polarizabilities as low as 1.05 Å^3. The polarizability distribution observed in the condensed phase is presented in (Fig. 1), which was estimated from M-site charges taken from a condensed-phase simulation in conjunction with Eq. 17. This distribution exhibits a width of observed molecular polarizabilities of about 0.20 Å^3.

Regarding electronegativity scaling, a final value of the p-parameter is determined to be $p = 0.80$; this value generates a condensed-phase dipole moment distribution with an average of 2.641($\pm$ 0.001) Debye, similar to that exhibited by the TIP4P-FQ model. For further comparisons, a model consisting of full scaling $p = 1.0$ was also developed and parameterized. The results of this model (referred to as QDP-P1) are also included in this work as a reference in order to more fully clarify differences between the TIP4P-FQ and TIP4P-QDP models; QDP-P1 notably higher dipole moments in the condensed phase ($\langle \mu \rangle \approx$ 2.75), an anticipated consequence of scaling electronegativity and hardness equivalently. Introduction of the scaling function and modification of the hardnesses further necessitated slight reparameterization of the remaining electrostatic and non-bonded parameters. The electronegativities of TIP4P-QDP and QDP-P1 were then reparameterized such that a single water molecule in vacuum minimizes to the experimental dipole moment of 1.85 D. Since the polarizabilities of QDP models in the condensed phase are higher than that of TIP4P-FQ, the Lennard–Jones parameters required minor modification to prevent over-polarization while still reproducing reasonable densities and energetics.

Table 1 Gas-phase, condensed-phase, and interfacial properties for TIP4P-FQ, TIP4P-QDP, and QDP-P1 water models

Property	TIP4P-FQ[a]	TIP4P-QDP	QDP-P1	Exper.
μ (debye)	1.85	1.85	1.85	1.85[b]
$\bar{\alpha}$ (Å^3)	1.12	1.4	1.4	1.47[c]
E_{dimer} (kcal/mol)	-4.50	-4.67	-4.44	-5.4 ± 0.7[d]
Dimer O–O length (Å)	2.92	2.91	2.98	2.98[d]
ρ_{liq} (g/cm^3)	1.0001 (0.0003)	0.9954 (0.0002)	0.9951 (0.0002)	0.997[e]
$\langle\mu_{liq}\rangle$ (Debye)	2.623 (0.001)	2.641 (0.001)	2.752 (0.001)	2.9 (0.6)[f]
ΔH_{vap} (kcal/mol)	10.49[g]	10.55 (0.12)	10.96 (0.12)	10.51[e]
$\langle\alpha_{iso,liq}\rangle$ Å^3	1.12[g]	1.309 (0.001)	1.323 (0.001)	1.34[h]
D_s (10^{-9} m^2/s) [m]	1.93 (0.05), 2.15	2.20 (0.04), 2.46	1.83(0.05), 2.04	2.30[i]
κ_T (10^{-10} Pa^{-1})	3.877 (0.098)	4.013 (0.062)	3.409 (0.051)	4.524[e]
C_p (cal/mol K)	21.0 (5.5)	16.4 (3.5)	18.5 (2.2)	18.0[e]
ε_∞	1.775, 1.592[g]	2.128	2.057	1.79[j]
ε	79. (8)[g]	85.8 (1.0)	97.6 (0.2)	78.[k]
$\Delta\Phi$ (kcal/mol)	-12.21 (0.05)	-11.98 (0.08)	-12.87 (0.05)	—
γ (dyne/cm)	72.7 (1.5)	71.0 (2.7)	81.2 (3.1)	71.9[l]

Values in parentheses denote the uncertainty in the property. Data from Bauer et al. [57]

[a] Reference [57], unless noted

[b] Reference [143]

[c] Reference [144]

[d] Reference [145]. Theoretical estimates from Mas et al. [146] suggest a value of 5.0 ± 0.1

[e] Reference [147]

[f] Reference [53]

[g] Reference [31]

[h] Estimated condensed-phase isotropic polarizability based on the gas-phase value of 1.47 Å^3 from reference [144] and assuming a 9% reduction in polarizability as deduced by Morita in reference [50]

[i] Reference [148]

[j] Reference [149]

[k] Reference [150]

[l] Reference [151]

[m] Values as calculated for a $N = 216$ system (left) and corrected for extrapolation to infinite system size (right)

The Lennard–Jones parameters were parameterized based on fitting to gas-phase water dimer binding energies and geometries (bond distances), condensed-phase density, and enthalpy of vaporization.

Condensed-phase properties computed using the TIP4P-QDP model include bulk liquid density from constant pressure MD simulations, enthalpy of vaporization, average condensed-phase molecular polarizability, diffusion constant, isothermal compressibility, heat capacity at constant pressure, and bulk dielectric constant. These properties are computed as discussed in Bauer et al. [57] and are shown in Table 1. We observe that without fitting explicitly to all of the properties in Table 1, the agreement between the predicted and experimental values is rather acceptable. We have applied this model to studies of the liquid-vapor coexistence curve [58] and ions at the liquid-vapor interface [59]. Studies using the Hirshfeld partitioning scheme [49, 52, 60–62] in conjunction with density functional theory have suggested the dynamic nature of polarizability for species such as methanol [62] and halides [52].

4 Development of CHEQ lipid bilayer force fields

Membranes and membrane-bound proteins are a vital part of biological systems. Membrane-bound proteins are involved in a number of physiological functions including but not limited to passive and active transport, signaling processes, and interfacial enzymatic processes [63]. Further emphasizing their importance, membrane-bound proteins have been found to make up approximately one-third of the human genome [64]. Equally important is the lipid environment within which these proteins function. Recent studies have explored structural properties of membranes as well as electrostatic properties such as the dielectric

Fig. 2 Schematic of the partitioning of DMPC lipid molecule into small-molecule analogs for parameterization of the CHEQ lipid force field

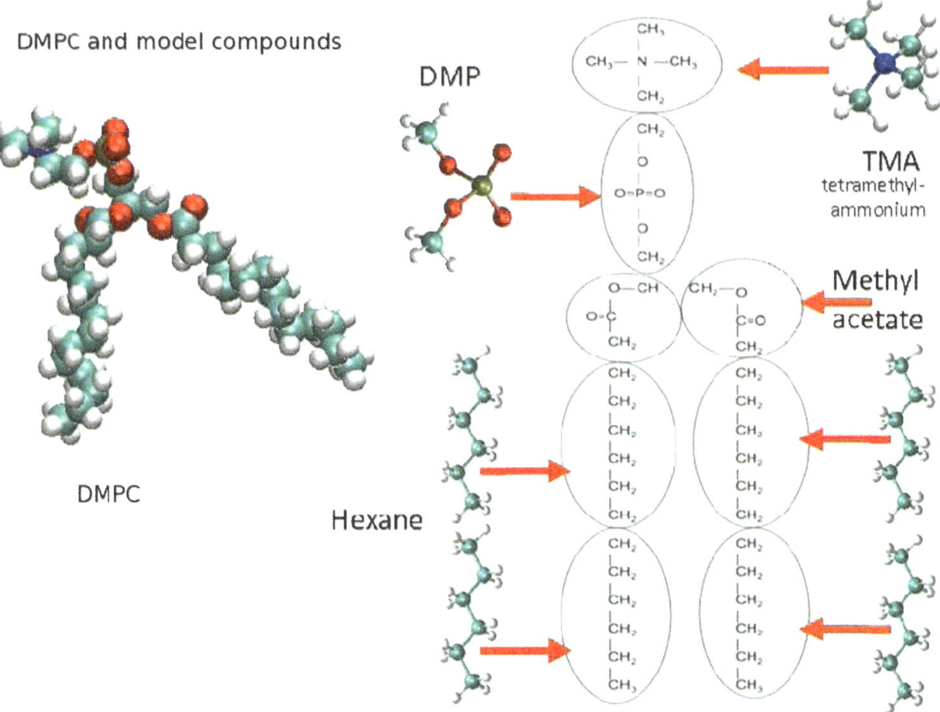

variation within a bilayer [65], the interfacial potential [66], and the interactions of polar or charged amino acid side chains with hydrocarbon tails [67]. Experimental studies have also probed these properties in recent years. For example, structural properties of bilayers have been determined by X-ray and neutron scattering [68, 69] and nuclear magnetic resonance (NMR) spectroscopy [70], while the water penetration into the bilayer interior has been investigated via electron spin spectroscopy [71, 72] and X-ray scattering techniques [73]. However, even current state-of-the-art experimental measurements are not always able to provide the type of detailed atomic level resolution that would provide significant insights into the mechanisms of membrane systems. To this end, computational methods such as molecular dynamics and Monte Carlo simulations have been employed to study properties and processes in such systems at the atomic level [63, 67, 74–85].

Recently, we have developed non-additive, charge equilibration (CHEQ) force fields for lipid bilayers based on DMPC and DPPC as model lipid molecules [86, 87]. Using an approach of "building up" a lipid force field based from small-molecule model compounds, we constructed the lipid force field for saturated chain lipids using linear alkanes, dimethylphosphate, tetramethyl-ammonium, and methyl acetate as model compounds Fig. 2. Properties of the small-molecule analogs that were targeted included gas-phase electrostatic properties (dipole moments, charges, and molecular polarizabilities), gas-phase interaction energies and geometries of small-molecule compounds with water

(using the TIP4P-FQ model as our solvent to which the force field is coupled), condensed-phase bulk liquid properties for alkanes (densities, vaporization enthalpies, self-diffusion constants, and relative torsional energetics), and hydration free energies of linear alkanes in water. The original CHEQ alkane force field [88] was refined using pair-specific Lennard–Jones interactions [89] in order to reduce the gain in free energy upon hydration predicted by the original force field. We discuss implications of this improvement in the next section. A range of lipid bilayer properties were computed following the parameterization process. These include bilayer structure (electron density profile, atomic number density profiles), deuterium order parameters, phosphate–nitrogen (P–N) vector orientation, and dipole potential.

4.1 Application of CHEQ lipid force fields

One interesting aspect of the CHEQ DMPC force field is increased water permeation into the hydrophobic core of the bilayer relative to the non-polarizable CHARMM27 and CHARMM27R force fields [86]. This is evidenced by an increased water number density beyond the carbonyl groups and reduced potential of mean force (PMF) for water entering the lipid tail region. The PMF calculated from water density profiles generated from long, unconstrained simulations of the solvated CHEQ DMPC bilayer suggests a 5-kcal/mol barrier for moving a water molecule from bulk to lipid interior [86]. Subsequent analysis using weighted histogram analysis method

(WHAM) suggests this barrier may as low as 4.5 kcal/mol [90].

Since the lipid groups were constructed from short chain alkane (hexane) parameters [43], the interactions of these alkanes with water have important implications for the interactions between water and lipids. Ongoing efforts towards refinement of the interactions of small molecules with water are of continued interest. One such study [89] presents refined interactions between small-chain alkanes and water, yielding alkane hydration free energies that are less favorable than those calculated using CHEQ Orig [86]. In the case of hexane, the reparameterization results in $\Delta G_{\text{hydration}} = 2.5 \pm 0.2$ kcal/mol (compared to the experimental value of $\Delta G_{\text{hydration}} = 2.550$ kcal/mol); this is approximately 0.7 ± 0.5 kcal/mol higher than the original parameterization [43]. The revised parameterization predicts a PMF barrier nearly 0.8 kcal/mol larger than the original parameterization. This is not surprising as qualitatively, an enhancement is expected from the less favorable interaction between water and the alkane. The range in values for the PMF barrier for different DMPC parameterizations (4.5–5.3 kcal/mol) is lower the values predicted from similar studies using non-polarizable models (5.4–13 kcal/mol) [86, 91–93].

5 Charge equilibration carbohydrate force fields

N-Acetyl-β -hexosaminides are common in nature. They constitute a broad spectrum of materials such as chitin [94] and extracellular matrix glycosaminoglycans [95, 96] and are an essential structural component of glycoproteins and glycolipids [97, 98]. They are also implicated in molecular recognition processes, acting as ligands for a variety of ribonucleases [99–101]. However, functions of poly-N-acetyl-glucosamine and glycosaminoglycans in cell signaling, development, and growth are not well understood, due to the lack of detailed information for the polymer solution [96, 102]. Vibrational spectroscopy [103–105], nuclear magnetic resonance (NMR) [106–110] spectroscopy, and X-ray crystallography [111–113] techniques are applied to investigate the solution structure of related polysaccharides. However, experimental approaches are often limited to certain molecular weight ranges and suffer from difficulties such as crystallization of glycoforms in polysaccharide solution systems.

Complementing experiment, computational methods are a powerful tool to study conformational properties of carbohydrates in solution [114]. Carbohydrate force fields developed for the CHARMM, AMBER, and GROMOS force fields [115, 116] enable MD simulations for polysaccharide solutions. Partly due to the sparsity of experimental data available for parameterization, development of

classical force fields for these types of molecules has been relatively slow. Current state-of-the-art force fields for carbohydrate molecular simulations are primarily the work of Woods and co-workers throughout the past several decades [115]. Ha et al. [117] parameterized classical, non-polarizable force fields for the GROMOS platform. Palma et al. [118] later corrected the O–C–C–O term to revise the barrier of hydroxyl rotations (PHLB model). Kuttel, Brady, and Naidoo modified the PHLB parameter set for lower primary alcohol rotational barriers and referred to the force field as the Carbohydrate Solution Force Field (CSFF) [119–120]. Recently, MacKerell et al. [116] have developed atomistic force fields for carbohydrates building upon the CHARMM force field platform.

Recently, we have developed a polarizable force field based on the charge equilibration force field for an aminoglycan; this will be applied in future studies of protein-ligand binding free energetics using statistical mechanical free energy methods in conjunction with polarizable protein and ligand force fields. Polarizable, or non-additive, force fields for modeling of carbohydrate interactions with polar biological components (in particular proteins) may prove influential within the context of quantifying binding free energetics and structures of complexes of carbohydrates and proteins. By analogy to protein-drug binding and structure research, the use of non-additive interactions may allow a broader range of interaction energies so as to provide a larger separation between biologically relevant complexes (based on free energetics and structure) and weakly interacting, less-relevant complexes. Development of polarizable force fields for carbohydrates in general will complement the current efforts in a number of groups working on non-additive models for proteins, nucleic acids, small molecules, and lipid membranes, as well as ions and small-molecule condensed-phase systems. Work to date on protein-ligand binding studies suggests that polarization (in the most general sense taken to be the variation of molecular mechanics partial charges or electrostatic multipole moments with relative orientations of ligand and substrate) is important in both polar systems as well as situations where ionic species are involved [121–122]. In our continuing future work, we plan to apply polarizable models for the NAG system and compare to non-polarizable (additive) force fields in the context of free energetics of binding in the hen egg-white lysozyme system.

5.1 Parameterization of a charge equilibration force field for NAG

The philosophy we employ for deriving the charge equilibration force field for the monomer NAG is to fit to gas-phase first-principles (ab initio and density functional theory (DFT)) calculation-based properties such as molecular polarizability, dipole moment, and interactions with water in various geometries. As a reference, gas-phase polarizability and dipole moment for NAG are calculated for the geometry optimized using density functional theory (DFT) with the B3LYP/6-311g(d,p) basis. Hardness parameters, η_i°, are determined by fitting to the molecular dipole polarizability (defined above). The augmented hardness matrix, $\mathbf{J}'$, is constructed using two charge normalization groups as shown in Fig. 3. Atomic electronegativities are then determined by fitting to DFT gas-phase charges and dipole moment according to the Merz–Singh–Kollman scheme. The force field gas-phase electrostatic properties are shown in Table 2 using the optimized electrostatic parameters (atom electronegativities and hardness parameters). Consistent with previous study [26], DFT method predicts higher polarizabilities than MP2 when using the same basis set. Higher level of theory predicts larger polarizability and DFT method shows similar polarizabilities when using aug-cc-pvDZ and aug-cc-pvTZ basis set. We note that the fitted gas-phase polarizability using the CHEQ model reproduces DFT and MP2 polarizability with the lower level of theory (6-311g(d,p)) and falls below the gas-phase reference values using larger basis set of aug-cc-pvDZ (14% for B3LYP and 15% for MP2). The absolute extent of reduction of molecular polarizabilities in the condensed phase still remains an ongoing avenue of research, with majority of scaling factors determined from empirical tuning of force field quality and associated stability of simulations using the force fields [36, 123].

After determining electrostatic parameters, solute–solvent interaction-derived non-bond repulsion–dispersion interaction parameters are determined. Solute–solvent non-bond interactions are determined by reproducing vacuum dimer energies and hydrogen binding distances. For reference, the optimized geometry of NAG–water heterodimer (Fig. 4) is obtained by DFT calculation using B3LYP/6-311g(d,p) level of theory. Table 3 shows optimized NAG–water interaction energy and geometry for the CHEQ force field, which is found to well-reproduce the reference first-principles results to at least comparable accuracy as the fixed-charge model. With this basic

Table 2 Electrostatic properties for NAG geometry optimized with B3LYP/6-311g(d,p) basis

	Polarizability/Å^3	Dipole moment/debye
B3LYP/6-311g(d,p)	17.29	5.1572
B3LYP/aug-cc-pvDZ	19.80	5.1604
B3LYP/aug-cc-pvTZ	19.82	5.1515
MP2/6-311g(d,p)	16.75	5.4377
MP2/aug-cc-pvDZ	19.46	5.4115
CHEQ	16.81	5.1160

NAG–water interaction model, we address in the next sections methods and results of condensed-phase molecular dynamics simulations of the NAG molecule (and its oligomers) in aqueous solution (Fig. 5).

5.2 Molecular dynamics simulations of polarizable protein-ligand complexes: hen egg-white lysozyme and NAG trimer

5.2.1 Molecular dynamics methodologies for simulations of hen egg-white lysozyme complexed with the NAG trimer

As an application, we performed molecular dynamics simulations of the complex between hen egg-white lysozyme (HEWL) and NAG trimer (NAG$_3$) with both the polarizable CHEQ and non-polarizable CHARMM carbohydrate [118] and protein [124] force fields. We used the CHEQ force field for proteins [36], water [31], and NAG$_3$ [125]. The simulations are the initial steps towards application of our models for protein-ligand absolute and relative binding free energy calculations using statistical mechanical approaches such as thermodynamic integration.

Two binding modes (ABD and BCD) of HEWL-(NAG)$_3$ complex are simulated independently. The ABC conformation is obtained from the X-ray crystal structure (PDB:1lzb). To construct the BCD structure, the carbohydrate residue at the A site is removed from the HEWL-(NAG)$_4$ X-ray structure bound in the ABCD site (PDB:1lzc). Each of the two starting structures is solvated in an octahedral water box. Bulk solution simulations were performed within the isothermal-isobaric ensemble (NPT) at 298 K and 1 bar. A constant temperature of 298 K was maintained with a Hoover thermostat [126–128], and constant pressure was maintained via a Langevin piston method [129]. Non-bonded interactions were switched to zero via a switching function from 10 to 12 Å. Conditionally convergent long-range electrostatic interactions were treated using the Smooth Particle Mesh Ewald [130] method with 24 grid points in each dimension and screening parameter of $\kappa = 0.32$. We performed ten

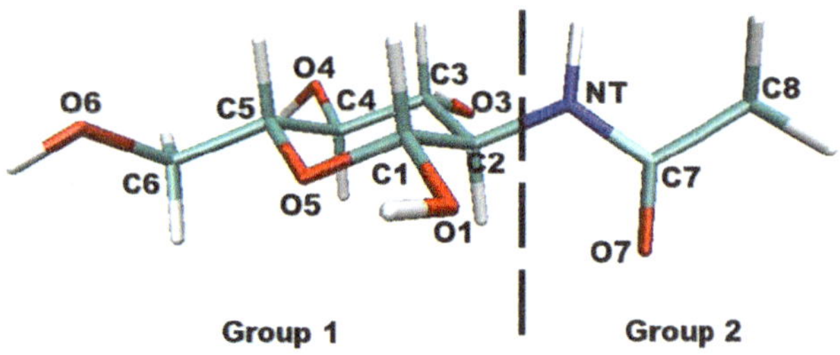

Fig. 3 Designation of charge conservation groups

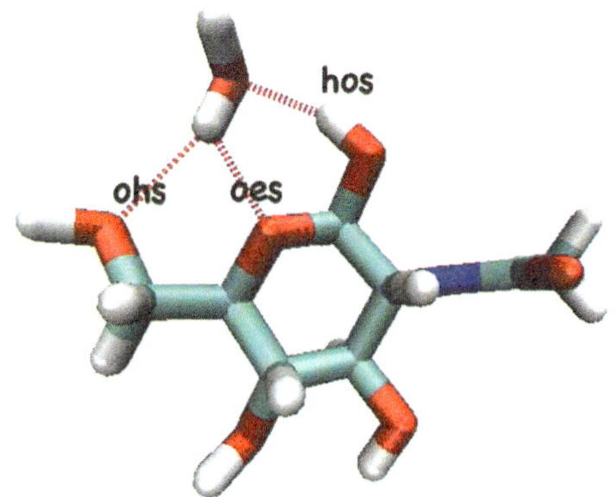

Fig. 4 Interactions between NAG and water in gas phase

Table 3 NAG–water gas-phase interactions and bond lengths

	E_{dimer} (kcal/mol)	R_{ohs} (Å)	R_{oes} (Å)	R_{hos} (Å)
B3LYP/6-311g(d,p)	−14.73	2.44	2.20	1.90
CHEQ	−14.10	2.85	2.18	1.99
PHLB	−9.55	2.82	1.98	2.06
PHLB-PSP	−13.51	2.87	1.60	1.81

Hydrogen bond types are described in Fig. 4

nanoseconds of molecular dynamics using a Verlet leap-frog integrator with timestep of 0.001 picoseconds (1 fs) using the CHARMM22 force field. Using the polarizable force field, we simulated the solvated X-ray structure (ligand at the ABC sites) for 8 ns and the one with $(NAG)_3$ bound to the BCD sites for 4 ns with timestep of 0.0005 ps (0.5 fs), the first 500 ps of which is considered as equilibration time. Charge degrees of freedom were maintained at 1K using a Nosé–Hoover bath; all charges were assigned to the same bath, and no local "hotspots" in charge were observed. Protein, water, and ligand charges degrees of freedom were assigned masses of 0.000095, 0.000069, and 0.0001 (kcal/mol ps^2)/e^2, respectively. Charge normalization was maintained by groups for the protein and ligand and over individual molecules for the water.

5.2.2 Analysis of MD simulations

HEWL X-ray crystallography studies suggest that $(NAG)_3$ binds at the ABC sites [131, 132] (Fig. 6). X-ray powder diffraction study [133] by Von Dreele, on the other hand, showed that $(NAG)_3$ occupies the ABC and BCD sites in the ratio of 35:65. Von Dreele [133] attributed the absence of BCD binding in the solid state to the hydrolysis at the CD site. We investigate the ABC and BCD complexes by two independent MD simulations starting from corresponding crystal binding structures as discussed above.

To evaluate the structural changes of the complex over the dynamic trajectory, the RMSD profiles for HEWL

backbone Fig. 7 and $(NAG)_3$ heavy atoms Fig. 7 are computed with respect to the starting ABC (panels a and c) and BCD (panels b and d) crystal structures. The results for the CHEQ force field are in panels c and d; those for the non-polarizable are shown in panels a and b. The protein backbone atom RMSD behaves similarly using both force fields, with slightly larger drift from the crystal structure with the CHEQ model, particularly when $(NAG)_3$ binds to the BCD sites (Fig. 7, panel d); this implies that the protein structure changes to accommodate the ligand. Such conformational relaxation is not found in the simulation with the CHARMM22 force field. Here, we emphasize that the starting BCD conformation is obtained from the crystal structure with $(NAG)_4$, which may deviate from the exact BCD structure and require longer relaxation time for protein as the CHEQ model suggests (Fig. 7d). On the other hand, the ligand RMSD profile with larger fluctuations and larger RMSD values indicates that $(NAG)_3$ at the ABC sites is relatively flexible and undergoes more structural changes from the starting crystal structure (Fig. 7 panels a and c).

6 Summary

With the continuing advances in computational hardware [134, 135] and novel force fields constructed using quantum mechanics, the outlook for non-additive force fields is promising. Our work in the past several years has slowly demonstrated the utility of polarizable force fields, particularly those based on the charge equilibration formalism, for a broad range of physical and biophysical systems. We have constructed polarizable force fields for small molecules [36, 42–44, 125, 136, 137], proteins [37], lipids, and lipid bilayers [43, 86, 138–140] and recently have begun work on carbohydrate force fields [125]. The latter area has been relatively untouched by force field developers with particular focus on polarizable, non-additive interaction potential models.

With respect to polarizable force field development, one aspect that has been difficult to address has been the absolute nature of the reduction in condensed-phase intrinsic molecular polarizability; this is particularly important for polarizable formalisms that are more classical in nature. We have recently explored the effects of condensed-phase polarizability reduction as well as developed novel water potentials that begin to consider explicitly the dynamical variation of molecular polarizability with the local environment [57]. This is an important area for further development. We also note that charge transfer effects have largely been neglected in classical formalisms; the implications of charge transfer are now becoming more apparent, and the development of interaction potentials capable of

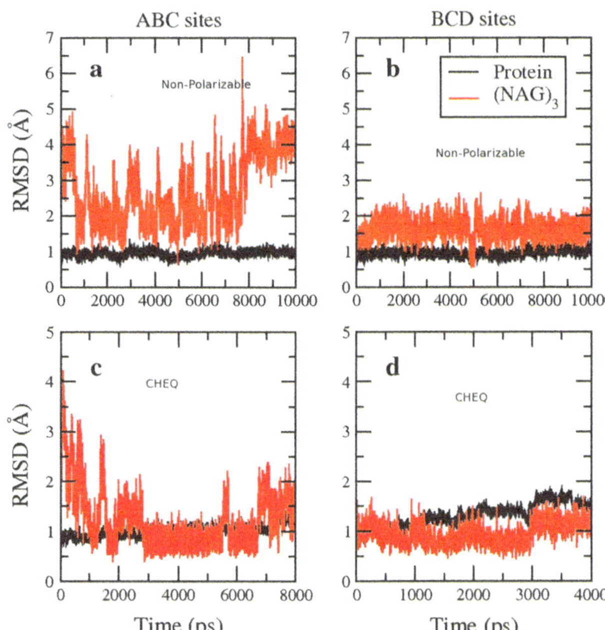

Fig. 7 Evolution of HEWL backbone and NAG₃ heavy-atom RMSD from crystal structure. *Left column* shows data for the ABC sites and *right column* for BCD sites. **a** and **b** Show results using the non-polarizable force field. **c** and **d** Show results using the CHEQ polarizable force field

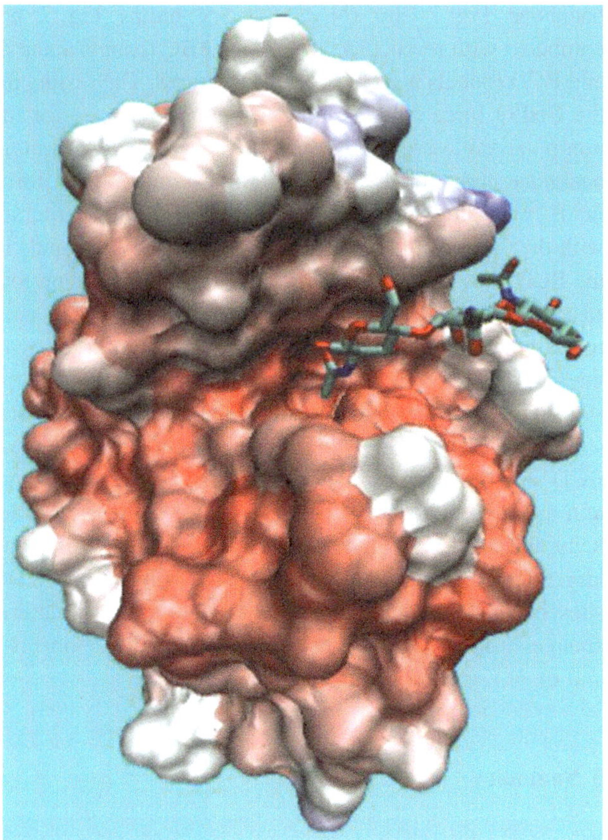

Fig. 5 Structure of the hen egg-white lysozyme complexed with NAG₃, taken from crystal structure PDBID:1LZB

Fig. 6 Binding site hydrogen bond topography of the hen egg-white lysozyme complexed with NAG₃ based on crystal structure PDBID:1LZB (*top panel* for sites ABC), and PDBID:1LZC (*bottom panel* for sites BCD)

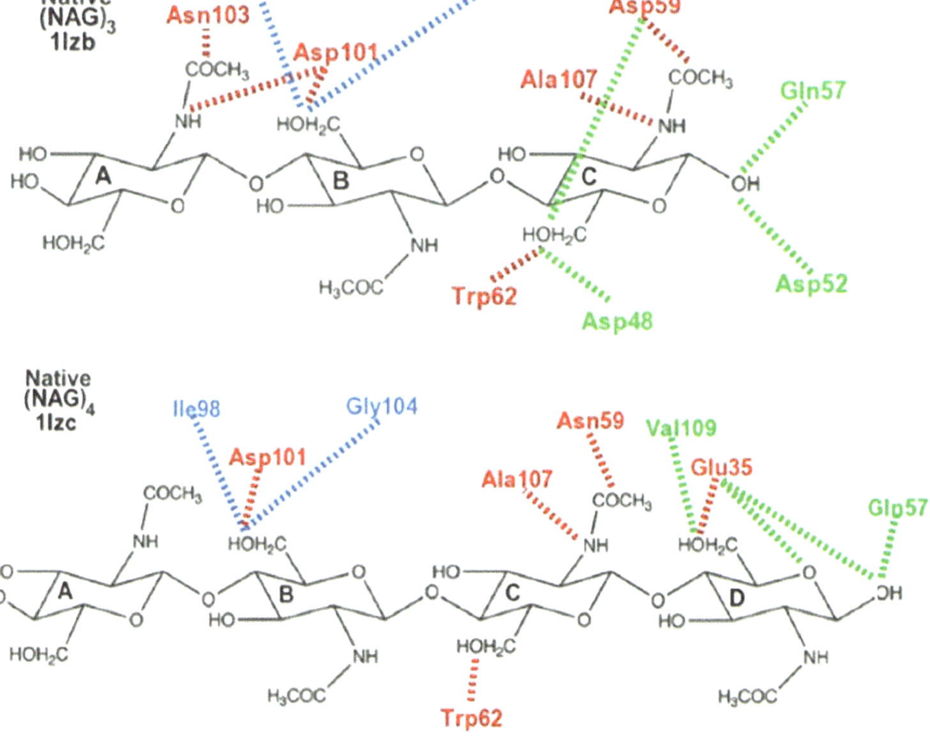

explicitly and dynamically including such effects is warranted, particularly for biophysical systems involving charge-dense ions.

With respect to biomacromolecular force fields, we have successfully constructed models for lipid molecules that we are now exploiting for molecular dynamics simulations of ion channels [138] and issues related to charge species in low-dielectric membrane environments [77, 141, 142]. Further development and refinement of such models continues aggressively in our lab.

We have also developed one of the first polarizable carbohydrate force fields for the NAG molecule and longer oligomers that are model systems for carbohydrate substrates to hen egg-white lysozyme, for instance. With the current efforts to include electronic polarization in classical force fields, we believe that this work establishes a platform on which to build more extensive polarizable force fields for an important class of biological macromolecules. We have successfully initiated studies of the hen egg-white lysozyme system complexed with its natural carbohydrate substrates and will progress to extended molecular dynamics simulations for absolute and relative binding free energy calculations in the near future.

References

1. Applequist J, Carl JR, Fung K (1972) J Am Chem Soc 94:2952–2960
2. Caldwell JW, Kollman PA (1995) J Phys Chem 99:6208–6219
3. Grossfield A, Ren P, Ponder JW (2003) J Am Chem Soc 125:15671–15682
4. Grossfield A (2005) J Chem Phys 122:024506
5. Jiao D, King C, Grossfield A, Darden TA, Ren P (2006) J Phys Chem B 110:18553–18559
6. Ponder JW, Wu C, Ren P, Pande VS, Chodera JD, Schnieders MJ, Haque I, Mobley DL, Lambrecht DS, DiStasio RA Jr, Head-Gordon M, Clark GNI, Johnson ME, Head-Gordon T (2010) J Phys Chem B 114:2549–2564
7. Ren P, Ponder JW (2002) J Comp Chem 23:1497–1506
8. Ren P, Ponder JW (2003) J Phys Chem B 107:5933–5947
9. Ren R, Ponder JW (2004) J Phys Chem B 108:13427–13437
10. Gresh N, Cisneros GA, Darden TA, Piquemal J-P (2007) J Chem Theory Comput 3:1960–1986
11. Thole BT (1981) Chem Phys 59:341–350
12. Anisimov VM, Lamoureux G, Vorobyov IV, Huang N, Roux B, MacKerell AD Jr (2005) J Chem Theory Comput 1:153–168
13. Lamoureux G, Roux B (2003) J Chem Phys 119:3025
14. Lamoureux G, MacKerell AD Jr, Roux B (2003) J Chem Phys 119:5185
15. Lamoureux G, Harder E, Vorobyov IV, Roux B, MacKerell AD Jr (2006) Chem Phys Lett 418:245–249
16. Lopes PEM, Lamoureux G, Roux B, MacKerell AD Jr (2007) J Phys Chem B 111:2873–2885
17. Vorobyov IV, Anisimov VM, AD MacKerell J (2005) J Phys Chem B 109:18988–18999
18. Vorobyov IV, Anisimov VM, Greene S, Venable RM, Moser A, Pastor RW, AD MacKerell J (2007) J Chem Theory Comput 3:1120–1133
19. Baker CM, Lopes PEM, Zhu X, Roux B, MacKerell AD Jr (2010) J Chem Theory Comput 6:1181–1198
20. Lamoureux G, Roux B (2006) J Phys Chem B 110:3308–3322
21. Yu H, Whitfield TW, Harder E, Lamoureux G, Vorobyov I, Anisimov VM, MacKerell AD Jr, Roux B (2010) J Chem Theory Comput 6:774–786
22. de Courcy B, Piquemal J-P, Garbay C, Gresh N (2010) J Am Chem Soc 132:3312
23. Devereux M, van Severen M-C, Parisel O, Piquemal J-P, Gresh N (2011) J Chem Theory Comput 7:138–147
24. Piquemal J-P, Chevreau H, Gresh N (2007) J Chem Theory Comput 3:824–837
25. Piquemal J-P, Williams-Hubbard B, Fey N, Deeth RJ, Gresh N, Giessner Prettre C (2003) J Comput Chem 24:1963
26. Piquemal J, Marquez A, Parisel O, Giessner-Prettre C (2005) J Comput Chem 26:1052
27. Piquemal J-P, Cisneros GA, Reinhardt P, Gresh N, Darden TA (2006) J Chem Phys 124:104101
28. Sanderson RT (1951) Science 114:670–672
29. Sanderson RT (1976) Chemical bonds and bond energy. Academic Press, New York
30. Rappe AK, Goddard WA (1991) J Phys Chem 95:3358–3363
31. Rick SW, Stuart SJ, Berne BJ (1994) J Chem Phys 101:6141–6156
32. Rick SW, Stuart SJ, Bader JS, Berne BJ (1995) J Mol Liq 65/66:31
33. Rick SW, Berne BJ (1996) J Am Chem Soc 118:672–679
34. Rick SW, Stewart SJ (2002) Reviews of computational chemistry. Wiley, New York, p 89
35. Rick SW (2001) J Chem Phys 114:2276–2283
36. Patel S, Brooks CL III (2004) J Comput Chem 25:1–15
37. Patel S, MacKerell AD Jr, Brooks CL III (2004) J Comput Chem 25:1504–1514
38. Patel S, Brooks CL III (2005) J Chem Phys 122:024508
39. Patel S, Brooks CL III (2006) Mol Simul 32:231–249
40. Olano LR, Rick SW (2005) J Comput Chem 26:699–707
41. Warren GL, Davis JE, Patel S (2008) J Chem Phys 128:144110
42. Zhong Y, Warren GL, Patel S (2007) J Comp Chem 29:1142–1152
43. Davis JE, Warren GL, Patel S (2008) J Phys Chem B 112:8298–8310
44. Bauer BA, Patel S (2008) J Mol Liq 142:32–40
45. Illingworth CJ, Domene C (2009) Proc R Soc A 465:1701–1716
46. Nalewajski RF, Korchowiec J, Zhou Z (1988) Int J Quant Chem 22:349–366
47. Chelli R, Procacci P, Righini R, Califano S (1999) J Chem Phys 111:8569–8575
48. Car R, Parrinello M (1985) Phys Rev Lett 55:2471–2474
49. Krishtal A, Senet P, Yang M, Van Alsenoy C (2006) J Chem Phys 125:034312
50. Morita A (2002) J Comput Chem 23:1466–1471
51. Schropp B, Tavan P (2008) J Phys Chem B 112:6233–6240
52. Bauer BA, Lucas TR, Krishtal A, van Alsenoy C, Patel S (2010) J Phys Chem A 114:8984–8992
53. Badyal YS, Saboungi M, Price DL, Shastri SD, Haeffner DR, Soper AK (2000) J Chem Phys 112:9206–9208
54. Kuo IW, Mundy CJ, Eggimann BL, McGrath MJ, Siepmann J, Chen B, Vicccli J, Tobias DJ (2006) J Phys Chem B 110:3738–3746
55. Silvestrelli PL, Parrinello M (1999) Phys Rev Lett 82:3308–3311
56. Sprik M (1991) J Chem Phys 95:6762–6769
57. Bauer BA, Warren GL, Patel S (2009) J Chem Theory Comput 5:359–373
58. Bauer BA, Patel S (2009) J Chem Phys 131:084709
59. Bauer BA, Patel S (2010) J Chem Phys 132:8107–8117

60. Hirshfeld FL (1977) Theor Chim Acta 44:129–138
61. Bultinck P, Van Alsenoy C, Ayers PW, Carbo-Dorca R (2007) J Chem Phys 126:144111
62. Krishtal A, Senet P, Van Alsenoy C (2008) J Chem Theory Comput 4:426–434
63. Roux B, Allen T, Berneche S, Im WQ (2004) Rev Biophys 37:15–103
64. Wallin E, Von Heijne G (1998) Protein Sci 7:1029–1038
65. Stern HA, Feller SE (2003) J Chem Phys 118:3401–3412
66. Wang L, Bose PS, Sigworth F (2006) J PNAS 103:18528–18533
67. MacCallum JL, Bennett WFD, Tielman DP (2008) Biophys J 94:3393–3404
68. Kučerka N, Liu Y, Chu N, Petrache HI, Tristram-Nagle S, Nagle JF (2005) Biophys J 88:2626–2637
69. Kučerka N, Nagle JF, Sachs JN, Feller SE, Pencer J, Jackson A, Katsaras J (2008) Biophys J 95:2356–2367
70. Petrache HI, Dodd SW, Brown MF (2000) Biophys J 79:3172–3192
71. Marsh D (2002) Eur Biophys J 31:559–562
72. Erilov DA, Bartucci R, Guzzi R, Shubin AA, Maryasov AG, Marsh D, Dzuba SA, Sportelli L (2005) J Phys Chem B 109:12003–12013
73. Mathai JC, Tristram-Nagle S, Nagle JF, Zeidel ML (2008) J Gen Physiol 131:69–76
74. Allen TW, Bastug T, Kuyucak S, Chung S-H (2003) Biophys J 84:2159–2168
75. Berneche S, Roux B (2002) Biophys J 82:772–780
76. Allen TW, Andersen OS, Roux B (2004) PNAS 101:117–122
77. Dorairaj S, Allen TW (2007) PNAS 104:4943–4948
78. Aliste MP, Tieleman DP (2005) BMC Biochem 6:30
79. Tieleman DP, MacCallum JL, Ash WL, Kandt C, Xu Z, Monticelli LM (2006) J Phys Condens Matter 18:S1221–S1234
80. Xu Z, Luo HH, Tieleman DP (2007) J Comput Chem 28:689–697
81. Feller SE (2000) Curr Opin Colloid Interface Sci 5:217–223
82. MacCallum JL, Tieleman DP (2006) J Am Chem Soc 128:125–130
83. Pandit SA, Bostick D, Berkowitz ML (2004) Biophys J 86:1345–1356
84. Berkowitz ML, Bostick DL, Pandit SA (2006) Chem Rev 106:1527–1539
85. Siu SWI, Vacha R, Jungwirth P, Bockmann RA (2008) J Chem Phys 128:125103
86. Davis JE, Rahaman O, Patel S (2009) Biophys J 96:385–402
87. Davis JE, Patel S (2009) J Phys Chem B 113:9183–9196
88. Patel S, Brooks CL III (2006) J Chem Phys 124:204706
89. Davis JE, Patel S (2010) Chem Phys Lett 484:173–176
90. Bauer BA, Lucas TR, Meninger D, Patel S (2011) Chem Phys Lett. doi:10.1016/j.cplett.2011.04.052
91. Orsi M, Sanderson WE, Essex JW (2009) J Phys Chem B 2009:12019–12029
92. Sugii T, Takagi S, Matsumoto Y (2005) J Chem Phys 123:184714
93. Jedlovszky P, Mezei M (2000) J Am Chem Soc 122:5125–5131
94. Austin PR, Brine CJ, Castle JE, Zikakis JP (1985) Science 212:749
95. Alberts B, Johnson A, Lewis J, Raff M, Roberts K, Walter P (2002) Molecular biology of the cell. Garland Science, New York
96. Fukuda M, Hindsgaul O (1994) Molecular glycobiology. IRL Press at Oxford University Press, Oxford
97. Varki A (2006) Cell 126:841
98. Weerapana E, Imperiali B (2006) Glycobiology 16:91R
99. Haltiwanger RS, Hill RL (1986) J Biochem 261:15696
100. Lew DB, Songu-Mize E, Pontow SE, Stahl PD, Rattazzi MC (1994) J Clin Invest 94:1855
101. Arkar K, Sarkar HS, Kole L, Das PK (1996) Mol Cell Biochem 156:109
102. Almond A (2005) Carbohydr Res 340:907
103. Kovacs A, Nyerges B, Izvekov V (2008) J Phys Chem B 112:5728
104. Paolantoni M, Sassi P, Morresi A, Santini S (2007) J Chem Phys 127:24504
105. Gallina ME, Sassi P, Paolantoni M, Morresi A, Cataliotti RS (2006) J Phys Chem B 110:8856
106. Bush CA, Martin-Pastor M, Imberty A (1999) Annu Rev Biophys Biomol Struct 28:269
107. Peters T, Binto BM (1996) Curr Opin Struct Biol 6:710
108. Almond A, DeAngelis PL, Blundell CD (2006) J Mol Biol 358:1256
109. Blundell CD, DeAngelis PL, Day AJ, Almond A (2004) Glycobiology 14:999
110. Prestegard JH, Al-Hashimi HM, Tolman JRQ (2000) Res Biophys 33:371
111. Kuroki R, Ito Y, Kato Y, Imoto T (1997) J Biol Chem 272:19976
112. Maenaka K, Matsushima M, Song H, Sunada F, Watanabe K, Kumagai I (1995) J Mol Biol 247:281
113. Ose T, Kuroki K, Matsushima M, Kumagai I (2009) J Biochem 146:651
114. Vilegenthart, JFG, Woods, RJ (eds) (2006) NMR spectroscopy and computer modeling of carbohydrates: recent advances. American Chemical Society, Washington
115. Woods RJ, Dwek RA, Edge CJ (1995) J Phys Chem 99:3832
116. Guvench O, Greene SN, Kamath G, Brady JW, Venable RM, Pastor RW, MacKerell AD Jr (2008) J Comp Chem 29:2543
117. Ha SN, Giammona A, Field M, Brady JW (1988) Carbohydr Res 180:207
118. Palma R, Himmel ME, Liang G, Brady JW (2001) In: ACS symposium series: glycosyl hydrolases in biomass conversion. American Chemical Society, Washington, p 112
119. Naidoo KJ, Kuttel MJ (2001) Comput Chem 22:445
120. M Kuttel JWB, Naidoo KJ (2002) J Comput Chem 23:1236
121. Jiao D, Golubkov PA, Darden TA, Ren P (2008) Proc Natl Acad Sci 105:6290–6295
122. Khoruzhii O, Donchev AG, Galkin N, Illarionov A, Olevanov M, Ozrin V, Queen C, Tarasov V (2008) Proc Natl Acad Sci 105:10378–10383
123. Patel S, MacKerell AD Jr, Brooks CL III (2004) J Comput Chem 25:1504
124. MacKerell AD Jr et al (1998) J Phys Chem B 102:3586–3616
125. Zhong Y, Bauer BA, Patel S (2011) J Comput Chem (in revision)
126. Nosé S (1984) J Chem Phys 81:511–519
127. Nosé S (1984) Mol Phys 52:255–268
128. Nosé S (1990) J Phys Condens Matter 2:SA115–SA119
129. Allen MP, Tildesley DJ (1987) Computer simulations of liquids. Clarendon, Oxford
130. Darden T, York D, Pedersen L (1993) J Chem Phys 98:10089–10092
131. Cheetham JC, Artymiuk PJ, Phillips DC (1992) J Mol Biol 224:613
132. Maenaka K, Mastushima M, Song H, Sunada F, Watanabe K, Kumagai I (1995) J Mol Biol 247:281
133. von Dreele RB (2004) Acta Cryst D 61:22
134. Bauer BA, Davis JE, Taufer M, Patel S (2011) J Comput Chem 32:375–385
135. Ganesan N, Bauer BA, Lucas TR, Patel S, Taufer M (2011) J Comput Chem (submitted, in revision)
136. Zhong Y, Patel S (2009) J Phys Chem B 113:767–778
137. Zhong Y, Patel S (2010) J Phys Chem B 114:11076–11092

138. Patel S, Davis JE, Bauer BA (2009) J Am Chem Soc 131:13890–13891
139. Patel S, Zhong Y, Bauer BA, Davis JE (2009) J Phys Chem B 113:9241–9254
140. Lucas TR, Bauer BA, Patel S (2011) J Comput Chem (in revision)
141. Vorobyov I, Allen TW (2010) J Chem Phys 132:185101
142. Schow EV, Freites JA, Cheng P, Bernsel A, von Heijne G, White SH, Tobias DJ (2011) J Membr Biol 239:35–48
143. Clough SA, Beers Y, Klein GP, Rothman LS (1973) J Chem Phys 59:2254–2259
144. Murphy WF (1977) J Chem Phys 67:5877–5882
145. Odutola JA, Dyke TR (1980) J Chem Phys 72:5062–5070
146. Mas EM, Bukowski R, Szalewicz K, Groenenboom GC, Wormer PES, van der Avoird A (2000) J Chem Phys 113:6687–6701
147. Lide, DR (eds) (1997) CRC handbook of chemistry and physics, 77th edn. CRC, Boca Raton
148. Krynicki K, Green CD, Sawyer DW (1978) Discuss Faraday Soc 66:199–208
149. Buckingham AD (1956) Proc R Soc Lond Ser A 238:235–244
150. Watanabe K, Klein ML (1989) Chem Phys 131:157–167
151. Cini R, Logio G, Ficalbi A (1972) J Colloid Interface Sci 41:287–297

Theor Chem Acc (2012) 131:1132
DOI 10.1007/s00214-012-1132-z

REGULAR ARTICLE

A distributed point polarizable force field for carbon dioxide

Fang-Fang Wang · Revati Kumar · Kenneth D. Jordan

Received: 14 June 2011 / Accepted: 19 August 2011 / Published online: 2 March 2012
© Springer-Verlag 2012

Abstract A distributed point polarizable model potential for carbon dioxide, with explicit terms for charge penetration and induction, is introduced. This model potential accurately describes the structures and interaction energies of small $(CO_2)_n$ clusters and also gives the second virial coefficients and radial distribution functions of supercritical CO_2 in excellent agreement with experiment.

Keywords Carbon dioxide · Force field · Charge penetration · Polarization · Symmetry-adapted perturbation theory

1 Introduction

Carbon dioxide (CO_2) has attracted considerable attention because of its use as a supercritical fluid for extraction processes and because of its role as a greenhouse gas. Not surprising, there have been many of computer simulations of CO_2 in the condensed phase as well as in a range of chemical environments. Most of these simulations have been carried out in model potentials, and, as a result, the

---

Published as part of the special collection of articles: From quantum mechanics to force fields: new methodologies for the classical simulation of complex systems.

F.-F. Wang · R. Kumar · K. D. Jordan (✉)
Department of Chemistry and Center for Molecular
and Materials Simulations, University of Pittsburgh,
Pittsburgh, PA 15260, USA
e-mail: jordan@pitt.edu

Present Address:
R. Kumar
Department of Chemistry, University of Chicago,
Chicago, IL 60637, USA

success of such simulations depends on the quality of the model potentials employed [1]. Several model potentials have been proposed to describe interactions between CO_2 molecules [2–17]. Most of these potentials employ a simple point-charge representation of the electrostatics and neglect many-body polarization effects.

Recently, our group introduced a distributed point polarizable (DPP2) water model that accurately describes the energetics of water clusters and is also quite successful for predicting properties of bulk water [18]. In the present study, we derive a DPP2-like force field for CO_2 and apply it to small CO_2 clusters and to supercritical CO_2.

2 Methodology

2.1 Ab initio calculations

High-level electronic structure calculations on clusters have become the method of choice for parameterizing and testing force fields for condensed phase simulations of molecular structures. The symmetry-adapted perturbation theory (SAPT) [19] method has proven particularly valuable in this regard as it allows for a separation of the net interaction energy into exchange-repulsion, electrostatics, induction, and dispersion components [18]. A combination of SAPT and coupled-cluster singles-plus-doubles and perturbative triples CCSD(T) [20] calculations on small water clusters were used in the parameterization and testing of the DPP2 water model [18], and a similar strategy is employed here for developing a DPP2-like rigid-monomer force field for CO_2.

The coordinate system used for generating the ab initio data for the CO_2 dimer is described in Fig. 1. The four degrees of freedom are chosen to be the distance R between

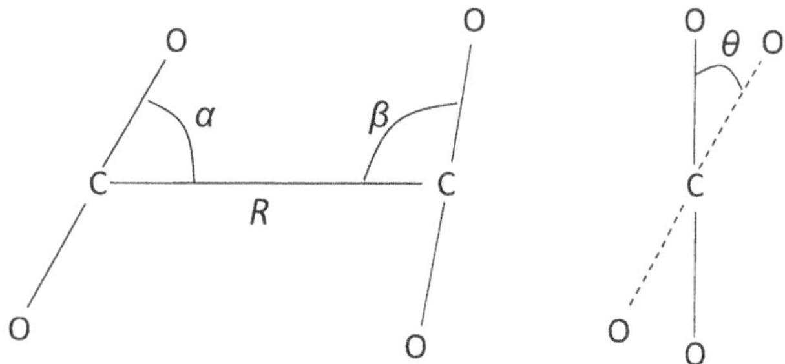

Fig. 1 Coordinates used to define the structure of the CO_2 dimers

the carbon atoms, the angles α and β between the molecular axes and the line connecting the two centers, and the dihedral angle θ. A grid of structures was generated by stepping R between 3.2 and 5.0 Å in steps of 0.3 Å and stepping each of the angles through 30 degree increments. Geometries with intermolecular O–O or C–O distances <2.7 and 2.5 Å, respectively, were excluded, leaving nearly 200 symmetry-independent dimer structures. For each of these structures, SAPT/aug-cc-pVTZ [21, 22] and CCSD(T)-F12/VTZ-F12 [23, 24] calculations were carried out. The F12 denotes the use of the explicitly correlated approach (the F12a variant) of Werner and coworkers [23], and the VTZ-F12 denotes the valence triple-zeta basis set of Peterson and coworkers [24], designed for use with the F12 method. The SAPT results were used to parameterize the electrostatics and dispersion terms in the model potential. The differences between the sum of the electrostatics, induction, and dispersion energies from the model potential and the CCSD(T)-F12/VTZ-F12 interaction energies for the dimer structures were used to fit the exchange-repulsion terms in the model potential.

The CCSD(T) three-body energies of three low-energy $(CO_2)_3$ structures shown in Fig. 2 were calculated using

$$\Delta E_3 = \sum_{i=1}^{N-2} \sum_{j=i+1}^{N-1} \sum_{k=j+1}^{N} \left[E(i,j,k) - E(i,j) - E(i,k) - E(j,k) \right. \\ \left. + E(i) + E(j) + E(k) \right], \tag{1}$$

where $E(i)$, $E(i,j)$, and $E(i,j,k)$ are, respectively, the energies of the monomer i, dimer (i,j), and trimer (i,j,k) cut out of the full cluster. As discussed below, the CCSD(T)-F12/VTZ-F12 three-body interaction energies were also used in parameterizing the polarization terms in the potential.

To test the DPP2 model for CO_2, CCSD(T)-F12/VTZ-F12 interaction energies were calculated for six low-lying potential energy minima of the CO_2 dimer, trimer, and tetramer (see Fig. 2). For the dimer and trimer species, the geometries were optimized at the MP2/aug-cc-pVTZ [21, 22] level, while for the tetramer the geometry was optimized at the MP2/aug-cc-pVDZ [21, 22] level. All optimizations were carried out under the constraint of rigid monomers, with the monomer CO bond lengths and OCO

angles fixed to their experimental values [25]. In addition to these minimum energy structures, 20 dimer and 20 trimer structures were randomly selected from a $T = 240$ K, $\rho = 14.9$ molecules/nm^3 *NVT* Monte Carlo simulation of supercritical CO_2 fluid to provide a further test of the model potential. These simulations were carried out using the Murthy[3] CO_2 potential, and only structures with CC distances <5 Å were retained.

The MP2 geometry optimizations and the CCSD(T)-F12 single-point calculations were performed using GAUSSIAN 03 [26] and MOLPRO [27], respectively. The SAPT calculations were carried out with the SAPT2008 [28] program interfaced with DALTON 2.0 [29]. The Monte Carlo simulations were carried out using in-house codes.

The SAPT calculations were wavefunction-based approach [19] rather than DFT based [30]. Jeziorski et al. [8] have found that the wavefunction-based SAPT procedure overestimates the dispersion of $(CO_2)_2$ near its equilibrium geometry by about 20%. However, this deficiency is largely remedied by our strategy of fitting the repulsive potential to the differences of the CCSD(T) and SAPT interaction energies.

2.2 Monte Carlo simulations

NVT Monte Carlo simulations of supercritical CO_2 as described by the DPP2 model potential were carried out for $T = 312$ K and a density of 11.4 molecules/nm^3 and for $T = 240$ K and a density of 14.9 molecules/nm^3. These conditions correspond closely to those for which radial distribution functions have been determined experimentally [31]. The simulations were carried out for 4×10^5 Monte Carlo moves following equilibration runs of 10^5 Monte Carlo steps. Each step consisted of a translation of the center of mass and a rotation of one of the monomers. The maximum values of the translational and rotational moves were optimized to give an acceptance rate of around 40%. Long-range interactions were treated with a spherical cutoff of 9.5 Å. (The interaction energy between two CO_2 molecules is negligible at distances >9.5 Å.) The same procedure was used in the Monte Carlo simulations with the Murthy potential (Table 1).

2.3 Parameterization of the model potential

2.3.1 Electrostatics

Most CO_2 model potentials represent the electrostatics by the use of three atom-centered point charges (Table 1), the values of which are chosen to reproduce the experimental quadrupole moment of the molecule. However, as seen from Table 2, such three-point-charge models give values of higher rank multipole moments very different from

Fig. 2 The MP2 and DPP2 optimized geometries of low-energy minima of $(CO_2)_n$, $n = 2$–4 (with the DPP2 values in parenthesis, distances in Å and angles in degrees). The D2 structure for $(CO_2)_2$ is a first-order saddle point at the MP2/aug-cc-pVDZ level, but a local minimum at the MP2/aug-cc-pVTZ level

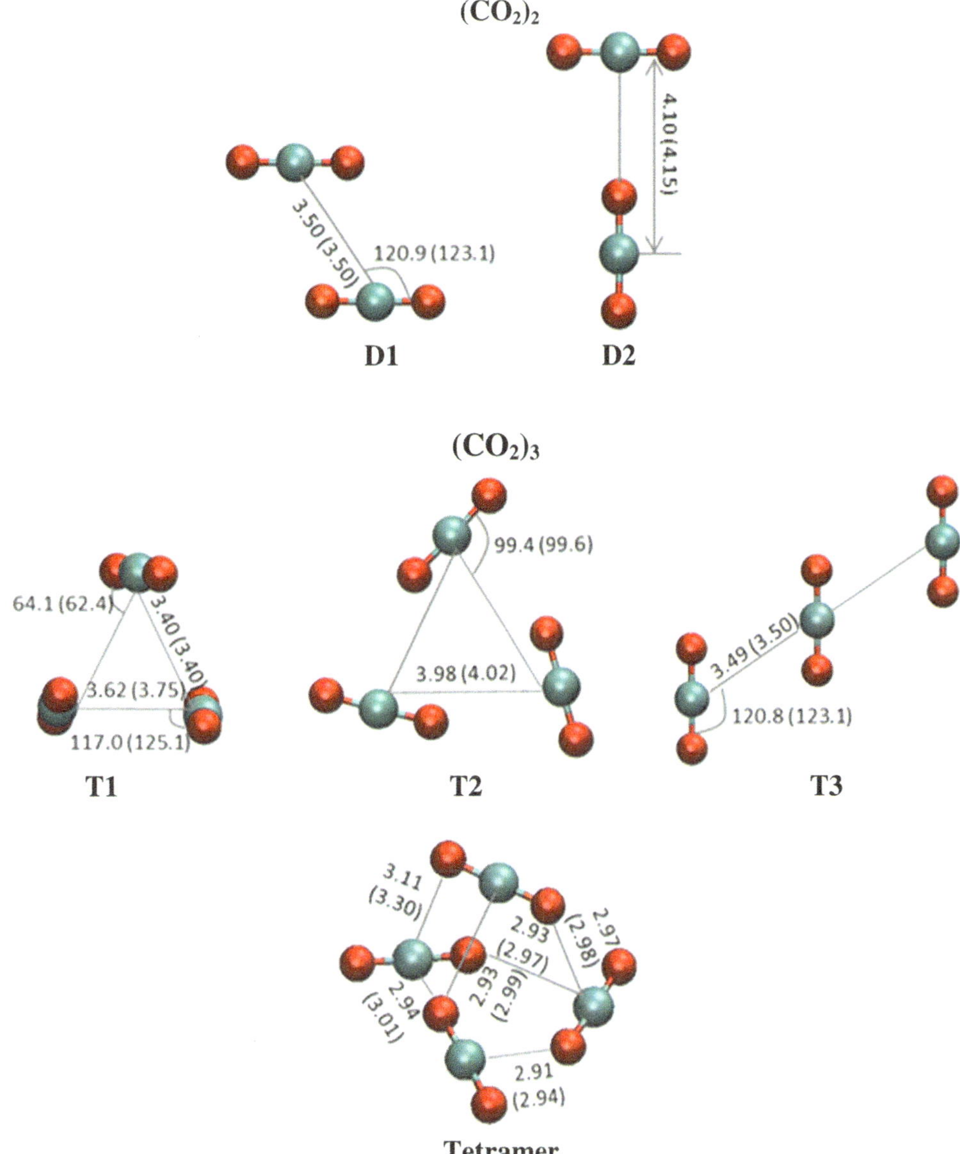

Table 1 Point charges (e), and their locations (Å) for the CO_2 models

	3-point	Murthy	DPP2
Charges			
qOn	−0.3321	−0.6418	−1.8861
qOp		0.1216	1.1971
qC	0.6642	1.0404	1.3781
Positions			
rOn	1.1599	1.0663	1.0483
rOp		1.5232	1.1599
rC	0.0	0.0	0.0

those obtained from a generalized distributed multipole analysis (GDMA) [32], analysis of the CCSD/aug-cc-pVTZ charge density of monomer. (The GDMA calculations were carried out with the ORIENT code of Stone [33].) For the minimum energy structure of the dimer, the electrostatic energy calculated using the moments through rank 4 is −0.87 kcal/mol, while that obtained from a three-point-charge model, with the charges chosen as described above, is only −0.49 kcal/mol. Clearly, the higher rank moments play an important role in the electrostatic interactions between CO_2 molecules.

Murthy's CO_2 model [3] employs five-point charges, one on the C atom and one on each side of each O atom, all located on the molecular axis. This leads to an improved description of the higher multipole moments and gives a value of −0.76 kcal/mol for electrostatic interaction energy of the dimer at its global minimum structure, in improved agreement with the GDMA result. Our DPP2-like model also employs five-point charges, but with three atom-

Table 2 Multipole moments (a.u.) of CO_2 from various models and from a GDMA analysis of the CCSD/aug-cc-pVTZ charge density and electrostatic energies (kcal/mol) of the $(CO_2)_2$ potential energy minimum

Moments	Approach			
	3-point	Murthy	DPP2	Ab initio
Rank2	−3.20	−3.20	−3.30	−3.30[a]
Rank4	−15.33	−4.47	−2.83	−2.85[a]
Rank6	−73.66	52.40	37.52	27.92[a]
Rank8	−353.88	797.20	380.88	387.85[a]
ES	−0.49	−0.76	−0.88	−0.87[b]

[a] Calculated with GDMA using CCSD/aug-cc-pVTZ charge density

[b] Calculated with ORIENT using atomic moments obtained by GDMA through rank 4. Thus this contribution does not include charge penetration

centered charges and additional off-atom charges located near each of the O atoms. The values of the charges and the locations of the off-atom charges were optimized by a least-square fitting to the electrostatic potential calculated at the CCSD/aug-cc-pVTZ level, constraining the quadrupole moment to the CCSD value of −3.30 a.u. The regions within 2.55 and 2.43 Å of the C and O nuclei, respectively, were excluded in the fitting. The results are summarized in Table 2, from which it is seen that the DPP2 model for CO_2 does significantly better than both the 3-point model and Murthy's 5-point model at reproducing the higher multipole moments and at representing the non-charge-penetration [34] portion of the electrostatic interaction energy in the CO_2 dimer.

To account for the short-range charge penetration portion of the electrostatics interaction between two CO_2 molecules, an approximation introduced by of Freitag et al. [35] is employed. This express the penetration energy as

$$E_{pen} = -\frac{1}{2R_{AB}}\left[q_A(q_B + 2Z_B)e^{-a_A R_{AB}} + q_B(q_A + 2Z_A)e^{-a_B R_{AB}} \right.$$
$$\left. + \frac{q_A q_B (a_A^2 + a_B^2)}{a_A^2 - a_B^2}(e^{-a_B R_{AB}} - e^{-a_A R_{AB}}) \right], \quad (2)$$

in the case that A and B refer to different types of charge sites, and

$$E_{pen} = -\frac{1}{R_{AB}}\left[q_A q_B\left(1 + \frac{a R_{AB}}{2}\right) + q_A Z_B + q_B Z_A \right]e^{-a R_{AB}}, \quad (3)$$

when A and B refer to the same type of site (i.e., C–C, O–O, or off-atom–off-atom).

In Eqs. 2 and 3, Z_A and Z_B are the numbers of valence electrons associated with atoms A and B, respectively, that is, six for O, four for C, and zero for off-atom sites, and q_A and q_B are the point charges employed in the model. The a parameters, obtained by least-squares fitting

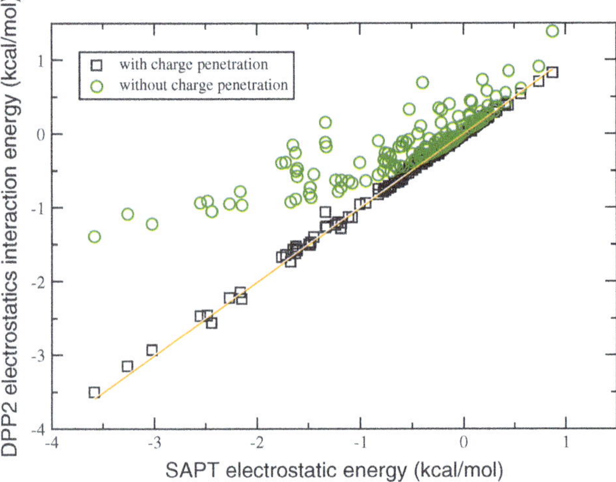

Fig. 3 Electrostatic interaction energies for the dimer training set

the electrostatic energy potential from the SAPT calculations on the ~ 200 CO_2–CO_2 dimer structures, are 0.35215, 0.40174, and 0.32296 Å^{-1} for C, O, and off-atom sites, respectively.

Figure 3 reports, for the CO_2 dimer training set, the electrostatic energies obtained using the DPP2 5-point charge model with and without charge penetration as well as the electrostatic energies from SAPT calculations. It is seen that the DPP2 5-point charge model with the charge penetration correction closely reproduces the electrostatic interaction energies from the SAPT calculations.

2.3.2 Polarization

The polarization interactions are modeled using mutually interacting atom-centered point polarizable sites, with Thole-type [36] damping between the charges and induced dipoles and between the induced dipoles. The procedure used to calculate the polarization energy is described in [18]. The atomic polarizabilities of CO_2 and the dipole–dipole damping coefficient were chosen so that the model gives $\alpha_\parallel$ and $\alpha_\perp$ polarizability components of CO_2 close to the experimental values [37]. The values of the charge-induced dipole damping coefficient was determined so that the model potential gives three-body energies for the three local minima of the $(CO_2)_3$ cluster close to those obtained from the CCSD(T) calculations. The resulting parameters are summarized in Table 3, and the resulting molecular polarizabilities and 3-body energies are reported in Tables 4 and 5, respectively. As noted by other researchers [8, 15], the 3-body interactions are quite small (≤ 0.1 kcal/mol in magnitude) for the CO_2 trimer.

Figure 4 compares, for the CO_2–CO_2 dimers in the training set, the polarization energies from the DPP2 model with the induction energies from the SAPT calculations, from which it is seen that for many of the sampled

Table 3 Atomic polarizabilities and damping parameters used for describing polarization in the DPP2 model of CO_2

Polarizability	
α_C (Å^3)	1.35
α_O (Å^3)	0.90
Damping parameters[a]	
a_{cd}	0.16
a_{dd}	0.65

[a] a_{cd} and a_{dd} refer to the damping parameters in the charge-induced dipole and induced dipole-induced dipole terms, respectively

Table 4 Molecular polarizability (Å^3) of the carbon dioxide monomer

	α_{xx}	α_{zz}	$\bar{\alpha}$
MP2[a]	1.91	4.04	2.65
CCSD(T)[a]	1.89	3.94	2.61
DPP2	1.93	3.88	2.58
Experiment[b]	1.91	3.94	2.59

[a] with aug-cc-pVTZ basis set

[b] [37]

Table 5 Three-body energies (kcal/mol) of local minima $(CO_2)_3$

Method	T1	T2	T3
DF-MP2/aug-cc-pV5Z	−0.04	−0.12	0.03
CCSD(T)-F12/VTZ-F12	0.01	−0.12	0.02
DPP2	−0.03	−0.12	0.02

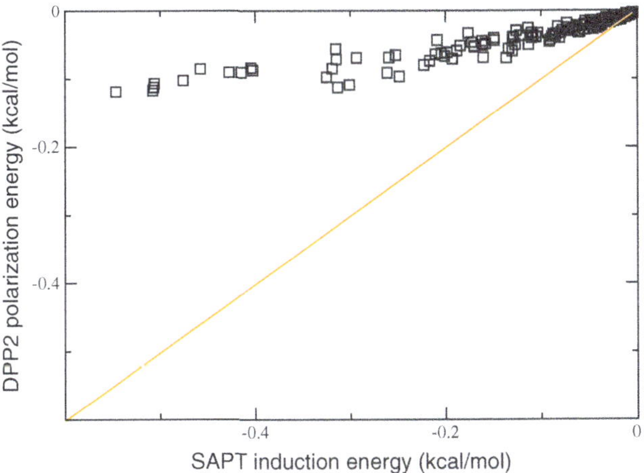

Fig. 4 Induction energies for CO_2 dimer training set calculated using the DPP2 model and the SAPT procedure

configurations the 2-body polarization energies from the DPP2 model significantly underestimate the 2-body induction energies from SAPT calculations. We note that the implementation of the DPP2 model for CO_2 neglects

charge-transfer, which an ALMO EDA [38] analysis indicates is in the order of −0.1 kcal/mol for the most stable dimer structure. In addition, the distributed inducible dipoles of the DPP2 model do a poor job at representing the quadrupole–quadrupole polarizability of the CO_2 monomer, which we estimate, if properly treated, contributes ~ -0.05 kcal/mol to the induction energy of the D1 dimer. We further note that charge penetration effects can enhance rather than attenuate induction energies [39]. In any case, the 2-body induction energies are far less important than the electrostatics and dispersion contributions, being at most −0.54 kcal/mol, and, to a large extent, the deficiency of the model in describing the induction is compensated for by the procedure used to fit the repulsion contributions.

2.3.3 Dispersion

The dispersion interactions between pairs of monomers are represented by damped r_{ij}^{-6} terms, that is, as

$$E_{disp} = \sum_{i,j} \frac{C_6^{ij} f(r_{ij}, \delta_{ij})}{r_{ij}^6}, \tag{4}$$

where i and j refer to atoms of different monomers, and $f(r_{ij}, \delta_{ij})$ is the Tang-Toennies damping function [40]. The C_6^{ij} and δ_{ij} parameters (summarized in Table 6) were obtained by fitting to the SAPT values of the dispersion energies (sum of dispersion plus exchange-dispersion) for the set of CO_2–CO_2 dimer structures. As seen from Fig. 5, the SAPT values of the dispersion energies for the CO_2–CO_2 dimers in the training set are well represented by the DPP2 model.

2.3.4 Exchange–repulsion

The 2-body exchange-repulsion interactions are represented as

$$E_{exch-rep} = \sum_{i,j} A_{ij} \exp(-B_{ij} r_{ij}), \tag{5}$$

where i and j refer to atoms associated with different monomers. The A_{ij} and B_{ij} parameters (summarized in Table 6) were determined by fitting for the CO_2–CO_2 dimer training set the residual energies obtained by subtracting for

Table 6 Parameters describing the dispersion and repulsion interactions in the DPP2 model for CO_2

Atoms	Dispersion		Repulsion	
	C_6^{ij} (kcal/mol Å^6)	δ_{ij} (Å^{-1})	A_{ij} (kcal/mol)	B_{ij} (Å^{-1})
C–C	−97.76	31.326	0.674764	0.67716
C–O	−288.57	33.307	6,727.82	3.0016
O–O	−497.23	28.026	240,059.0	4.22009

✎ Springer

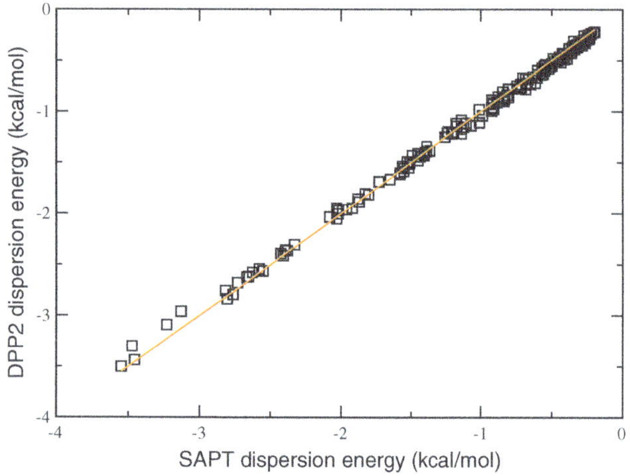

Fig. 5 Dispersion interaction energies for CO_2 dimer training set calculated using the model potential and the SAPT procedure

Table 7 Interaction energies (kcal/mol) of selected stationary points $(CO_2)_n$, $n = 2–4$

Method	$(CO_2)_2$		$(CO_2)_3$			$(CO_2)_4$
	D1	D2	T1	T2	T3	
MP2-F12[a]	−1.46	−1.20	−3.97	−3.83	−2.93	−7.26
CCSD(T)-F12[a]	−1.50	−1.23	−3.99	−3.94	−3.03	−7.33
DPP2[a]	−1.52	−1.13	−4.05	−3.73	−3.07	−7.26
Murthy[a]	−1.34	−1.03	−3.57	−3.25	−2.73	−6.36
DPP2 (opt)[b]	−1.56	−1.14	−4.17	−3.74	−3.15	−7.41

[a] Results for MP2 optimized geometries

[b] Results for DPP2 optimized geometries

each structure the sum of the DPP2 values of the electrostatics, polarization, and dispersion energies from the CCSD(T)-F12 interaction energies. As noted above, this approach ensures that deficiencies in the model for describing the electrostatics, induction, and dispersion energies are partially compensated for by the so-called exchange–repulsion term. It is seen from Fig. 6 that the DPP2 model accurately reproduces the CCSD(T)-F12 values of the net interaction energies of the $(CO_2)_2$ training set.

3 Testing the DPP2 model

3.1 $(CO_2)_n$ ($n = 2–4$)

The geometries of the stationary points of the $(CO_2)_n$, $n = 2–4$ clusters obtained from MP2 and DPP2

optimizations are shown in Fig. 2, and the calculated interaction energies are summarized in Table 7. From Fig. 2 it is seen that the DPP2 model gives geometries very close to those from the MP2 optimizations. As seen from Table 7, the DPP2 model also closely reproduces the CCSD(T)-F12 interaction energies of these clusters with the mean absolute error being <0.1 kcal/mol. In contrast, for the Murthy model the mean absolute error in the interaction energies for the $(CO_2)_n$, $n = 2–4$ clusters is nearly 0.5 kcal/mol.

Figure 7 reports for dimer and trimer structures sampled in the $T = 240$ K Monte Carlo simulation, the interaction energies obtained using DPP2 and Murthy models as well as from the CCSD(T)-F12 method. The DPP2 model accurately reproduces the interaction energies from the CCSD(T)-F12 calculations, with the maximum error being about 0.2 kcal/mol. The Murthy model also performs well for structures with large intermolecular separations; however, it is less successful for structures with small intermolecular separations.

One question that can be raised about the DPP2 model is the appropriateness of including many-body polarization

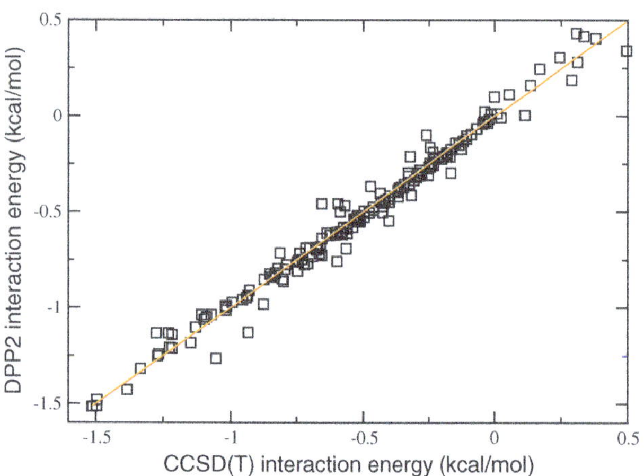

Fig. 6 Net interaction energies for the CO_2 dimer training set calculated using the DPP2 model potential and the CCSD(T)-F12 method

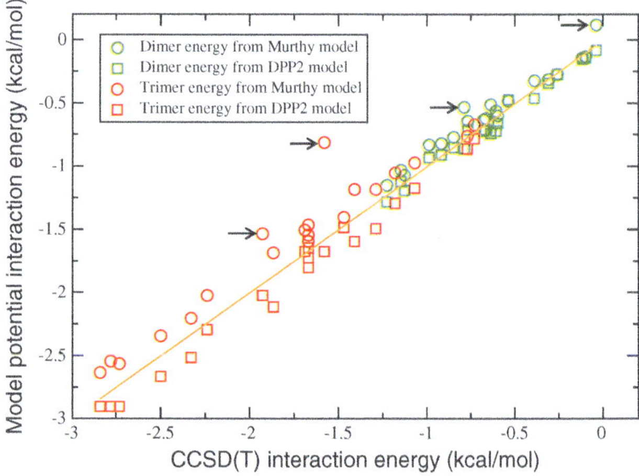

Fig. 7 Net interaction energies for a set of dimer and trimer structures sampled in a $T = 240$ K Monte Carlo simulation. *Arrows* indicate structures with intermolecular O–O distances shorter than 2.9 Å

effects while ignoring many-body dispersion interactions. To address this issue, we have carried out 3-body SAPT [41] calculations on the global minimum structure of $(CO_2)_3$. These calculations give the following three-body contributions, exchange (-0.04 kcal/mol), induction, including the δ(HF) contribution (-0.01 kcal/mol), and dispersion (0.24 kcal/mol) for a net 3-body contribution of 0.20 kcal/mol, compared to the CCSD(T) value of 0.01 kcal/mol. Thus, although the dispersion contribution is largest in magnitude, all these contributions to the 3-body energy are small compared to the net interaction energy. In addition, it should be noted that the DPP2 model gives a 3-body interaction energy of -0.03 kcal/mol, very close to the CCSD(T) value. Indeed, the way the DPP2 model was parameterized the so-called 3-body polarization contribution can be thought of as effectively incorporating exchange and dispersion interactions as well.

3.2 Second virial coefficient

The second virial coefficient is a basic thermodynamic quantity that represents the departure from ideality due to interactions between pairs of molecules. The second virial coefficient as a function of temperature for the DPP2-like CO_2 model was calculated by numerically integrating the Mayer function [42]:

$$B_2(T) = -2\pi \int_0^\infty \left[\langle e^{-\beta u(r)} \rangle - 1 \right] r^2 dr, \tag{2}$$

where the average of $e^{-\beta u(r)}$ was taken over randomly sampled molecular orientations at a fixed intermolecular distance r. Calculated and measured [43] B_2 versus T curves are reported in Fig. 8. It is seen that the DPP2 force field for CO_2 gives values of the second virial coefficients in excellent agreement with experiment.

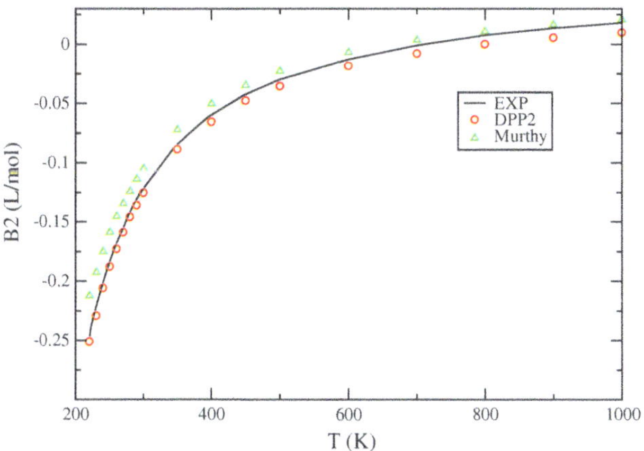

Fig. 8 Calculated and experimental [43] second virial coefficient for CO_2

3.3 Supercritical CO_2

Radial distribution functions (RDF) available for supercritical CO_2 are available from neutron diffraction studies [31]. These experientially determined RDFs do not resolve the contributions for the various atom pairs. Thus to enable comparison with experiment we have calculated atom–atom partial radial distribution functions using the expression

$$g_m(r) = 0.403 g_{oo}(r) + 0.464 g_{co}(r) + 0.133 g_{cc}(r) \tag{6}$$

from [44]. This expression accounts for the difference in the neutron scattering cross sections of C and O. As seen from Fig. 9, for both sets of conditions ($T = 240$ K, $\rho = 14.9$ molecules/nm^3 and $T = 312$ K, $\rho = 11.4$ molecules/nm^3), the radial distribution functions calculated using the DPP2 force field for CO_2 are in excellent agreement with those determined experimentally.

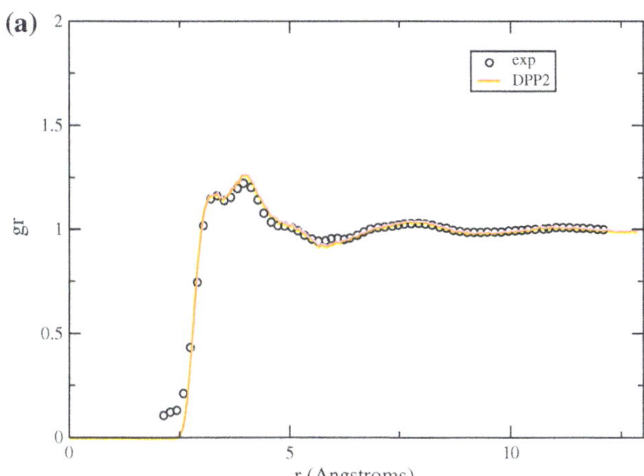

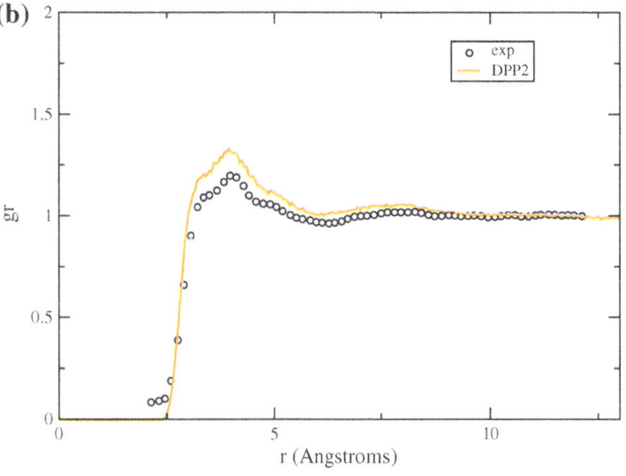

Fig. 9 Radial distribution function of supercritical CO_2. **a** calculated and experimental results for $T = 240$ K and $\rho = 14.9$ molecules/nm^3 **b** calculated and experimental results for $T = 312$ K and $\rho = 11.4$ molecules/nm^3. The experimental results are from [31]

4 Conclusions

In this study, we have constructed a CO_2 force field using the same strategy as used previously to develop the DPP2 water model, that is, using a combination of SAPT and CCSD(T) calculations to fit individual electrostatic, induction, dispersion, and exchange–repulsion contributions. The resulting DPP2-like force field of CO_2 is shown to accurately reproduce the MP2 geometries and the CCSD(T)-F12 interaction energies of small $(CO_2)_n$ clusters and also to give second virial coefficient values and radial distribution functions of supercritical CO_2 in excellent agreement with experiment. In future work we plan to extend the DPP2 model to describe CO_2-water interactions, to enable its use in simulations of CO_2-liquid water mixtures and of CO_2 hydrates.

Acknowledgments This research was carried out with the support of the National Science foundation, grant CHE 518253. The calculations were carried out on computers in the University of Pittsburgh's Center for Simulation and Modeling.

References

1. MacKerell AD Jr (2001) Atomistic models and force fields. In: Becker OM, MacKerell AD Jr, Roux B, Watanabe M (eds) Computational biochemistry and biophysics. Marcel Dekker Inc, New York
2. Murthy CS, Singer K, McDonald IR (1981) Mol Phys 44:135
3. Murthy CS, O'Shea SF, Mcdonald IR (1983) Mol Phys 50:531
4. Böhm HJ, Meissner C, Ahlrichs R (1984) Mol Phys 53:651
5. Zhu SB, Robinson GW (1989) Comput Phys Commun 52:317
6. Etters RD, Kuchta B (1989) J Chem Phys 90:4537
7. Möller D, Fischer J (1994) Fluid Phase Equilib 100:35
8. Bukowski R, Sadlej J, Jeziorski B, Jankowski P, Szalewicz K, Kucharski SA, Williams HL, Rice BM (1999) J Chem Phys 110:3785
9. Bock S, Bich E, Vogel E (2000) Chem Phys 257:147
10. Potoff JJ, Siepmann JI (2001) AIChE J 47:1676
11. Vrabec J, Stoll J, Hasse H (2001) J Phys Chem B 105:12126
12. Zhang Z, Duan Z (2005) J Chem Phys 122:214507
13. Zhu A, Zhang X, Liu Q, Zhang Q (2009) Chin J Chem Eng 17:268
14. Merker T, Engin C, Vrabec J, Hasse H (2010) J Chem Phys 132:234512
15. Oakley MT, Wheatley RJ (2009) J Chem Phys 130:034110
16. Persson RAX (2011) J Chem Phys 134:034312
17. Yu K, McDaniel JG, Schmidt JR (2011) J Phys Chem B 115:10054
18. Kumar R, Wang F–F, Jenness GR, Jordan KD (2010) J Chem Phys 132:014309
19. Jeziorski B, Moszynski R, Szalewicz K (1994) Chem Rev 94:1887
20. Purvis GD, Bartlett RJ (1982) J Chem Phys 76:1910
21. Kendall RA, Dunning TH, Harrison RJ (1992) J Chem Phys 96:6796
22. Woon DE, Dunning TH (1994) J Chem Phys 100:2975
23. Adler TB, Knizia G, Werner H-J (2007) J Chem Phys 127:221106
24. Peterson KA, Adler TB, Werner H-J (2008) J Chem Phys 128:084102
25. Eggenberger R, Gerber S, Huber H (1991) Mol Phys 72:433
26. Frisch MJ, Trucks GW, Schlegel HB, Scuseria GE, Robb MA, Cheeseman JR, Montgomery Jr. JA, Vreven T, Kudin KN, Burant JC, Millam JM, Iyengar SS, omasiJT, aroneVB, Mennucci B, Cossi M, Scalmani G, Rega N, Petersson GA, Nakatsuji H, Hada M, Ehara M, Toyota K, Fukuda R, Hasegawa J, Ishida M, Nakajima T, Honda Y, Kitao O, Nakai H, Klene M, Li X, Knox JE, Hratchian HP, Cross J B, Bakken V, Adamo C, Jaramillo J, Gomperts R, Stratmann RE, Yazyev O, Austin AJ, Cammi R, Pomelli C, Ochterski JW, Ayala PY, Morokuma K, Voth GA, Salvador P, Dannenberg JJ, Zakrzewski VG, apprichSD, Daniels AD, Strain MC, Farkas O, Malick DK, Rabuck AD, Raghavachari K, Foresman JB, Ortiz JV, Cui Q, Baboul AG, Clifford S, Cioslowski J, Stefanov BB, Liu G, Liashenko A, Piskorz P, Komaromi I, Martin RL, Fox DJ, Keith T, Al-Laham MA, Peng CY, Nanayakkara A, Challacombe M, Gill PMW, Johnson B, Chen W, Wong MW, Gonzalez C, Pople JA (2004) Gaussian 03, Revision C.02, Gaussian, Inc, Wallingford, CT
27. Werner H-J, Knowles PJ, Lindh R, Manby FR, Schütz M, Celani P, Korona T, Rauhut G, Amos RD, Bernhardsson A, Berning A, Cooper DL, Deegan MJO, Dobbyn AJ, Eckert F, Hampel C, Hetzer G, Lloyd AW, McNicholas SJ, Meyer W, Mura ME, Nicklass A, Palmieri P, Pitzer R, Schumann U, Stoll H, Stone AJ, Tarroni R, Thorsteinsson T, MOLPRO, version 2006.1, a package of ab initio programs. http://www.molpro.net
28. Bukowski R, Cencek W, Jankowski P, Jeziorska M, Jeziorski B, Kucharski SA, Lotrich VF, Misquitta AJ, Moszyński R, Patkowski K, Podeszwa R, Rybak S, Szalewicz K, Williams HL, Wheatley RJ, Wormer PES, Żuchowski PS, SAPT2008: An Ab Initio Program for Many-Body Symmetry-Adapted Perturbation Theory Calculations of Intermolecular Interaction Energies, University of Delaware and the University of Warsaw
29. DALTON, a molecular electronic structure program, Release 20 (2005). http://www.kjemi.uio.no/software/dalton/dalton.html
30. Misquitta AJ, Podeszwa R, Jeziorski B, Szalewicz K (2005) J Chem Phys 123:214103
31. Cipriani P, Nardone M, Ricci FP (1998) Phys B 241–243:940
32. Stone AJ (2005) J Chem Theory Comput 1:1128
33. Stone AJ, Dullweber A, Engkvist O, Fraschini E, Hodges MP, Meredith AW, Nutt DR, Popelier PLA, Wales DJ (2002) 'Orient: a program for studying interactions between molecules, version 45,' University of Cambridge, Enquiries to Stone AJ, ajs1@cam.ac.uk
34. Stone AJ (1996) The theory of intermolecular forces. Clarendon, Oxford
35. Freitag MA, Gordon MS, Jensen JH, Stevens WJ (2000) J Chem Phys 112:7300
36. Thole BT (1981) Chem Phys 59:341
37. Maroulis G, Thakkar AJ (1990) J Chem Phys 93:4164
38. Khaliullin RZ, Cobar EA, Lochan RC, Bell AT, Head-Gordon M (2007) J Phys Chem A 111:8753
39. Wang B, Thuhlar DG (2010) J Chem Theory Comput 6:3330
40. Tang KT, Toennies JP (1984) J Chem Phys 80:3726
41. Podeszwa R, Szalewicz K (2007) J Chem Phys 126:194101
42. McQuarrie DA (2000) Statistical mechanics. University Science Books
43. Span R, Wagner W (1996) J Phys Chem Ref Data 25:1509
44. Ishii R, Okazaki S, Odawara O, Okada I, Misawa M, Fukunaga T (1995) Fluid Phase Equilibria 104:291

Theor Chem Acc (2012) 131:1198
DOI 10.1007/s00214-012-1198-7

REGULAR ARTICLE

Toward accurate solvation dynamics of lanthanides and actinides in water using polarizable force fields: from gas-phase energetics to hydration free energies

Aude Marjolin · Christophe Gourlaouen · Carine Clavaguéra · Pengyu Y. Ren · Johnny C. Wu · Nohad Gresh · Jean-Pierre Dognon · Jean-Philip Piquemal

Received: 8 September 2011 / Accepted: 25 February 2012 / Published online: 17 March 2012
© Springer-Verlag 2012

Abstract In this contribution, we focused on the use of polarizable force fields to model the structural, energetic, and thermodynamical properties of lanthanides and actinides in water. In a first part, we chose the particular case of the Th(IV) cation to demonstrate the capabilities of the AMOEBA polarizable force field to reproduce both reference ab initio gas-phase energetics and experimental data including coordination numbers and radial distribution functions. Using such model, we predicted the first polarizable force field estimate of Th(IV) solvation free energy, which accounts for −1,638 kcal/mol. In addition, we proposed in a second part of this work a full extension of the SIBFA (Sum of Interaction Between Fragments Ab initio computed) polarizable potential to lanthanides (La(III) and Lu(III)) and to actinides (Th(IV)) in water. We demonstrate its capabilities to reproduce all ab initio contributions as extracted from energy decomposition analysis computations, including many-body charge transfer and discussed its applicability to extended molecular dynamics and its parametrization on high-level post-Hartree–Fock data.

Published as part of the special collection of articles: From quantum mechanics to force fields: new methodologies for the classical simulation of complex systems.

Electronic supplementary material The online version of this article (doi:10.1007/s00214-012-1198-7) contains supplementary material, which is available to authorized users.

A. Marjolin · C. Gourlaouen · J.-P. Dognon (✉)
Laboratoire de Chimie de Coordination des Eléments-f,
CEA, CNRS UMR 3299, CEA Saclay,
91191 Gif-sur Yvette Cedex, France
e-mail: jean-pierre.dognon@cea.fr

A. Marjolin · J.-P. Piquemal (✉)
Laboratoire de Chimie Théorique, UMPC, CNRS UMR 7616,
CC 137 4 Place Jussieu, 75252 Paris Cedex 05, France
e-mail: jpp@lct.jussieu.fr

C. Clavaguéra (✉)
Laboratoire des Mécanismes Réactionnels, Département
de Chimie, Ecole Polytechnique, CNRS,
91128 Palaiseau Cedex, France
e-mail: carine.clavaguera@dcmr.polytechnique.fr

P. Y. Ren · J. C. Wu
Department of Biomedical Engineering, University of Texas
at Austin, Austin, TX 78712-1062, USA

N. Gresh
Laboratoire de Chimie et Biochimie Pharmacochimiques
et Toxicologiques, UMR 8601 CNRS, UFR Biomédicale,
Université Paris Descartes, 45 rue des Saints-Pères,
75270 Paris Cedex 06, France

Keywords Lanthanides · Actinides · Energy decomposition analysis · Polarizable force field · Charge transfer · Hydration free energy

1 Introduction

The fields of interest in lanthanide(III) ions have been extended in the recent years. The main applications cover medical diagnosis (contrast agents in magnetic resonance imaging and luminescent probes for proteins) [1–3], catalysis and organic synthesis [4], organic light-emitting diodes, and nuclear chemistry. The study of the coordination of lanthanide(III) ions and the water exchange in aqueous solutions is of particular importance for the understanding of the chemical processes in which these ions are involved.

Concerning the actinides, the safe management of highly radioactive spent fuel is currently a major challenge for the nuclear energy industry. Fundamental researches are carried out in such areas as nuclear waste management and nuclear toxicology [5–7]. Consequently, there is a great interest in the knowledge of the chemistry of actinide species. In the reprocessing strategy for nuclear waste, chemical processes are involved; at present, these are based on solvent extraction processes. Such a complex chemical process clearly relies on a delicate balance between the different interactions that govern the system, particularly extractant molecule-cation and water-cation interactions. In this context, the water molecule not only acts as solvent, but it also plays a non-negligible role in complex formation. One of the preliminary steps in understanding the solvent extraction process consists of the study of the lanthanide and actinide cation hydration, and more particularly of the organization of the first coordination sphere around these cations.

Molecular modeling can help to better understand the local complexation properties of the ions and to support the design of new extractant molecules. Classical molecular dynamics (MD) simulations using explicit solvent are essential to investigate statistical and dynamical properties that can be related to rather rare events taking place over hundreds of picoseconds.

In addition, many recent studies highlighted the failure of traditional fixed charge force fields to capture the main physical effects that govern interaction for highly charged ions in polar solvents [8–19].

Consequently, several polarizable models are available in the literature, mainly applied to the study of lanthanide hydration [20–22]. It was stressed [13] that some of them are not general due to the non-transferability of their parameters leading to debatable results including predicted shorter water residence times for the heavier lanthanides than for the lighter ones, in contradiction with well-established experimental data [22]. Moreover, the trends in the water exchange rate are shifted to heavier ions with a maximum at Tb^{3+} instead of Gd^{3+}, in contradiction with the well-known gadolinium break along the lanthanide series [22].

In our previous work on the uranyl ion [23], we pointed out that only an explicit high-level treatment for non-covalent interactions is able to reproduce experimental data. This approach was also pioneered by Hagberg et al. on UO_2^{2+} [11] and Cm^{3+} hydration [12]. The NEMO force field was used including multipolar electrostatic expansions and complete many-body effects, i.e., polarization and charge transfer, but within a rigid molecular framework [24]. The parameters were extracted from a metal–water curve interaction computed at the relativistic multiconfigurational level of theory followed by perturbation theory (CASSCF/CASPT2) and validated by molecular dynamics simulations in water. Two more recent studies were reported on the hydration of actinide ions using polarizable force fields. The EXAFS spectra of a Cf^{3+}

aqueous solution were successfully explained thanks to the design of two specific polarizable intermolecular potentials for eight and nine coordination of the cation and Monte Carlo simulations [16]. Réal et al. performed MD simulations on the Th(IV) in aqueous solution using two different polarizable force fields and including explicit ad hoc Th^{4+}–water charge-transfer term [17]. The influence of the parameter set on the structuration of the first shells was explored, and we will discuss their results later in this paper.

Thanks to high-level electrostatics and full treatment for polarizable effects, MD simulations with AMOEBA [25, 26] were able to reproduce structural and dynamical experimental properties of hydrated mono and divalent metal cations [27–30]. In our previous studies, the AMOEBA polarizable and flexible force field was extended to the lanthanide(III) ions [8–10]. We present here an extension to actinide ions in the same framework.

In addition, we proposed in a second part of this work a full extension of the SIBFA (Sum of Interaction Between Fragments Ab initio computed) [31] approach to trivalent lanthanides and tetravalent actinides. Indeed, the anisotropic polarizable SIBFA force field has been developed to provide an accurate description of divalent metal cations [31–33] in large biological binding sites [34, 35]. SIBFA is able to treat "explicitly" both polarization and charge-transfer contributions and therefore can be used as a reference to understand the many-body effects in large systems.

The parametrization of these two advanced force field requires the use of accurate ab initio reference data. In the case of heavy elements, especially for actinides, relativistic effects and multiconfigurational wave functions are essential to provide reliable reference data. A complete study on the Th(IV)–water dimer has been carried out to state that MRCI level is the best reference to obtain dissociation curves for force field development [36].

We will present first the ab initio methods used for the parameterizations. Then, the second part will be devoted to the extension of AMOEBA to the Th(IV) ion, to the validation of its energetics on Thorium–water clusters and to MD simulations for the determination of structural and thermodynamical data (hydration free energy) in aqueous solution. In the last part, we will present the extension of SIBFA to closed-shell trivalent lanthanides and tetravalent actinides.

2 Procedures

2.1 Interaction energy calculations and levels of ab initio computations

Due to the complex physical and chemical characteristics of lanthanide and actinide systems, high-level ab initio calculations were mandatory, so as to generate reliable data for the parameterization of the AMOEBA and SIBFA force fields.

For example, in the dissociation of the $[Th(H_2O)]^{4+}$ complex, several electronic states which may cross along the minimum energy path are involved. The $[Th(H_2O)]^{4+}$ adiabatic dissociation curves computed using a MRCI wave function are given in Fig. 1. At about 2.6 Å, the S_0 state crosses a higher lying S_1 state. An avoided crossing results, which causes strong orbitals, changes. This leads to a net charge transfer that transforms a Th^{4+}–H_2O complex into a Th^{3+}–H_2O^+ as the ligand is ionized. In order to compute the required dissociation curve (see curve S_d in Fig. 1) to develop polarizable ab initio-based force fields, a state of the art diabatic representation is mandatory. The full methodology used is described in Gourlaouen et al. [36]. The MRCI calculations [37] were performed with the MOLPRO [38] program package.

The interaction energy is calculated as the difference between the dimer energy and that of the separate fragments for varying M–O_w distances.

$$E_{int} = E[M - (OH_2)]^{m+} - E(M^{m+}) - E(OH_2)$$

The reference geometry was one in which the water molecule was kept frozen in the experimental geometry (bond length $d_{O-H} = 0.957$ Å and bond angle HOH $= 104.5°$ [39]), so that the dimer system interaction energy was calculated in the C_{2v} group. In this manner, the active space used for the multireference calculations consisted for all three systems of the oxygen's highest occupied molecular orbital ($2p_y$) as well as the first virtual orbitals of symmetry 2 of each cation, i.e., two $4f$ orbitals and a $5d$ orbital for La(III), two $5d$ and a $6p$ orbital for Lu(III), and two $5f$ and a $6d$ orbital for Th(IV). Dunning's augmented triple zeta basis set [40] was used on the O and H atoms, while the Stuttgart quasi-relativistic small-core pseudopotentials and associated basis sets as in MOLPRO's internal library were, respectively, used for La, Lu, and Th. On the other hand, calculations at lower levels of theory (HF and MP2) with both small-core and large-core pseudopotentials were carried out with the Gaussian03 [41] package. The geometry was also kept partially frozen in the experimental geometry except for the Ln/An–O_w bond length. All interaction energy values were corrected of the basis set superposition error (BSSE) by the counterpoise method.

2.2 Energy decomposition analyses

Energy decomposition analyses (EDA) were then carried out to compute the different contributions to the interaction energy of the dimer systems for the parameterization of the AMOEBA and SIBFA force fields. The constrained space orbital variation [42] (CSOV) decomposition at the Hartree–Fock level as implemented in our modified version of HONDO95.3 [43, 44] was carried out for the lanthanide systems, while the restricted variational space [45] (RVS) analysis was carried out with the GAMESS [46] package on the Thorium–water dimer as well as on the different $[M-(OH_2)_n]^{m+}$ clusters. Both approaches are related to the Morokuma procedure [47] in which the Hartree–Fock interaction energy is separated in four physically meaningful contributions, namely the Coulomb electrostatic energy (ES), the Pauli exchange-repulsion term (REP) (which adds up to give the "frozen core" energy), and POL and CT, which are the polarization energy and the charge-transfer term, respectively; those latter contributions sum up to give the non-frozen energy resulting of the interaction between the two approaching monomers. One advantage of CSOV and RVS is that such methods are able to maintain

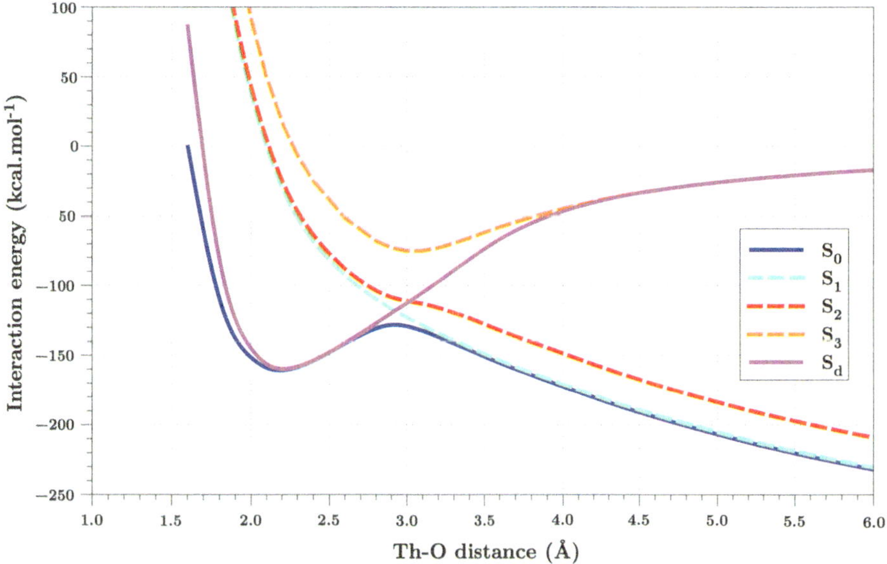

Fig. 1 Adiabatic (*blue, cyan, red*, and *orange*) and diabatic (*magenta*) MRCI dissociation curves for $[Th(H_2O)]^{4+}$ complex

the correct antisymmetry of the wave function leading to the correct evaluation of polarization and charge-transfer energies [43, 48].

2.3 Sum of interactions between fragments ab initio computed (SIBFA)

SIBFA [31] is a polarizable force field, based on the decomposition of the interaction energy in such a way that its equations are strongly analogous to the EDA description of the energy terms. The interaction energy E_{int} is therefore calculated as follows, and a brief outline of the different contributions is given. However, an extended description of the force field can be found in more detailed works [31, 38, 48].

$$E_{\text{int}} = E_{\text{mtp}*} + E_{\text{rep}} + E_{\text{pol}*} + E_{\text{ct}} + E_{\text{disp}}$$

$E_{\text{mtp}*}$ is the Coulomb electrostatic contribution, computed with distributed multipoles (up to quadrupoles) derived from the ab initio HF wave function of the monomer fragments and distributed on the atoms and bond barycenters using the procedure developed by Vigné-Maeder and Claverie [49]. This electrostatic energy includes a correction, the penetration energy, accounting for the overlap of orbitals at short range [50]. Second, E_{rep} is the Pauli short-range repulsion between pairs of parallel spin electrons, given as the sum of the different bond–bond, bond–lone pair, and lone pair–lone pair interactions [51]. $E_{\text{pol}*}$ is the polarization energy contribution, computed with distributed, anisotropic polarizabilities on the individual fragments. The polarizabilities are distributed on the centroids of the localized orbitals (lone pairs and bond barycenters) using the procedure of Garmer and Stevens [52]. The field polarizing each monomer is computed with both the permanent multipoles, already derived for the electrostatic contribution, hence no additional cost of calculation, and the dipoles induced on all the other monomers, in an iterative way. A gaussian function is used for the screening of the polarizing field. In addition to $E_{\text{pol}*}$, the polarization energy based only on the permanent dipoles, E_{pol} can also be calculated. E_{ct} is the charge-transfer contribution which includes all the effects brought about by the permanent and induced dipoles, therefore leading to a coupling with the polarization term. And last, E_{disp} is the dispersion energy originating from the Van der Waals interaction between induced dipoles. Throughout the different terms, SIBFA takes into accounts many physical properties that are yet inaccessible to most force fields, such as the many-body effects of non-additivity, namely in the polarization energy, in the "explicit" many-body charge-transfer and exchange-repulsion [53] terms as well as anisotropy, accounted for in the short-distance electrostatics, repulsion, dispersion, and the charge-transfer contribution.

For consistency purpose within this work, we propose here a reparametrization of the SIBFA water potential at the aug-cc-pVTZ/HF level based on the reference CSOV values.

2.4 The AMOEBA force field

The AMOEBA (Atomic Multipoles Optimized Energetics for Biomolecular Applications) polarizable force field [25] has already been tested in molecular dynamics (MD) simulations of various systems including metal cation hydration [27–29]. The electrostatic component of the model accounts for permanent charges on each atom as well as dipole and quadrupole moments, all of which are derived from quantum chemical calculations. Furthermore, polarization effects are also explicitly included in this electrostatic component via atomic dipole induction as shown in the equation below:

$$\mu_{i,\alpha}^{\text{ind}} = \alpha_i \left(\sum_j T_\alpha^{i,j} M_j + \sum_{j'} T_{\alpha\beta}^{i,j'} \mu_{j'\beta}^{\text{ind}} \right) \quad \text{for } \alpha, \beta = 1, 2, 3$$

with α_i the atomic polarizability of the considered cation, T the usual interaction matrix for sites i and j, and M_j the permanent multipole components. The first term in the equation represents the dipole on site i induced by the permanent multipoles, while the second term corresponds to the dipole on site i induced by the induced dipoles produced at the other atoms. A polarization-damping scheme is used via a smeared charge distribution as proposed by Thole [54]:

$$\rho = \frac{3a}{4\pi} \exp(-au^3)$$

with

$$u = R_{ij}/(\alpha_i \alpha_j)^{1/6}$$

where u is the effective distance between atoms i and j as a function of atomic polarizabilities between them, a is a dimensionless width parameter of the smeared charge distribution, which controls the strength of damping. This parameter was set to 0.39 for water and monovalent ions; however, the value was adjusted to smaller values in the case of divalent cations such as Ca^{2+}, Mg^{2+}, and Zn^{2+} that need a wider charge distribution. Consequently, the value for Th^{4+} was also adjusted using the procedure described below. Repulsion–dispersion interactions (van der Waals) between pairs of non-bonded atoms are represented by a buffered 14–7 potential.

$$U_{ij}^{\text{buff}} = \varepsilon_{ij} \left(\frac{1+\delta}{\rho_{ij}+\delta} \right)^{n-m} \left(\frac{1+\gamma}{\rho_{ij}^m+\delta} - 2 \right)$$

where ε_{ij} is the potential well depth. In addition, ρ_{ij} is R_{ij}/R_{ij}^0 where R_{ij} is the separation distance between atoms i and j, and R_{ij}^0 is the minimum energy distance. Following Halgren [55], we used fixed values of $n = 14$, $m = 7$, $\delta = 0.07$, and $\gamma = 0.12$. The polarizable water model as developed by Ren and Ponder [25] is employed in this study, and the values for R_{ij} and ε_{ij} are parametrized for the Th^{4+} cation. If charge transfer is not explicitly taken into account, it is nevertheless implicitly included via the van der Waals parameters that are derived from the ab initio calculations that contain all these effects. [27, 28] Such strategy has been shown to be robust enough to perform accurate cluster [8, 9] and condensed phase simulations [27, 28].

3 Results

3.1 AMOEBA: from gas-phase clusters to free energy

3.1.1 Extending AMOEBA to the Thorium (IV) ion

The parameters for Th^{4+} were derived from the ab initio diabatic dissociation curve of the $[Th(H_2O)]^{4+}$ complex obtained at the MCSCF/MRCI level, using Dolg et al. quasi-relativistic effective core potential for Thorium and aug-cc-pVTZ Dunning basis sets for H and O [40]. The $6s$, $6p$ Thorium electrons, and all the valence H_2O electrons were correlated. The dipole polarizability α_{Th} of Th^{4+} was previously determined at 1.143 Å^3 [56]. The repulsion–dispersion parameters were fitted upon 60 various configurations chosen to sample the potential energy surface by varying the Th–O distance and the orientation of the water molecule, so that $\varepsilon_{Th} = 2.50$ kcal/mol and $R_{Th} = 3.90$ Å. In addition, the damping factor "a" was adjusted to 0.20 for Th^{4+}, so that the AMOEBA polarization energy matched the CSOV polarization values as much as possible following the procedure detailed in previous studies [27, 28].

3.1.2 Simulation details: from clusters to periodic boundary conditions

In order to be consistent with our previous studies and to examine the effect of periodic boundary conditions on Th(IV) solvation, we performed both cluster and periodic boundary conditions simulations. Molecular dynamics simulations were then performed on the $[Th(H_2O)_{214}]^{4+}$ cluster at constant temperature with a Nosé-Hoover [57, 58] thermostat. The Beeman [59] algorithm was used for the propagation of dynamical trajectories. The water–Thorium cluster was confined by spherical boundary

conditions with a van der Waals soft wall characterized by a 12–6 Lennard-Jones potential which was set to a fixed buffer distance of 2.5 Å outside the specified radius of 15 Å. This value was optimized after several tests to probe the role of the size of the radius sphere. All molecular dynamics simulations were carried out with the TINKER [60] software package at 298 K with a 1 femtosecond time step, for a total simulated time of 1 ns per trajectory. The data are accumulated from four trajectories. Furthermore, simulations using periodic boundary condition (PBC) were also performed to compare the dynamics of the Th^{4+} solvation with the cluster simulations. The long-range electrostatics was modeled using the smooth Particle-Mesh Ewald [61] summation for atomic multipoles with a cutoff of 7 Å in real space, and the convergence criterion for induced dipole computation was set to 10^{-6} D. The temperature was maintained at 298 K using the Berendsen [62] weak coupling method. Two systems were studied, one containing 215 water molecules and the Th^{4+} ion to match the cluster conditions, and the other 511 water molecules and the Th^{4+} ion, the unit box having 18.643 and 24.857 Å side length, respectively.

3.1.3 Validation of the AMOEBA potential on Thorium–water clusters: gas-phase energetics

Total interaction energies are reported in Table 1 for the Th–water clusters with 8, 9, and 10 water molecules. These geometries were chosen as they are the representative of the final condensed phase coordination of Th(IV) in water. We propose to test AMOEBA against 3 different geometries of Th(IV) interacting with 10 water molecules within the first and second shells, respectively: $9 + 1$, $8 + 2$, and 10. Due to the size of the basis set for the initial reference level of the AMOEBA model, we had to restrict ourselves to a slightly smaller basis set, namely cc-pVTZ. The given values are corrected of the BSSE. All geometries were optimized at both the MP2 level and with the AMOEBA force field starting from the same structures. As can be seen from Table 1, the AMOEBA absolute interaction energies are in very good agreement with the MP2 ones and the global error on the interaction energy by comparison with

Table 1 Interaction energies (in kcal/mol) of the Th^{4+} clusters calculated at the MP2 level and with the AMOEBA force field

$[Th–(H_2O)_n]^{4+}$	MP2	AMOEBA	% Error
8	−740.04	−749.85	1.3
9	−786.56	−790.31	0.5
10	−819.26	−818.30	−0.1
$9 + 1$	−832.90	−835.04	0.3
$8 + 2$	−834.81	−843.29	1.0

MP2 is smaller than 1.5%. It is important to point out that through the geometry optimization steps, the spatial organization of the water molecules around the ion are preserved going from size 8 to size 10. Beyond this essential validation, the force field is also able to reproduce energetic order of clusters of a specific size, i.e., the relative energies for the 8 + 2, 9 + 1, and 10 clusters. The transferability of the Th(IV) parameters is then assumed from gas-phase clusters to condensed phase.

3.1.4 Molecular dynamics structural results: AMOEBA versus experiments

Table 2 presents the main results concerning the structure of the solvation shells around the Th^{4+} ion extracted from an analysis of the molecular dynamics trajectories in both cluster and PBC simulations. Since the results are very similar for the four independent trajectories in cluster and the two PBC simulations with different water box sizes, these

Table 2 Molecular dynamics results for the Th^{4+}–water cluster and from PBC simulations with 215 and 511 water molecules

Properties	MD cluster	MD PBC 215 H_2O	MD PBC 511 H_2O	Experimental [64]
1st shell				
Average CN	9	9	9	9
Th^{4+}–O_w distance (Å)	2.4	2.41	2.4	2.462
2nd shell				
Number of H_2O	18.1	17.7	18.1	18
Th^{4+}–O_w distance (Å)	4.59	4.57	4.63	4.657

structural data are considered to be converged. The average coordination number (CN) in the first sphere was found to be exactly 9.0 from the integration of the narrow first peak observed on the radial distribution function g(r) for Th–O pairs (see Fig. 2). The number of experimental structural investigations of the hydrated Thorium(IV) ion in aqueous solution is limited. Recently, Wilson et al. [63] reported a mean Th^{4+}–O_w bond distance of 2.46 Å with a coordination number of 10 from wide angle X-ray scattering (WAXS) study in highly concentrated hydrobromic acid. On the other hand, Torapava et al. [64] performed X-ray absorption fine structure LAXS experiments in aqueous solution. As their results have been obtained in lower concentration than the previous authors, they are, therefore, more comparable to our simulations. Consequently, their derived coordination number of 9 and the Th^{4+}–O_w distance of 2.462 Å have been taken as reference and directly compared with our calculated values. In these conditions, the computed CN of 9.0 is in very good agreement with experiments. No water exchange was observed between the first and the second coordination shells as expected from experimental observations of a residence time of about 20 ns [65]. The maximum peak for the first shell is located at 2.40 Å in good agreement with experimental data. The second sphere is well resolved with a larger peak centered at 4.59 Å, which corresponds to about 18 water molecules. These values match the experimental data [64]. Beyond this distance, no specific geometric arrangement can be pointed out anymore both in cluster and PBC simulations. Figure 3 shows the distribution of the cosines of the angle between the HOH bisector and the Th–O axis for both the first and second shells. A strong radial alignment is obtained with an angle lower than 20° and 25° for the first and second shells, respectively. This confirms the strong influence of the ion on the two first shells.

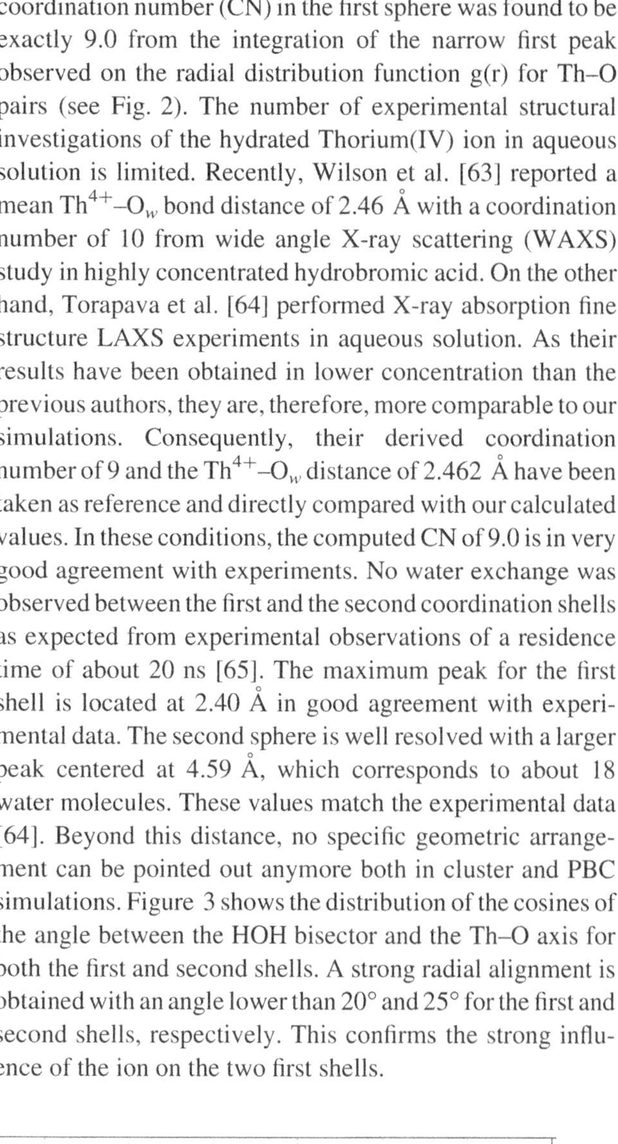

Fig. 2 Radial distribution function g(r) of Th^{4+}–O and integrated curve for the Th^{4+}–water cluster

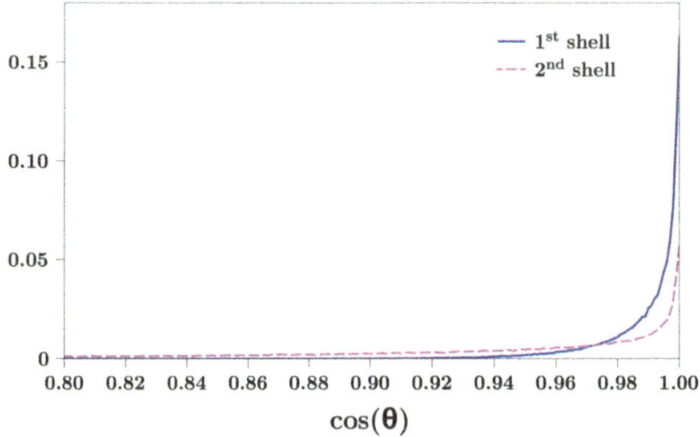

Fig. 3 Probability distribution of the cosines of the tilt angle (between the HOH bisector and the Th–O axis) for the water molecules in the first and second coordination shells in the Th^{4+}–water cluster case

In a recent study, the solvation of Th(IV) was investigated by molecular dynamics using a polarizable force field [17] with the water geometry constrained to the experimental one. The force fields parameters were adjusted from ab initio calculations, and it was found that the organization of the two first coordination spheres around the cation strongly depends on the parameter sets, especially on the parameters of the charge-transfer term. These simulations led to a coordination number between 8.05 and 8.45, lower than the recent experimental results of Torapova et al. An equilibrium was found between 8- and 9-fold coordination that implies water exchange between the two first coordination spheres. This was never observed experimentally [63].

3.1.5 Hydration free energy of Th(IV)

Molecular dynamics simulations were performed to compute the solvation free energy of Th(IV). Fourteen independent simulations were first performed to "grow" the Th VdW particle by setting the charge and polarizability to zero and gradually varying R as $R(\lambda) = \lambda(R\text{final})$ and ε as $\varepsilon(\lambda) = \lambda(\varepsilon\text{final})$, where $\lambda = (0.0, 0.0001, 0.001, 0.010, 0.1, 0.2, 0.3, 0.4, 0.5, 0.6, 0.7, 0.8, 0.9, 1.0)$. Twenty-one further simulations were then performed to "grow" the +4 charge q of Thorium along with its polarizability α such that $q(\lambda') = \lambda'(q\text{final})$ and $\alpha(\lambda') = \lambda'(\alpha \text{ final})$, where $\lambda' = (0.0\text{–}1.0$ with a fixed 0.05 increment). Each simulation ran for 500 ps with a 1.0 fs timestep and the same previously described conditions as the periodic boundary conditions computation of the 511 water box. The absolute free energy calculation was carried out on each of the frames saved every 0.1 ps after the first 50 ps equilibration period using the Bennet acceptance ratio [66] (BAR), a free energy calculation method that utilizes forward and reverse perturbations to minimize variance.

Twenty additional simulations of 400 ps each were also performed to sample tighter frames for values of λ' between 0.000 and 0.100 and between 0.900 and 1.000 with 0.01 steps. However, the difference between the preliminary results and the final value is negligible and would not have required any extra steps as the variation of the ΔG value was quite constant between each pair of frames. Conclusively, the accumulated absolute free energy values for the different frames sum up to $-1,635 (\pm18)$ kcal/mol. Despite a significantly different coordination number, this value is in the range of the available reference (1,458 kcal/mol) value due to David et al. [67] Such value was computed upon EXAFS data using an empirical hydration model based on five basic characteristics of the hydrated ions: crystallographic ionic radius, the coordination number, the cation–oxygen distance (first hydration shell), the number of water molecules in the second hydration shell, and the cation effective charge Z_{eff}. This result is also consistent with the somewhat older results from Marcus [68] who found a value of $-1,391$ kcal/mol with a less refined empirical model also using partial experiment data. Moreover, the values previously obtained with the AMOEBA force field for the divalent cations Ca^{2+}, Mg^{2+}, and Zn^{2+} [27] (see also Supplementary Informations) show that the relative free energies computed with polarizable molecular dynamics simulations are in line with the experimental literature values, including the one obtained for Th^{4+}.

3.1.6 Conclusions

The AMOEBA force field has yet again proved to be robust enough to handle dynamic simulations and yield structural and energetics data for cations. However, the non-explicit inclusion of charge transfer in the model can lead to inaccurate modeling in the particular case of lanthanide and actinide systems where water exchange and charge transfer are key phenomena. Being smaller in magnitude compared to polarization, a good percentage of it (namely the two-body part) could be accurately included within AMOEBA's van der Waals term when the many-body charge transfer is not the driving force of the solvation dynamics. Therefore, as Th(IV) retains its first shell water molecules throughout the whole dynamics, the remaining induction contribution (charge transfer) can be said to be included in the Van der Waals term as a result of matching the total binding energy of AMOEBA to that of quantum mechanics, regardless of the different individual contributions. While the present study does not call for the explicit treatment for charge transfer, other systems will require the computation of both induction terms so as to capture the overall many-body effects, and hence, a project aiming to include an explicit contribution of the charge transfer in the

AMOEBA force field is underway. Nonetheless, we still aimed to investigate the variation of the charge-transfer energy in the Thorium–water clusters as well as in the Lanthanum– and lutetium–water clusters, which are, respectively, the first and last of the lanthanide series using the SIBFA force field as described in the next section.

3.2 Extending the SIBFA force field to lanthanides and actinides

As we have seen, in SIBFA, the intermolecular interaction is expressed under the form of distinct contributions. Each contribution is calibrated in order to closely reproduce its ab initio counterpart as obtained from energy decomposition analyses on monoligated $(M–H_2O)^{n+}$ metal cation–water complexes.

3.2.1 Interaction energies

The aim of working at different levels of calculation was to determine whether the HF/large-core pseudopotential/aug-cc-pVTZ level as in the energy decomposition scheme and consequently for the parameterization of the SIBFA force field was relevant. The comparable profiles of the interaction energy curves in the considered range for the parameterization (1.5–3.5 Å) in all cases show that the chosen level can safely be used for the energy decomposition within the CSOV and RVS frameworks for the development of the SIBFA potential (Fig. 4). The difference between the post-HF and HF energies at the minimum of the curves is the missing correlation energy in the Hartree–Fock approach and can be partially recovered in the force field through the dispersion energy as a rough approximation. Namely, this correlation energy accounts for nearly 10% of the total energy at MP2 level and is due to the high number of electrons in the lanthanides and actinides that are not taken into account explicitly in the large-core pseudopotentials. Therefore, the correlation calculated at the MP2 level recovers the missing energy. The BSSE is negligible in all cases (less than 2% in the considered range) at the HF level.

3.2.2 Energy decomposition analyses

Detailed numerical values about EDA contributions can be found in several Tables in the Supplementary Information section.

Table 3 shows the energy decomposition analysis around the optimized distances of 2.4, 2.1, and 2.2 Å, respectively, for the $[La–OH_2]^{3+}$ $[Lu–OH_2]^{3+}$ and $[Th–OH_2]^{4+}$ dimers. All energies are given in kcal/mol. In both cases of the lanthanide cations bearing the same +3 charge, the contribution of the first-order energies amount to the

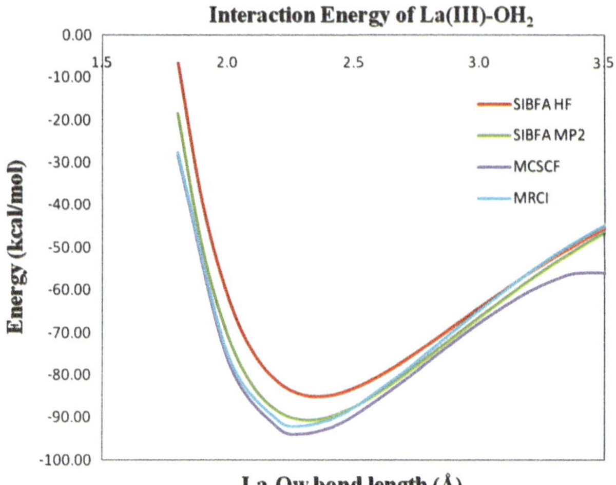

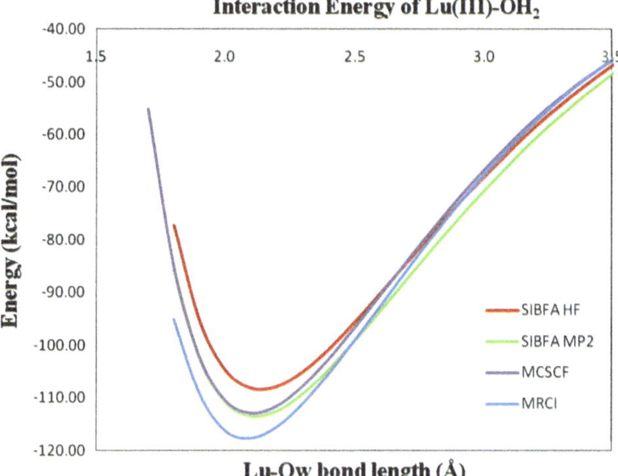

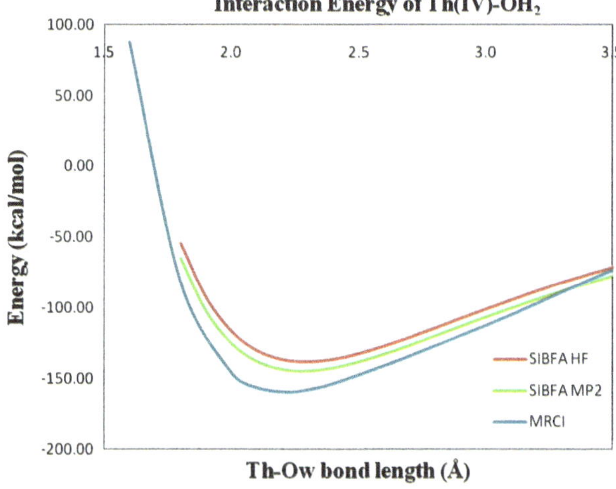

Fig. 4 Ab initio interaction energy curves at different levels of calculation for all three dimers using the procedure described in Sect. 1

same value even though the separate energies of electrostatics (ES) and exchange repulsion (REP) are significantly higher in the Lutetium–water dimer. This is of course due

Table 3 The different energy contributions, in kcal/mol, as given by the CSOV and RVS decomposition at the HF level, as well as the global interaction energy (E_{int}) at the equilibrium distances for the $[La–OH_2]^{3+}$, $[Lu–OH_2]^{3+}$, and the $[Th–OH_2]^{4+}$ dimers

Dimer	ES	REP	E1	POL	CT	E_{int}
$[La–OH_2]^{3+}$	−67.56	35.50	−32.07	−40.17	−10.40	−82.64
$[Lu–OH_2]^{3+}$	−85.43	52.84	−32.59	−55.29	−16.92	−104.79
$[Th–OH_2]^{4+}$	−102.55	83.13	−19.42	−92.80	−30.45	−145.26

to the increased number of electrons from the Lanthanum $4f^0$ configuration to the completely filled $4f^{14}$ shell of Lutetium, even though they are not accounted for explicitly in the pseudopotentials. However, both the polarization energy (POL) and the charge transfer (CT) are higher in the Lutetium system and are therefore responsible for the global increase in the interaction energy value, of 22.15 kcal/mol, from $[La–OH_2]^{3+}$ to $[Lu–OH_2]^{3+}$. In the case of the $[Th–OH_2]^{4+}$ dimer, the Coulomb energy is as expected, significantly higher than in the trivalent systems and so is the repulsion energy, thus resulting in a decreased first-order energy (E1) with respect to the Lanthanum and Lutetium systems. Conversely, both the polarization energy and charge-transfer contributions have increased, with POL accounting for more than 60% of the interaction energy of the Thorium system and $\sim 50 \pm 2\%$ in the case of the two lanthanide systems. It is clearly shown here that, despite the high positive charge of the cations, the main contribution to the interaction energy is the polarization term when compared to E1. In addition to polarization, the charge-transfer term appears to be clearly non-negligible and therefore hints to more covalent interactions in such systems. Therefore, the degree of covalency in those bonds is linked to the second-order energies and will be clearly dependent on the type of cation (and on the involved orbitals and presence and number of f electrons in the system). As a result, we pointed out that neglecting either one or the other of the second-order contributions will induce a shift in the equilibrium distance and corresponding energy as shown in Fig. 5 for lanthanide and actinide systems. It is important to note that the use of a EDA formalism such as CSOV is important for such ab initio decomposition. Indeed, CSOV generates consistent polarization data as the HF wave functions are kept strictly antisymmetric. That way, here, we do not have any overestimation of polarization/charge transfer due to such lack of antisymmetry (i.e., exchange polarization is included, see references 40 and 45 for details). In any case, these

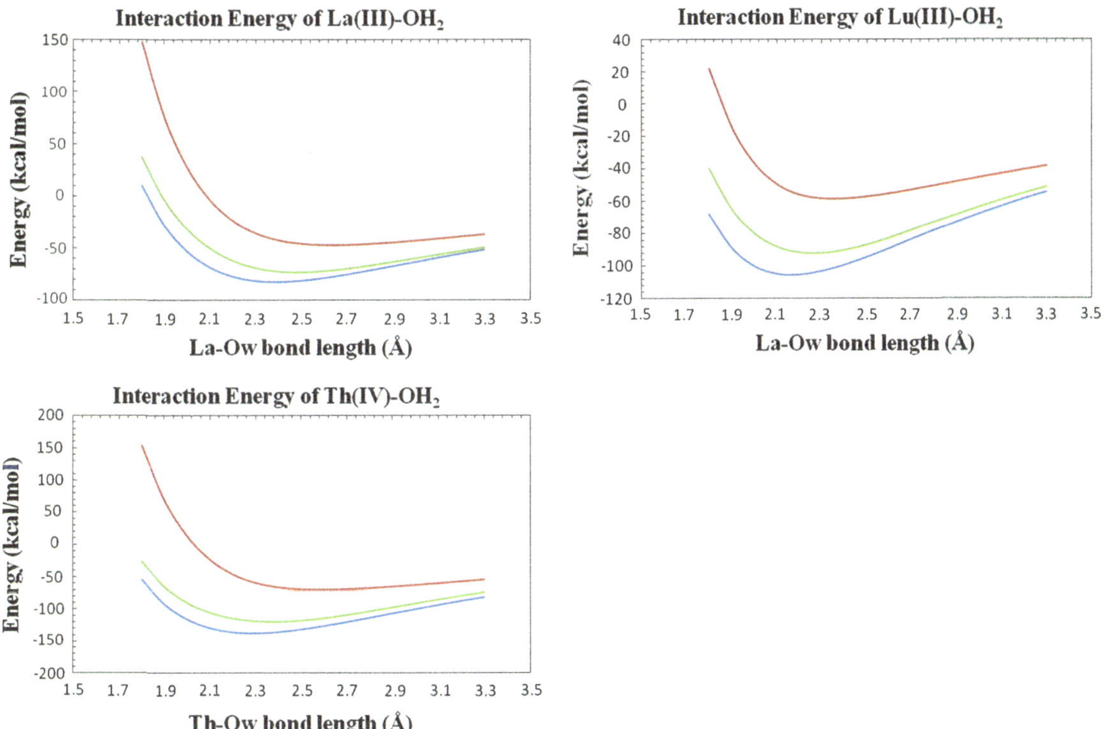

Fig. 5 Ab initio energy curves excluding either the charge-transfer term (*green*) or the polarization energy (*red*) as compared to the interaction energy (*blue*) for all three dimers

results clearly rise key issue toward force fields development that should either: (a) embody an explicit functional form for charge transfer or (b) at least follow a consistent parametrization strategy to incorporate charge transfer in van der Waals based on EDA as some of us proposed [27].

3.2.3 Extension of the SIBFA potential to lanthanides and actinides

At the end of this work, a full SIBFA potential was obtained for all three considered cations, with a very good description of each energy contribution by the force field with respect to the reference data as shown in Fig. 6 for Th(IV) –(La(III) and Lu(III) Figures are available as SI: see Figures SI7 and SI8). It can be noted here that SIBFA's first handling of trivalent and tetravalent cations shows a very good reproduction of the different ab initio contribution curves. Moreover, concerning the electrostatic contribution, the inclusion of a correction of the quantum penetration energy through an overlap term prevented any divergence at short range. Also, all pair approximations made for the repulsion term are entirely transferable up to the heavy elements as previously shown for the Pb(II) cation [69], with correct behavior of the SIBFA force field at both short- and long range. Concerning the second-order energies, results are satisfactory. At very short distances, the SIBFA polarization energy diverges from the CSOV curve, indicating that the damping is not sufficient since the

CSOV polarization term grows more rapidly than its SIBFA counterpart as the Ln/An–O_w distance decreases. Such behavior which is nevertheless not critical will be avoided in a near future through the use of a Gaussian electrostatic potential based on density fitting such as the Gaussian electrostatic model (GEM) [70]. This approach which is being implemented within SIBFA [71] will allow for a better reproduction of the short-range curvature of the polarization and thus avoid the polarization catastrophe. Lastly, the charge transfer gives satisfying agreement with ab initio, namely around the optimized distance with a slight divergence however at very short and very long distances. Overall, the agreement between SIBFA's global interaction energy with the ab initio value obtained from the quantum EDA is also very satisfactory (Fig. 7) and indicates SIBFA's ability to reproduce the ab initio interaction energy for trivalent lanthanide–water and tetravalent actinide–water dimers.

3.2.4 Transferability of the potential

The transferability of the different parameters optimized for the dimer systems was tested for several $[Ln–(OH_2)_n]^{3+}$ and $[Th–(OH_2)_n]^{4+}$ clusters, with $n = 4$, 6, 8, and 9 (structures for which RVS computations were technically possible). The geometries of the tetra-coordinated clusters were optimized at the HF level using aug-cc-pVTZ basis sets for the water molecules and a small-core

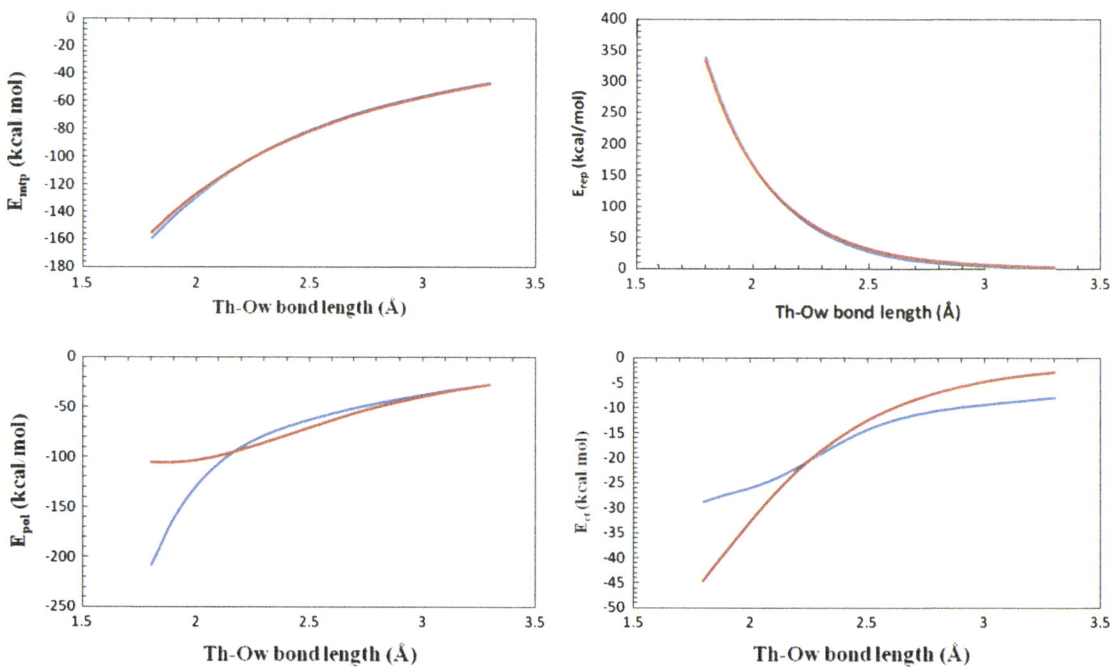

Fig. 6 CSOV (*blue*) versus SIBFA (*red*) electrostatic (*top left*), repulsion (*top right*), polarization (*bottom left*), and charge-transfer (*bottom right*) energies as a function of the Th–Ow distance in Th(IV)–OH$_2$

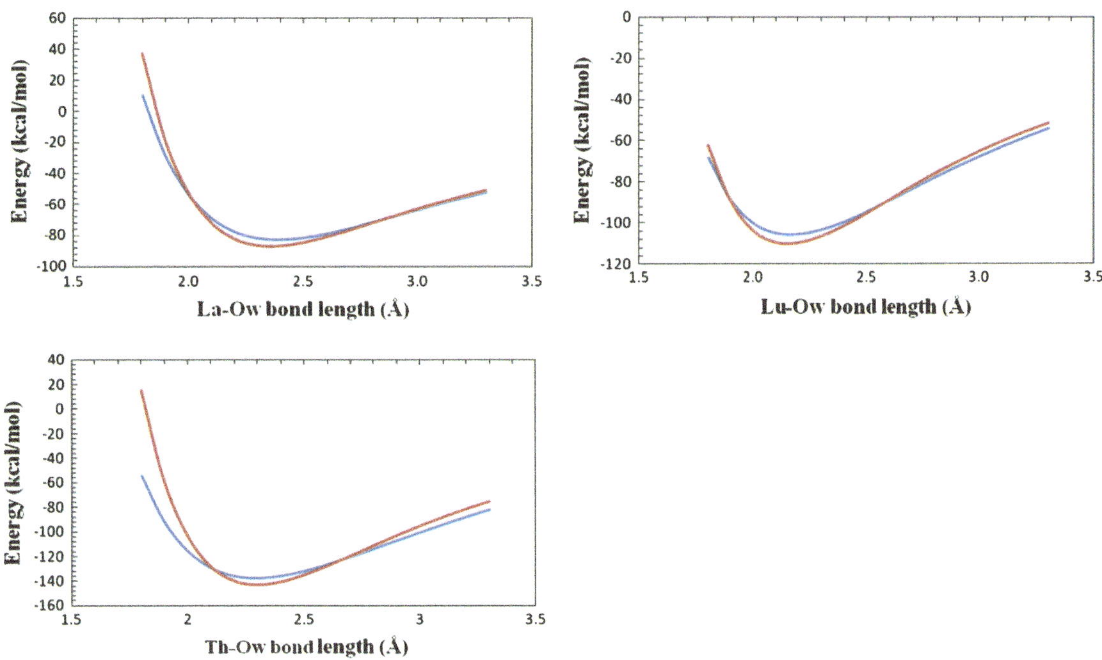

Fig. 7 CSOV (*blue*) versus SIBFA (*red*) total interaction energies as a function of metal M–Ow distances in M($n+$)–OH$_2$ at the HF level

pseudopotential for Th. The structure of the complex with eight-coordinated water molecules is square antiprismatic (SAP), while the structure of the complex with nine water molecules is tricaped trigonal prismatic (TTP). As SIBFA uses a rigid fragment approximation, initial geometries were projected within the usual SIBFA model water internal geometry. The Ln/An–O$_w$ distances only were then fixed at three different distances around the optimized bond length. The hexa-coordinated structures however were standard, symmetrical complexes in which only the Ln/ An–O$_w$ distances were varied. EDA were then carried out with the GAMESS software to yield the different contributions to the interaction energy of the systems. We have to point out here that while the RVS scheme worked fine for all tetracoordinated clusters, we have not been able to obtain a converged decomposition in the case of the different hexacoordinated clusters. The energy decomposition was therefore carried out in the Kitaura–Morokuma (KM) framework, with as a result, a global second-order energy that was compared to the SIBFA $E2 = E_{int} - E_{mtp} - E_{rep}$ values instead of the separate polarization and charge-transfer contributions. The comparable energies namely around the optimized distances for each of the tetra- and hexa-coordinated clusters account for the transferability of the parameters, with less than 2% error in the global interaction energy around the considered distances.

Lastly, the parameters were tested on more relevant structures of eight- and ninefold coordination for all three systems. However, due to computational limits, only the interaction energy at the Hartree–Fock level calculated

with the Gaussian program was used for comparison against the SIBFA interaction energy on the same structures. Once again, the error is kept under the aforementioned 2% in all three Lanthanum, Lutetium, and Thorium systems as shown in Tables 4, 5, 6. We have therefore obtained a fully transferable, polarizable force field, including charge transfer, for trivalent closed-shell lanthanide and tetravalent actinide cations.

3.2.5 The issue of dispersion: toward correlated molecular dynamics

The last contribution taken into account by the SIBFA force field is the dispersion energy, i.e., the energy arising from the dynamically induced dipoles. Per se, this contribution cannot be derived from a Hartree–Fock energy decomposition analysis, and therefore, it has been taken in this study as the difference between an MP2 dissociation curve and the HF curve used to parameterize the SIBFA force field. Despite the approximate fit of the SIBFA E_{disp} component on the Δ(MP2–HF) curve, the overall MP2 interaction energy curve matches that obtained with the SIBFA potential corrected with the estimated dispersion component from the fit, referred to as SIBFA + d, as it can be seen in Supplementary Informations (SI: Figure 9). While the results for the cation-monoaqua systems feature less than 5% error, the extension to the different clusters is still out of reach because the nature of the dispersion energy is not fully accounted for in the Δ(MP2–HF) approximation as correlation is not restricted to the sole

Table 4 Interaction energies in kcal/mol of the $[La–(OH_2)_n]^{3+}$ clusters as calculated at the Hartree–Fock (HF) level and with the SIBFA force field for several ion–water distances

n	d (La–O$_w$) (Å)	HF	SIBFA	Δ	% Error
4	2.3	−268.66	−263.38	−5.28	2.0
	2.4	−279.69	−277.45	−2.24	0.8
	2.5	−281.39	−280.13	−1.26	0.5
6	2.1	−189.21	−166.13	−23.08	12.2
	2.2	−277.58	−264.90	−12.68	4.6
	2.3	−330.42	−323.53	−6.89	2.2
	2.4	−358.71	−354.76	−3.95	1.1
	2.5	−370.24	−367.49	−2.75	0.7
8	2.60	−419.71	−415.11	−4.6	1.1
9	2.63	−451.24	−447.5	−3.73	0.8

The relative difference Δ between each pair of values is given as well as the corresponding percentage error

Table 5 Interaction energies in kcal/mol of the $[Lu–(OH_2)_n]^{3+}$ clusters as calculated at the Hartree–Fock (HF) level and with the SIBFA force field for several ion–water distances

n	d (Lu–O$_w$) (Å)	HF	SIBFA	Δ	% Error
4	2.2	−351.36	−344.07	−7.29	2.1
	2.3	−350.65	−347.05	−3.60	1.0
	2.4	−342.44	−340.00	−2.44	0.7
6	2.1	−419.09	−389.94	−29.15	7.0
	2.2	−448.29	−432.33	−15.96	3.6
	2.3	−457.28	−448.88	−8.40	1.8
	2.4	−453.47	−448.63	−4.84	1.1
	2.5	−441.26	−437.78	−3.48	0.8
8	2.42	−505.72	−496.18	−9.54	1.89
9	2.43	−536.21	−533.26	−2.95	0.55

The relative difference Δ between each pair of values is given as well as the corresponding percentage error

Table 6 Interaction energies in kcal/mol of the $[Th–(OH_2)_n]^{4+}$ clusters as calculated at the Hartree–Fock (HF) level and with the SIBFA force field for several ion–water distances

n	d (Th–O$_w$) (Å)	HF	SIBFA	Δ	% Error
4	2.3	−469.22	−470.13	0.91	−0.2
	2.4	−470.46	−477.44	6.98	−1.5
	2.5	−462.10	−469.25	7.15	−1.6
6	2.1	−475.89	−390.34	−85.55	18.0
	2.2	−555.23	−518.73	−36.5	6.6
	2.3	−598.21	−588.47	−9.74	1.6
	2.4	−615.69	−618.50	2.81	−0.5
	2.5	−615.57	−622.26	6.69	−1.1
8	2.50	−693.80	−701.81	8.01	−1.2
9	2.56	−735.70	−743.22	7.52	−1.0

The relative difference Δ between each pair of values is given as well as the corresponding percentage error

dispersion term. Indeed, preliminary calculations have shown that the parameters derived for dispersion in SIBFA do not exhibit the same transferability as that obtained for the other contributions. A means of accessing a more physically meaningful dispersion contribution is through post-HF EDA, such as correlated CSOV or by performing EDA on correlated energy curves. This however implies intensive studies so as to determine which functional is more appropriate to each of the investigated systems while remaining cautious on the ability of the density functional theory to assess lanthanide and actinide systems.

4 Conclusion

The parameterization of the SIBFA force field for the first and last cations of the lanthanide series and the closed-shell Thorium(IV) cation was undertaken. It was found that SIBFA, which includes many-body charge transfer, is capable of handling heavy elements from dimers to ninefold coordinated hydrated complexes with around 2% error with respect to the ab initio Hartree–Fock interaction energy. Furthermore, the fact that the force field is based on EDA has led us to investigate the separate contributions to the interaction energy of the considered dimers. The sizeable values of the polarization and charge-transfer energies secures the idea that only a polarizable force field including charge transfer can be used to correctly model lanthanide and actinide systems. Nevertheless, through its careful parametrization on the Th(IV)–water dimer and without including any experimental data, the AMOEBA force field leads to very accurate results. The transferability of the parameters has been demonstrated since the force field was able to reproduce gas-phase reference energetic, structural, and thermodynamic experimental quantities, including Th(IV) solvation free energies. Such results are encouraging and should lead us to improved modeling of lanthanides and actinides within complex environments beyond solvation. To conclude, based on the SIBFA results, explicit charge transfer will be added to AMOEBA as full SIBFA cluster MD simulations will be undergone. Moreover, as our approaches are not limited to treat closed-shell systems, [32] future work will also deal with open-shell actinides systems (Marjolin et al. in preparation).

Acknowledgments Two of the authors, C. G. and J.-P. D., thank the direction of simulation and experimental tools of the CEA nuclear energy division CEA/DEN/RBPCH for financial support. This work was granted access to the HPC resources of [CCRT/CINES/IDRIS] under the allocation x2011086146 made by GENCI (Grand Equipement National de Calcul Intensif).

References

1. Merbach AE, Toth E (2001) The chemistry of contrast agents in medical magnetic resonance imaging. Wiley, Chichester
2. Nicholas LK, Long NJ (2006) Chem Soc Rev 35:557
3. Moore EG, Samuel APS, Raymond KN (2009) Acc Chem Res 42:542
4. Hou ZM, Wakatsuki Y (2002) Coord Chem Rev 231:1
5. Silva R, Nitsche H (1995) Radiochim Acta 70:377
6. Grenthe I, Fuger J, Konings R, Lemire R, Muller A, Nguyen-Trung C, Wanner H (1992) In: Wanner H, Forest I (eds) Chemical thermodynamics of Uranium. North-Holland, Amsterdam
7. Nash KL, Madic C, Mathur JN (2006) Actinide separation science and technology. In: Morss LR, Edelstein NM, Fuger J (eds) The chemistry of the actinide and transactinide elements. Springer, Dordrecht
8. Clavaguéra C, Calvo F, Dognon J-P (2006) J Chem Phys 124:074505
9. Clavaguéra C, Sansot E, Calvo F, Dognon J-P (2006) J Phys Chem B 110:12848
10. Clavaguéra C, Pollet R, Soudan JM, Brenner V, Dognon J-P (2005) J Phys Chem B 109:7614
11. Hagberg D, Karlström G, Roos BJ, Gagliardi L (2005) J Am Chem Soc 127:14250
12. Hagberg D, Bednarz E, Edelstein NM, Gagliardi L (2007) J Am Chem Soc 129:14136
13. Villa A, Hess B, Saint-Martin H (2009) J Phys Chem B 113:7270
14. Duvail M, Vitorge P, Spezia R (2009) J Chem Phys 130:104501
15. Beuchat C, Hagberg D, Spezia R, Gagliardi L (2010) J Phys Chem B 114:15590
16. Galbis E, Hernández-Cobos J, den Auwer C, Le Naour C, Guillaumont D, Simoni E, Pappalardo RR, Sánchez ME (2010) Angew Chem Int Ed 49:3811
17. Réal F, Trumm M, Vallet V, Schimmelpfennig B, Masella M, Flament J-P (2010) J Phys Chem B 114:15913
18. Duvail M, Spezia R, Vitorge P (2008) Chem Phys Chem 9:693
19. Duvail M, Martelli F, Vitorge P, Spezia R (2011) J Chem Phys 135:044503
20. Kowall Th, Foglia F, Helm K, Merbach AE (1995) J Am Chem Soc 117:13790
21. D'Angelo P, Zitolo A, Migliorati V, Chillemi G, Duvail M, Vitorge P, Abadie S, Spezzia R (2011) Inorg Chem 50:4572
22. Helm L, Merbach AE (2005) Chem Rev 105:1923
23. Clavaguéra-Sarrio C, Brenner V, Hoyau S, Marsden CJ, Millié P, Dognon J-P (2003) J Phys Chem B107:3051
24. Engkvist O, Åstrand PO, Karlström G (2000) Chem Rev 100:4087
25. Ren P, Ponder JW (2003) J Phys Chem B 107:5933
26. Shi Y, Wu C, Ponder JW, Ren P (2011) J Comput Chem 32:967
27. Wu J, Piquemal J-P, Chaudret R, Reinhardt P, Ren P (2010) J Chem Theory Comput 6:2059
28. Piquemal J-P, Perera L, Cisneros GA, Ren P, Pedersen LG, Darden TA (2006) J Chem Phys 125:054511
29. Jiao D, King C, Grossfield A, Darden TA, Ren P (2006) J Phys Chem B 110:18553
30. Grossfield A, Ren P, Ponder JW (2003) J Am Chem Soc 125:15671
31. Gresh N, Cisneros GA, Darden TA, Piquemal J-P (2007) J Chem Theory Comput 3:1960
32. Piquemal J-P, Williams-Hubbard B, Fey N, Deeth RJ, Gresh N, Giessner-Prettre C (2003) J Comput Chem 24:1963
33. Gresh N, Piquemal J-P, Krauss M (2005) J Comput Chem 26:1113
34. Roux C, Gresh N, Perera L, Piquemal J-P, Salmon L (2007) J Comput Chem 28:938
35. Jenkins LMM, Hara T, Durell SR, Hayashi R, Inman JK, Piquemal J-P, Gresh N, Appella E (2007) J Am Chem Soc 129:11067
36. Gourlaouen C, Clavaguéra C, Piquemal J-P, Dognon J-P (2012) submitted
37. Simah D, Hartke B, Werner H-J (1999) J Chem Phys 111:4523
38. Werner HJ, Knowles PJ with contributions from Almlof J, Amos RD, Bernhardsson A, Berning A, Cooper DL, Deegan MJO, Dobbyn AJ, Eckert F, Hampel C, Lindh R, Lloyd AW, Meyer W, Mura ME, Nicklass A, Peterson K, Pitzer R, Pulay P, Rauhut G, Schutz M, Stoll H, Stone AJ, Taylor PR, Thorsteinsson T (2008) Molpro, version 2008.1. University of Stuttgart and Birmingham, Germany and Great Britain
39. Benedict WS, Gailar N, Plyler EK (1956) J Chem Phys 24:1139
40. Dunning TH Jr (1989) J Chem Phys 90:1007
41. Frisch MJ, Trucks GW, Schlegel HB, Scuseria GE, Robb MA, Cheeseman JR, Montgomery JJA, Vreven T, Kudin KN, Burant JC, Millam JM, Iyengar SS, Tomasi J, Barone V, Mennucci B, Cossi M, Scalmani G, Rega N, Petersson GA, Nakatsuji H, Hada M, Ehara M, Toyota K, Fukuda R, Hasegawa J, Ishida M, Nakajima T, Honda Y, Kitao O, Nakai H, Klene M, Li X, Knox JE, Hratchian HP, Cross JB, Bakken V, Adamo C, Jaramillo J, Gomperts R, Stratmann RE, Yazyev O, Austin AJ, Cammi R, Pomelli C, Ochterski JW, Ayala PY, Morokuma K, Voth GA, Salvador P, Dannenberg JJ, Zakrzewski VG, Dapprich S, Daniels AD, Strain MC, Farkas O, Malick DK, Rabuck AD, Raghavachari K, Foresman JB, Ortiz JV, Cui Q, Baboul AG, Clifford S, Cioslowski J, Stefanov BB, Liu G, Liashenko A, Piskorz P, Komaromi I, Martin RL, Fox DJ, Keith T, Al-Laham MA, Peng CY, Nanayakkara A, Challacombe M, Gill PMW, Johnson B, Chen W, Wong MW, Gonzalez C, Pople JA (2009) Gaussian 09. Gaussian Inc., Wallingford, CT
42. Bagus PS, Hermann K, Bauschlicher CW Jr (1984) J Chem Phys 81:1966
43. Piquemal J-P, Marquez A, Parisel O, Giessner-Prettre C (2005) J Comput Chem 26:1052
44. Dupuis M, Marquez A, Davidson ER HONDO95.3, Quantum Chemistry Program Exchange (QCPE). Indiana University, Bloomington, IN 47405
45. Stevens WJ, Fink W (1987) Chem Phys Lett 139:15–22
46. Schmidt MW, Baldridge KK, Boatz JA, Elbert ST, Gordon MS, Jensen JH, Koseki S, Matsunaga N, Nguyen KA, Su S, Windus TL, Dupuis M, Montgomery JA Jr (1993) J Comput Chem 14:1347
47. Kitaura K, Morokuma K (1976) Int J Quantum Chem 10:325
48. Cisneros GA, Darden TA, Gresh N, Reinhardt P, Parisel P, Pilmé J, Piquemal J-P (2009) In: York DM, Lee T-S (eds) Multi-scale quantum models for biocatalysis: modern techniques and applications, for the book series: challenges and advances in computational chemistry and physics. Springer, Berlin, pp 137–172
49. Vigné-Maeder F, Claverie P (1988) J Chem Phys 88:4934
50. Piquemal J-P, Gresh N, Giessner-Prettre C (2003) J Phys Chem A 107:10353
51. Piquemal J-P, Chevreau H, Gresh N (2007) J Chem Theory Comput 3:824
52. Garmer DR, Stevens WJ (1989) J Chem Phys 93:8263
53. Chaudret R, Gresh N, Parisel O, Piquemal J-P (2011) J Comput Chem 7:618
54. Tholé BT (1981) Chem Phys 59:341
55. Halgren TA (1992) J Am Chem Soc 114:7827
56. Réal F, Vallet V, Clavaguéra C, Dognon JP (2008) Phys Rev A 78:052502
57. Nose S (1984) J Chem Phys 81:511–519
58. Hoover WG (1985) Phys Rev A 31:1695
59. Beeman D (1976) J Comp Phys 20:130

60. Ponder JW (2009) TINKER: software tools for molecular design, version 6.2. Washington University School of Medicine, Saint Louis
61. Essmann U, Perera L, Berkowitz ML, Darden T, Lee L, Pedersen LG (1995) J Chem Phys 101:8577
62. Berendsen HJC, Postma JPM, van Gunsteren WF, DiNola A, Haak JR (1984) J Chem Phys 81:3684
63. Wilson RE, Skanthakumar S, Burns PC, Soderholm L (2007) Angew Chem 119:8189
64. Torapava N, Persson I, Eriksson L, Lundberg D (2009) Inorg Chem 48:11712
65. Farkas I, Grenthe IJ (2000) Phys Chem A 104:1201
66. Bennett CH (1976) J Comput Phys 22:245
67. David FH, Vokhmin V (2003) New J Chem 27:1627
68. Marcus YA (1994) Biophys Chem 51:111
69. Devereux M, van Severen M-C, Parisel O, Piquemal J-P, Gresh N (2011) J Chem Theory Comput 7:138
70. Piquemal J-P, Cisneros GA, Reinhardt P, Gresh N, Darden TA (2006) J Chem Phys 124:104101
71. Chaudret R, Gresh N, Darden TA, Parisel O, Cisneros GA, Piquemal J-P (2011) J Chem Phys 2012 (in press)

Theor Chem Acc (2012) 131:1138
DOI 10.1007/s00214-012-1138-6

REGULAR ARTICLE

Automation of AMOEBA polarizable force field parameterization for small molecules

Johnny C. Wu · Gaurav Chattree · Pengyu Ren

Received: 12 May 2011 / Accepted: 29 August 2011 / Published online: 26 February 2012
© Springer-Verlag 2012

Abstract A protocol to generate parameters for the AMOEBA polarizable force field for small organic molecules has been established, and polarizable atomic typing utility, Poltype, which fully automates this process, has been implemented. For validation, we have compared with quantum mechanical calculations of molecular dipole moments, optimized geometry, electrostatic potential, and conformational energy for a variety of neutral and charged organic molecules, as well as dimer interaction energies of a set of amino acid side chain model compounds. Furthermore, parameters obtained in gas phase are substantiated in liquid-phase simulations. The hydration free energy (HFE) of neutral and charged molecules have been calculated and compared with experimental values. The RMS error for the HFE of neutral molecules is less than 1 kcal/mol. Meanwhile, the relative error in the predicted HFE of salts (cations and anions) is less than 3% with a correlation coefficient of 0.95. Overall, the performance of Poltype is satisfactory and provides a convenient utility for applications such as drug discovery. Further improvement can be achieved by the systematic study of various organic compounds, particularly ionic molecules, and refinement and expansion of the parameter database.

Keywords AMOEBA · Polarizable force field · Small molecule modeling · Poltype · Atomic typer · Molecular dynamics

Abbreviations

AMOEBA	Atomic multipole optimized energetics for biomolecular applications
DMA	Distributed multipole analysis
HFE	Hydration free energy
ESP	Electrostatic potential

Published as part of the special collection of articles: From quantum mechanics to force fields: new methodologies for the classical simulation of complex systems.

Electronic supplementary material The online version of this article (doi:10.1007/s00214-012-1138-6) contains supplementary material, which is available to authorized users.

J. C. Wu · G. Chattree · P. Ren (✉)
Department of Biomedical Engineering, University of Texas at Austin, Austin, TX 78712-1062, USA
e-mail: pren@mail.utexas.edu

1 Introduction

The classical fixed-charge molecular mechanics force field has been the conventional model to study biological macromolecular systems. However, Buckingham [1] pointed out as early as 1967 that the intermolecular forces are electrostatic in nature and can be modeled with electric multipole moments and induction. Despite investigations of various polarizable force fields [2–8], few large-scale macromolecular simulations have taken advantage of these models. Furthermore, the need for a polarizable force field has been widely acknowledged [9, 10]. Studies identify that fixed-charge force fields have difficulties with calculating the solvation free energy of polar small molecules, particular those containing hydroxyl groups [11], which are common in carbohydrates. Patel et al. [12] studied polarization in further detail and proposed the TIP4P-QDP charge-dependent polarizability water model. This work revealed the effects of polarization variability on the enhanced structure at liquid–vapor interface. Additionally, the energetics [13] as well as thermodynamics of single-atom monovalent [14, 15] and divalent [16] ions in solvent have been studied with polarizable models. The importance of polarization in protein–ligand recognition [17–19] has been identified. Although other classical force fields have been extended to include polarization interactions, such as PIPF-CHARMM [20], most polarizable force fields lack higher-order electronic moments.

The AMOEBA (Atomic multipole optimized energetics for biomolecular applications) polarizable force field is an effort to achieve chemical accuracy of conformational and interaction energies to quantum mechanical models. Moreover, AMOEBA addresses differences in systems where assuming an averaged polarization is not adequate. Its permanent point multipole up to quadrupole is capable of describing the intricate electrostatic potential surfaces. Polarization is represented with polarizable point dipoles that can fully describe the directionality of electron redistribution without the need for fictitious particles. While small molecule parameters can be obtained relatively easily for fixed-charge force fields [21, 22], adoption of a multipole-based force field has been limited due to the lack of automation for parameterization. Higher-order multipole moments require the choice of a local frame, and the assignment of atomic polarizabilities is necessary as well. This work articulates a procedure to generate the AMOEBA force field parameters for small molecules such as protein ligands. Additionally, a utility, Poltype, has been implemented to fully automate this procedure and is available at http://water.bme.utexas.edu/wiki/index.php/Software:Poltype. The parameters obtained from this procedure are substantiated via comparisons with quantum mechanics calculations, other molecular mechanics simulations, and experimental measurements for a range of properties.

2 Methods

2.1 AMOEBA force field

The AMOEBA force field is a polarizable molecular mechanics model that treats electrostatic interactions with higher-order moments up to quadrupoles. The potential energy model has been described previously [23–25] and is briefly explained here for reference. The potential energy function comprises bonded and non-bonded interactions. Bonded interactions include bond stretching, angle-bending, bond-angle stretch-bending, out-of-plane bending, and

rotation about torsion. Non-bonded interactions include van der Waals, permanent and induced electrostatics.

$$U = U_{\text{bond}} + U_{\text{angle}} + U_{b\theta} + U_{\text{oop}} + U_{\text{torsion}} + U_{\text{vdW}} + U_{\text{ele}}^{\text{perm}} + U_{\text{ele}}^{\text{ind}}$$

Bonded interactions in the AMOEBA force field adopt non-harmonic functional forms. Bond stretch energies utilize the fourth-order Taylor expansion of the Morse potential. Bond-angle bend and torsion energies utilize a sixth-order potential and a six-term Fourier series expansion, respectively. These valence functional forms are the same as those used by the MM3 [26] classical molecular mechanics potential. Additionally, out-of-plane bending was restrained at sp^2-hybridized trigonal centers with a Wilson–Decius cross-function [27].

$$U_{\text{bond}} = K_b(b - b_0)^2[1 - 2.55(b - b_0) + 3.793125(b - b_0)^2$$

$$U_{\text{angle}} = K_\theta(\theta - \theta_0)^2[1 - 0.014(\theta - \theta_0) + 5.6 \times 10^{-5}(\theta - \theta_0)^2 - 7.0 \times 10^{-7}(\theta - \theta_0)^3 + 2.2 \times 10^{-8}(\theta - \theta_0)^4]$$

$$U_{b\theta} = K_{b\theta}[(b - b_0) + (b' - b'_\theta)](\theta - \theta_0)$$

$$U_{\text{torsion}} = \sum_n K_{n\phi}[1 + \cos(n\phi \pm \delta)]$$

$$U_{\text{oop}} = K_\chi \chi^2.$$

Bond lengths, bond/torsion phase angles, and energies are in units of Å, degrees, and kcal/mol, respectively. The repulsion–dispersion interactions are represented with a buffered 14-7 potential [28].

$$U_{\text{vdw}}(ij) = \varepsilon_{ij}\left(\frac{1.07}{\rho_{ij} + 0.07}\right)^7\left(\frac{1.12}{\rho_{ij}^7 + 0.12} - 2\right).$$

The potential is a function of separation distance, R_{ij}, between atoms i and j $\rho_{ij} = R_{ij}/R_{ij}^0$ where R_{ij}^0 is the minimum energy distance and is combined for heterogeneous atom pairs $R_{ij}^0 = \frac{(R_{ii}^0)^3 + (R_{jj}^0)^3}{(R_{ii}^0)^2 + (R_{jj}^0)^2}$. In addition, the potential minimum in kcal/mol is combined for heterogeneous atom pairs $\varepsilon_{ij} = \frac{4\varepsilon_{ii}\varepsilon_{jj}}{(\varepsilon_{ii}^{1/2} + \varepsilon_{jj}^{1/2})^2}$.

Permanent electrostatic interactions are computed with higher-order moments where

$$M_i = [q_i, d_{ix}, d_{iy}, d_{iz}, Q_{ixx}, Q_{ixy}, Q_{ixz}, Q_{iyx}, Q_{iyy}, Q_{iyz}, Q_{izx}, Q_{izy}, Q_{izz}]^T$$

is a multipole composed of charge, q_i, dipoles, $d_{i\alpha}$, and quadrupoles, $Q_{i\beta\gamma}$. The interaction energy between two multipole sites is

$$
U_{\text{emp}}^{\text{perm}} =
\begin{bmatrix} q_i \\ d_{ix} \\ d_{iy} \\ d_{iz} \\ Q_{ixx} \\ \vdots \end{bmatrix}^{T}
\begin{bmatrix}
1 & \frac{\partial}{\partial x_j} & \frac{\partial}{\partial y_j} & \frac{\partial}{\partial z_j} & \cdots \\
\frac{\partial}{\partial x_i} & \frac{\partial^2}{\partial x_i \partial x_j} & \frac{\partial^2}{\partial x_i \partial y_j} & \frac{\partial^2}{\partial x_i \partial x_j} & \cdots \\
\frac{\partial}{\partial y_i} & \frac{\partial^2}{\partial y_i \partial x_j} & \frac{\partial^2}{\partial y_i \partial y_j} & \frac{\partial^2}{\partial y_i \partial y_j} & \cdots \\
\frac{\partial}{\partial z_i} & \frac{\partial^2}{\partial z_i \partial x_j} & \frac{\partial^2}{\partial z_i \partial y_j} & \frac{\partial^2}{\partial z_i \partial z_j} & \cdots \\
\vdots & \vdots & \vdots & \vdots & \ddots
\end{bmatrix}
\frac{1}{R_{ij}}
\begin{bmatrix} q_j \\ d_{jx} \\ d_{jy} \\ d_{jz} \\ Q_{jxx} \\ \vdots \end{bmatrix}.
$$

Multipoles are defined at atomic centers in relation to a local frame defined by other atoms that are bonded to it. A triplet of atoms is used to specify a local frame following the z-then-x convention. Figure 1a illustrates an example of an asymmetric local frame for atom A defined by atoms A, B, and C. The vector created by AB is the direction of the positive z-axis. The positive x-axis lies on the plane created by ABC and creates an acute angle with AC. The positive y-axis is defined to create a cubic right-handed coordinate system. Figure 1b illustrates an example of a local frame in which B and C are symmetric with respect to A, such as a water molecule. The z-axis is defined as the bisector of $\angle$BAC. The x-axis is defined as the vector along the aforementioned plane that creates an acute angle with AB and is orthogonal to the z-axis. As with the former case, the y-axis is defined to create a right-handed coordinate system.

Electronic polarization describes the redistribution of electron density due to an external field. Polarizable point dipoles are utilized by AMOEBA at atomic centers to describe this effect. An iterative induction approach originally developed by Thole [29] is adopted in which an induced dipole at site i continues to polarize all other sites until convergence is achieved at all induced dipole sites. This method imposes a damped polarization interaction at very short range in order to avoid a well-known artifact of point polarizability models by smearing one of the atomic multipole moments in each pair of interaction sites [30]. The damping functions for charge, dipole, and quadrupole interactions have been derived previously [24]. The smearing function of a charge has the form

$$
\rho = \frac{3a}{4\pi} \exp\left(-au^3\right)
$$

and $u = r_{ij}/\left(\alpha_i \alpha_j\right)^{1/6}$ where r_{ij} is the linear separation between sites i, j and α_i, α_j are their corresponding atomic

polarizabilities. The factor "a" is a dimensionless width parameter that determines the damping strength.

2.2 Protocol

Given the structure, net charge, and multiplicity of a molecule, all parameters can be systematically determined. The AMOEBA force field requires parameters for atomic charges, dipoles, quadrupoles, polarizabilities, damping coefficients (for high valence ions only), van der Waals diameters, and well depths. Valence parameters include force constants and equilibrium values for bond lengths, angles, and torsion force constants of up to sixfold. Figure 2 depicts an overview of the parameterization process. This procedure is implemented by the Poltype polarizable atomic typing utility.

Prior to parameterization, the *molecule with coordinates* (Fig. 2) is needed. Rotatable bonds are identified about four heavy atoms in which the second and third atoms share a single bond. Bond types can be provided in the structure given to Poltype. If not assigned, atom and bond perception are performed by taking advantage of the mechanism offered by The Open Babel Package, version 2.3.0. available at http://openbabel.sourceforge.net. Symmetric *multipoles are classified* (Fig. 2) based on an iterative algorithm to identify graph invariant indices [31, 32] based on the maximum graph theoretical distance, heavy valence, aromaticity, ring atom, atomic number, heavy bond sum, and formal charge of an atom.

Multipoles are obtained with Stone's distributed multipole analysis [33] and then refined via electrostatic potential fitting. All quantum mechanics (QM) calculations are performed with Gaussian 09 [34]. The structure is first optimized at the HF/6-31G* level. The *initial QM single-point calculation* (Fig. 2) then computes the electron density matrix using the MP2/6-311G** level of theory and basis set. Additionally, a grid of electrostatic potentials is populated from *high-level basis set single-point calculations* (Fig. 2), which may be either the MP2/6-311++G(2d, 2p) or MP2/aug-cc-pVTZ basis sets. The electrostatic potential is computed for a grid around each molecule. Grid points of four shells of increasing distance around a molecule with an offset of 1 and 0.35 Å apart are generated. The GDMA program [35] implements distributed multipole analysis [33]. It arranges multipole sites at atomic centers and analytically *assigns initial multipoles* (Fig. 2) based on the density matrix. In GMDA v2.2, "Switch 0" and "Radius H 0.65" are set to access the original DMA procedure. Atomic polarizabilities are assigned based solely on the element type of each atom. Polarization groups are partitioned between rotatable bonds. The *final multipole parameters* (Fig. 2) are further optimized by fitting to electrostatic potentials with a

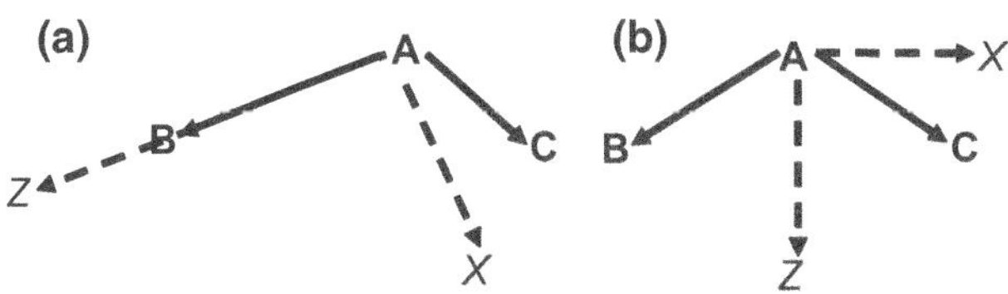

Fig. 1 Given atoms A, B, and C, the local frame is identified by the z-axis and x-axis as shown. The y-axis is defined to create a right-handed coordinate system with the existing axes

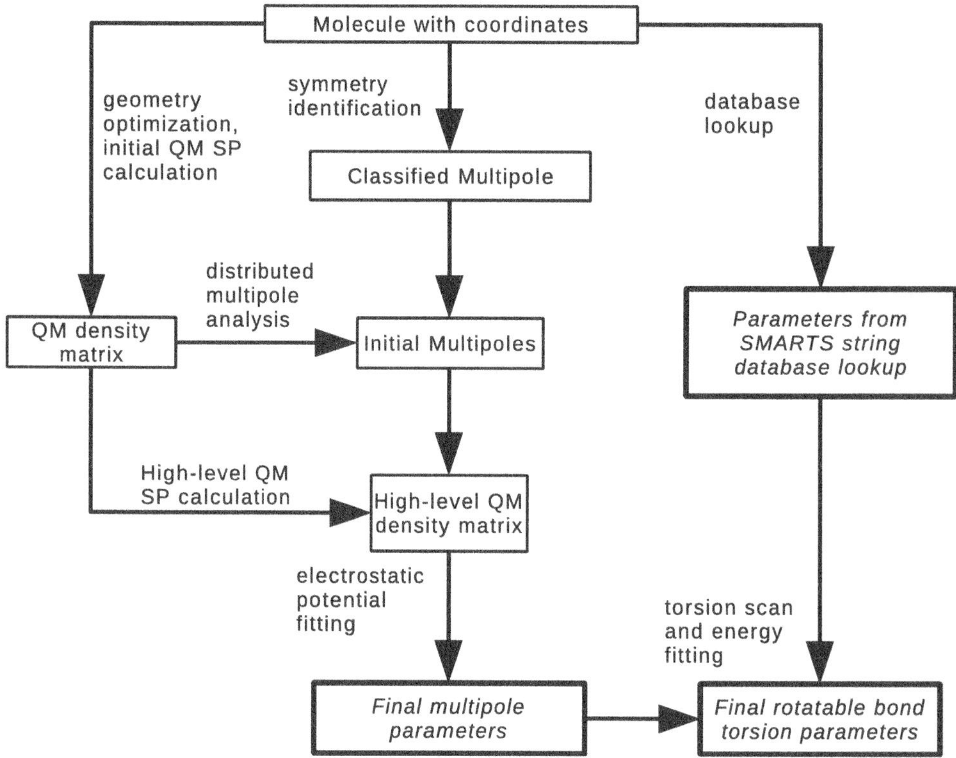

Fig. 2 Overview of the parameterization procedure for Poltype

0.1 kcal mol^{-1} electron^{-2} gradient convergence criteria. When there is intramolecular polarization, the electrostatic potential around the molecule is calculated from the permanent multipoles with fully induced dipoles added. Potential fitting and multipole assignment are currently based on modified utilities available in TINKER 5.1, and standalone versions of Poltype will be developed within the Force Field X (FFX) platform available at http://ffx. kenai.com/. In accordance with the solvation study conducted by Shi et al. [36], quadrupoles of hydroxyl groups are scaled by 60% after electrostatic potential fitting.

Diameter and well-depth values for van der Waals are assigned based on elements and their valence orbitals. A SMARTS string pattern was used to search for bond orders with its neighbors and assigned after a *database lookup* (Fig. 2). Hydrogen atoms also have a reduction factor that is based on the valence orbital of the atom to which it is bonded. Force constants for bond length, angle-bend, stretch-bend, out-of-plane bend, and torsions about non-rotatable bonds are similarly obtained from a *database lookup* (Fig. 2). Equilibrium values are taken from the QM optimized geometry.

Torsional parameters about rotatable bonds (Fig. 2) are obtained by comparing the conformational energy profile calculated from QM with the AMOEBA model that includes electrostatics, vdW, bonds, angles, parameters. A set of torsion parameters is identified by 4 atom classes that surround the rotatable bond and are composed of force constants

for each periodicity (1–6). The dihedral angle is scanned by minimizing all torsions about the rotatable bond of interest at 30° intervals with restraints. The sixth-order Fourier series is then fit to the difference between the QM conformational energy and AMOEBA's energy without the rotatable-bond torsion term. The QM conformational energy was obtained at the M06L/6-31G** level. Since torsion scanning gives 12 data points, no more than 8 parameters may be used to fit to the conformational energy profile. Torsions about the same central bond that are also in-phase are collapsed into one set of parameters for the fitting, and the contributions are distributed evenly among the parameters. Additionally, if a torsional parameter is greater than the difference between the maximum and minimum energy, then that parameter was omitted, and the rest of the parameters were fit again. However, if all parameters are removed after the magnitude test, the torsion parameters of only the atoms used to restrain the torsions were fitted. If more than one rotatable bond contains the same classes, the force constants of all classes are averaged.

3 Results and discussion

3.1 Monomeric comparisons

Quantum mechanics calculations provide molecular properties such as dipole moments, optimized structures, conformational energies, and electrostatic potentials of a grid around a molecule. The parameters of a diverse set of small organic molecules have been obtained using the Poltype polarizable atomic typing utility. A representative set of amino acid side-chain model compounds obtained from the Atlas of Protein Side-Chain Interactions [37, 38] was parameterized. Additionally, the parameters for a subset of small molecules, for which experimental hydration free energies are available [39, 40], were obtained as well. The full listing of dipole moments, electrostatic potential room mean square deviation, optimized structure, and conformation energies computed with Poltype/AMOEBA parameters and quantum mechanics are provided in the supplementary material. The molecular dipole moment of the optimized geometry computed using the Poltype/AMOEBA parameters compared with quantum mechanics calculations is shown in Fig. 3. The RMS error of all molecular dipoles is 0.16 Debye, and the correlation coefficient is 0.998. Molecules with particularly large dipole moment errors are anionic molecules such as CH3S- and C6H5S- with errors of 0.62 and 0.42 Debye, respectively. The molecular dipole moment from quantum mechanics calculations are 6.71 and 3.40 Debye, respectively. Although C6H5S- may not have a large relative error, it poses one of the largest absolute errors. The electrostatic potential (ESP) RMS difference of a grid of point charges around a molecule is 0.16 kcal/mol, and the molecules with the largest errors follow the same trend as that for dipoles. The average RMS distance between optimized geometries from Poltype/AMOEBA molecules and those optimized from quantum mechanics is 0.08 Å. For conformational energies, the correlation of all

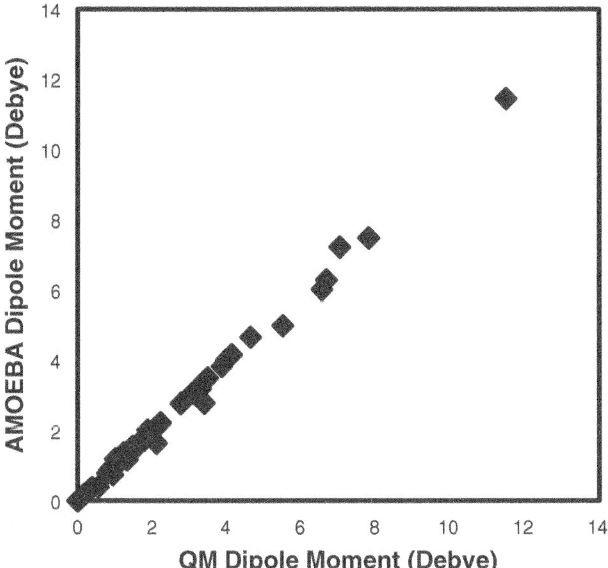

Fig. 3 Molecular dipole moment computed from AMOEBA parameters and quantum mechanical calculations

conformations prior to torsional fitting about rotatable bonds yielded a 3.6 kcal/mol RMS deviation from QM and a 0.13 correlation coefficient. After fitting, the RMS deviation decreased to 1.24 kcal/mol with a 0.91 correlation coefficient.

3.2 Dimer calculations

The packing of side chains makes significant contribution to protein stability. Investigation of interactions between side-chain model compounds is commonly used for evaluating the potential energy models. Typically, fixed-charge potential energy models are not able to accurately reproduce both gas-phase and solution-phase properties. However, AMOEBA aims to capture the energetics in different environments by including explicit polarization effects. The aforementioned Atlas of Protein Side-Chain Interactions [37, 38] were compiled by clustering interacting side chain pair conformations in 2,548 non-homologous protein structures from the Protein Data Bank. The geometry of the top cluster from the side chain pairs was used for dimer calculations. As the atlas only contains the conformation of heavy atoms, the systems were prepared [41] by adding hydrogen atoms to each model compound, and then each pair was optimized at the DFT/TZVP level. Heavy atoms were held fixed during optimization. Table 1 shows the interaction energy of the most common dimer configurations calculated with AMOEBA compared with other QM and molecular mechanics methods. Amino acids are identified by their standard abbreviations. Charged residues that are neutralized have an "(N)" designation. Interactions calculated with CCSD(T)/CBS [42–44] are considered reference energies. The OPLS-AA/L [45] and Amber parm03 [46] are energies computed with fixed-charge force fields and were taken directly from a study by Berka and coworkers [41]. Note that typically fixed-charge force fields use "enhanced" atomic charges for condensed-phase modeling such that comparison with gas-phase QM is not entirely useful. The DFT/TZVP and RI-MP2/aVTZ are quantum mechanics calculations used for comparison.

Overall, the mean relative error (MRE) and maximal relative error (MRX) of interaction energies computed with parameters from Poltype for the AMOEBA force field are lower than errors of other force fields as well as DFT, but comparable to RI-MP2 results. The mean absolute error (MAE), maximal absolute error (MAX), and root mean square error (RMS) are also lower than other molecular mechanics methods, but worse than DFT. Interestingly, the MRX of all molecular mechanical methods perform better than DFT, as the latter shows significant "relative" errors for weak associating dimers. Similarly, AMOEBA and the fixed-charge force fields yield a lower MAE compared to DFT. However, when an empirical dispersion function was

Table 1 Interaction energies (kcal/mol) for amino acid pairs (identified by their standard abbreviations) calculated using several approaches in the gas phase

Dimer[a]	CCSD(T) CBS[b]	Poltype AMOEBA	OPLS-AA/L[c]	parm03[d]	DFT TZVP	RI-MP2 aVTZ
RD	−110.80	−100.33	−105.71	−90.37	−110.60	−110.21
KE	−108.40	−104.86	−106.02	−103.57	−108.27	−107.75
DH(N)	−30.64	−28.15	−12.20	−22.36	−28.83	−30.91
D(N)H(N)	−17.97	−15.10	−10.90	−7.80	−16.26	−17.94
R(N)D(N)	−16.32	−12.38	−8.94	–	−14.71	−15.92
K(N)E(N)	−10.76	−9.54	−8.80	−9.11	−9.81	−10.65
QN	−7.37	−4.86	−8.61	−8.84	−5.66	−6.92
TT	−6.50	−7.99	−7.96	−6.83	−4.81	−6.28
YY	−4.66	−5.41	−3.84	−3.62	1.35	−5.51
TS	−4.50	−5.15	−4.38	−4.40	−3.36	−4.30
LW	−4.04	−4.16	−3.46	−3.46	1.00	−4.74
YP	−3.79	−3.83	−3.05	−3.09	0.44	−4.11
FF	−2.33	−2.41	−1.97	−2.26	1.11	−3.04
MM	−2.03	−1.95	−3.14	−2.35	1.22	−2.01
LY	−1.72	−1.72	−1.86	−1.52	0.96	−1.66
LL	−1.62	−1.60	−1.40	−1.66	0.00	−1.60
MC	−1.46	−1.28	−2.01	−1.20	0.25	−1.43
VV	−1.39	−1.52	−1.36	−1.43	0.44	−1.28
IL	−1.39	−1.36	−1.19	−1.41	0.06	−1.35
II	−1.24	−1.22	−1.13	−1.20	0.62	−1.11
LT	−1.09	−1.11	−0.91	−1.05	0.02	−1.02
VL	−1.08	−1.06	−0.81	−1.11	0.11	−1.01
AL	−1.07	−1.11	−1.00	−0.94	0.71	−0.93
LG	−0.77	−0.75	−0.75	−0.53	−0.09	−0.71
MRE [%]		8.69	19.54	13.55	83.61	6.52
MRX [%]		34.01	60.19	56.58	166.28	−30.62
MAE		1.28	2.11	2.22	2.03	0.26
MAX		10.47	18.44	20.43	6.01	−0.85
RMS		2.61	4.16	4.78	1.4	0.36

Interaction energies computed with the CCSD(T) level of theory extrapolated to the complete basis set limit (CBS) is used as the reference method. Interaction energies calculated with the AMOEBA force field parameterized with Poltype are performed as a part of this work. The DFT method was carried out with the TPSS functional and TZVP basis. The aug-cc-pVTZ basis set and resolution of identity approximation was used for the MP2 method. MRE is the unsigned mean relative error (%), MRX is the signed maximal relative error (%), MAE is the unsigned mean absolute error, MAX is the signed maximal absolute error, and RMS is the signed root mean square error

[a] Other than Poltype/AMOEBA, interaction energies were calculated by (Berka [41])

[b] Reference calculation (Tsuzuki [42]; Sinnokrot [43]; Hobza [44])

[c] Interaction energy computed using OPLS-AA/L force field (Kaminski [45])

[d] Interaction energy computed using parm03 force field (Duan [46])

incorporated in the DFT method [47, 48], the interaction energy prediction improves significantly [41]. RI-MP2 performs remarkably well in producing accurate interaction energies when compared to CCSD(T)/CBS.

The charged pairs arginine–aspartate and lysine–glutamate (RD and KE) seem to be the source of the largest absolute error for all molecular mechanics methods. However, the relative error of the RD pair was less than

10% for Poltype/AMOEBA and OPLS-AA/L with a larger error for Amber parm03. The relative error for the KE pair was less than 5% for AMOEBA and OPLS force fields. Additionally, an SAPT decomposition of the KE interaction reveals that higher-order energy beyond first-order electrostatics and repulsion and second-order induction and dispersion stabilizes the pair by about 3 kcal/mol. Conversely, higher-order energy stabilizes the RD pair by more than 6 kcal/mol and suggests that the difficulty with this pair may be due to interactions not captured by the energy function of molecular mechanics models.

Pairs with polar residues yield lower absolute error for Poltype/AMOEBA. However, it should be noted that the conformation of residues such as aspartic acid may be artificial due to the system preparation described above. Since all geometries chosen for the aspartic acid have C–O bond lengths in a narrow range between 1.24 and 1.25 Å, this geometry only corresponds to the COO–carboxylate ion [49, 50]. Typically, protonated carboxylic acid exhibits asymmetric bond lengths of 1.31 and 1.2 Å. Since geometries in the current test set are obtained from PDB structures and then minimized with heavy atoms fixed and hydrogen atoms added, pairs with artificially protonated carboxylic acid such as D(N)H(N) do not accurately describe electron distributions of charged carboxylate nor neutral carboxylic acid. Reassignment of multipoles with the artificial structure indeed yields an interaction energy of the D(N)H(N) pair that more closely matches the reference energy of the structure. The R(N)D(N) pair exhibits a similar sensitivity to geometry in which an assignment of multipoles with the given structure yields the error in Table 1, but the assignment of multipoles with a monomer-optimized structure further underestimated the interaction energy by ~ 3 kcal/mol. These examples suggest that the electron distribution of unphysical structures, particularly protonation states that are incompatible with its heavy atom conformation, cannot be captured by molecular mechanical models including AMOEBA. The parameterization of molecular mechanics models is based on minimum energy structures as it is unlikely for simple classical mechanical model to capture the complete potential energy surface especially when the structures deviate significantly from the local minima and "chemical" changes are involved. Nonetheless, dimer interaction energy calculations provide insight into the non-covalent interactions of a system and are conducive to the development of a force field. This is particularly true for AMOEBA since polarization allows parameters to be transferable between gas- and condensed-phase without the need to "pre-polarize" and scale up partial charges. Moreover, other workers [51] support the proficiency of AMOEBA in predicting interaction energies of fragment pairs decomposed from the HIV-II protease crystal structure and show improvement over other classical molecular mechanics models.

3.3 Solvation

The thermodynamic properties of molecules developed with Poltype are studied and compared with experimental values. The parameters of several families of small molecules containing functional groups in drug-like molecules

were obtained for AMOEBA using Poltype, and their hydration free energies (HFE) are computed with the Bennett acceptance ratio (BAR) [52]. In a similar procedure as a previous AMOEBA HFE study [36], perturbations of the solute required the decoupling of electrostatic and van der Waals interactions. The perturbation of electrostatic atomic multipoles and polarizabilities was scaled down linearly with $\lambda = (1.0, 0.9, 0.8, 0.7, 0.6, 0.5, 0.4, 0.3, 0.2, 0.1,$ and $0.0)$. We also scale down the radius and well depth of vdW interactions linearly with $\lambda = (1.0, 0.9, 0.8, 0.75, 0.7, 0.65, 0.6, 0.5, 0.4, 0.2,$ and $0.0)$. Molecular dynamics in solvent were carried out by placing the solute molecule at the origin of a cubic, pre-equilibrated, 28.78-Å periodic box containing 800 water molecules. The system was then equilibrated for 50 ps at 298 K. For each perturbation step, 500-ps molecular dynamics simulations were performed with 1 fs time steps and vdW cutoff of 12 Å at 298 K constant temperature using the Berendsen thermostat [53]. The long-range electrostatics for all the systems were treated using particle mesh Ewald (PME) summation [54–56]. The atomic coordinates at every 500 fs were used for post-analysis except for first 100 ps simulation. Gas-phase simulations were run on the single solute molecule for 50 ps with a time step of 0.1 ps at 298 K using a stochastic thermostat. Atomic coordinates at every 100 fs were used for post-analysis. Previously, Mobley et al. [57] conducted a study to compute HFE for a larger set of molecules with the fixed-charge general Amber force field (GAFF) [58] by assigning AM1-BCC partial charges [8, 59]. Hydration free energies with the AMOEBA force field, GAFF, and experimental results [39, 40] are listed in Table 2. Included in the table is also the free energy difference observed, while electrostatics and van der Waals interactions are perturbed. The RMS error of HFE with Poltype/AMOEBA for the set of molecules in this study is 0.75 kcal/mol. Although previous work with a larger set of molecules with GAFF yielded a lower RMS error, the error for the set in this study is 1.56 kcal/mol. When families of molecules are considered, alkenes have errors of ~ 0.6 kcal/mol, while the errors for GAFF are ~ 1 kcal/mol. Similarly, HFE predicted by Poltype/AMOEBA of nitro-containing molecules consistently yield lower errors than GAFF.

However, GAFF had errors of lower magnitude than Poltype/AMOEBA for two alkanes (22-dimethylbutane and n-octane). It should be noted that the free energy differences observed for these molecules due to van der Waals perturbations are consistent between AMOEBA and GAFF. It seems that electrostatics is too "attractive" in AMOEBA. The source of error in AMOEBA electrostatics is likely due to the ESP optimization procedure. For example, in the fitting process of n-octane, there are significantly changes in the quadrupoles of the terminal hydrogen atoms, which

Table 2 Hydration free energies (kcal/mol) of small molecules obtained from experiment, Poltype/AMOEBA, and general Amber force field (GAFF) are shown in bold. The free energy differences as a result of scaling electrostatics and van der Waals, and the errors from the experimental value for Poltype/AMOEBA and GAFF are shown as well

Molecule name	Exp[a]	Poltype/AMOEBA				GAFF[b]			
	ΔG_{exp}	ΔG_{ele}	ΔG_{vdw}	ΔG_{AMOEBA}	Error $_{AMOEBA}$	ΔG_{ele}	ΔG_{vdw}	ΔG_{GAFF}	Error $_{GAFF}$
2 Methylbut-2–ene	**1.31**	−1.78	2.51	**0.72**	−0.59	−0.55	2.83	**2.28**	0.97
But-1-ene	**1.38**	−1.65	2.48	**0.83**	−0.55	−0.37	2.85	**2.48**	1.10
1-Nitrobutane	**−3.09**	−4.50	1.95	**−2.55**	0.54	−2.43	0.92	**−1.51**	1.58
2-Nitrophenol	**−4.58**	−5.43	1.24	**−4.19**	0.39	−5.40	0.06	**−5.34**	−0.76
4-Nitrophenol	**−10.64**	−11.09	1.49	**−9.61**	1.03	−8.04	−0.18	**−8.22**	2.42
22-Dimethylbutane	**2.51**	−0.75	2.44	**1.70**	−0.81	0.01	2.52	**2.53**	0.02
n-Octane	**2.88**	−1.37	3.11	**1.74**	−1.14	0.01	3.12	**3.13**	0.25
23-Dimethylphenol	**−6.16**	−7.24	2.28	**−4.96**	1.20	−6.49	1.82	**−4.67**	1.49
Tert-butylbenzene	**−0.44**	−3.54	2.21	**−1.33**	−0.89	−2.98	2.56	**−0.42**	0.02
3-Chloropyridine	**−4.01**	−5.42	1.40	**−4.02**	−0.01	−3.78	1.28	**−2.50**	1.51
Di-n-butylamine	**−3.24**	−6.86	3.28	**−3.58**	−0.34	−4.71	3.08	**−1.63**	1.61
Di-n-propyl ether	**−1.16**	−5.41	2.71	**−2.70**	−1.54	−2.67	2.88	**0.21**	1.37
Methyl isopropyl ether	**−2.01**	−5.37	2.62	**−2.75**	−0.74	−2.89	2.14	**−0.75**	1.26
Di-n-propyl sulfide	**−1.28**	−4.16	2.35	**−1.81**	−0.53	−2.15	2.64	**0.49**	1.77
Dimethyl disulfide	**−1.83**	−3.22	1.99	**−1.23**	0.60	−0.72	2.20	**1.48**	3.31
Isobutyraldehyde	**−2.86**	−5.29	2.17	**−3.12**	−0.26	−4.98	2.05	**−2.93**	−0.07
Propionaldehyde	**−3.43**	−5.26	1.85	**−3.41**	0.02	−5.06	1.98	**−3.08**	0.35
Methanethiol	**−1.24**	−2.85	2.02	**−0.83**	0.41	−2.25	1.99	**−0.26**	0.98
n-Butanethiol	**−0.99**	−3.77	2.10	**−1.68**	−0.69	−2.39	2.27	**−0.12**	0.87
Methyl acetate	**−3.13**	−6.20	2.01	**−4.19**	−1.06	−5.44	1.71	**−3.73**	−0.60
Oct-1-yne	**0.71**	−2.40	2.94	**0.54**	−0.17	−0.83	3.29	**2.46**	1.75
Pent-1-yne	**0.01**	−2.37	2.39	**0.02**	0.01	−0.81	2.74	**1.93**	1.92
Octan-1-ol	**−4.09**	−7.87	2.68	**−5.20**	−1.11	−5.13	2.48	**−2.65**	1.44
p-Dibromobenzene	**−2.30**	−3.7	2.22	**−1.48**	0.82	−1.70	1.69	**−0.01**	2.29
Tribromomethane	**−2.13**	−3.82	2.45	**−1.37**	0.76	−0.70	1.58	**0.88**	3.01

[a] Experimental HFE (Abraham et al. [39] and Chambers et al. [40])

[b] HFE calculated from general Amber force field (Mobley [57] and Wang [58])

are shared by 6 atoms. Large deviations from those obtained from DMA may result in unphysical electrostatic parameters. Relaxing the convergence criteria of ESP fitting from 0.1 to 0.5 kcal mol^{-1} electron^{-2} gradient convergence criteria or moving the grid points away from the vdW surface may prevent unphysical multipoles resulting from the optimization. We are currently investigating this procedure especially for large linear molecules. Additionally, care must be taken when defining polarization groups. Some families such as aldehydes cannot be partitioned between carbonyl C=O and its neighboring heaving atom. When the groups are inappropriately partitioned across the bond, errors in HFE prediction increase to 1.53 and 1.55 kcal/mol for isobutyraldehyde and propionaldehyde, respectively.

Additionally, the hydration of ionic molecules is studied. For this preliminary study, a generic scaling factor for formally charged atoms was applied. For hydrogen atoms bonded to atoms with positive formal charge, their vdW diameters were scaled down by 10% from their original parameters. Conversely, the vdW diameters of atoms with a formal negative charge were scaled up by 10%. These scaling methods will be further investigated and refined. Simulations details are the same as those of neutral solutes. It should be noted, though, that difficulties arise in the comparison with experiments of homogenous ions as they are not directly accessible and must be conducted in salt solutions. The contributions of anions and cations then must be determined through various schemes such as self-consistent thermodynamic analysis [60], the TATB assumption [61], or the cluster-pair approximation [62]. The experimental hydration data that we compare with here apply the cluster-pair approximation, which is based on the correlation between ion–water clustering data and aqueous solvation free energies of neutral ion pairs. However, when comparing ion solvation quantities, differences between

Table 3 Hydration free energy (kcal/mol) of ionic molecules and their corresponding salt

Molecule	ΔG exp	ΔG salt exp	ΔG AMOEBA	ΔG salt AMOEBA
(CH3)2PH2+	−57	−131.5	−47.87	−134.37
CH3PH3+	−63	−137.5	−51.80	−138.30
(CH3)2SH+	−64.5	−139	−51.53	−138.03
CH3SH2+	−74	−148.5	−57.54	−144.04
CH3NH3+	−76.4	−150.9	−68.16	−154.66
(CH3)2NH2+	−68.6	−143.1	−63.01	−149.51
HC(OH)NH2+	−78	−152.5	−67.31	−153.81
C6H5NH3+	−72.4	−146.9	−63.15	−149.65
C5H5NH+	−58	−132.5	−52.19	−138.69
imidazoleH+	−64	−138.5	−54.45	−140.95
CH3C(OH)CH3+	−77.1	−151.6	−53.57	−140.07
(CH3)2OH+	−79.7	−154.2	−59.01	−145.51
CH3CH2OH2+	−88.4	−162.9	−68.32	−154.82
CH3OH2+	−93	−167.5	−73.75	−160.25
CH3O−	−95	−198.2	−105.67	−197.47
HCO2−	−76.2	−179.4	−91.27	−183.07
CH3S−	−73.8	−177	−85.13	−176.93
C6H5S−	−63.4	−166.6	−67.25	−159.05

experimental methods and measurements should be taken into consideration. Therefore, the comparison of an anion–cation salt pair is more appropriate than single ions alone, if the HFE of the pair has been determined in a consistent manner. Experimental hydration free energy of single ions and salt and corresponding energies using Poltype/AMOEBA are shown in Table 3. The salt HFE of an anionic molecule is taken here to be the sum of HFE of the molecule and the sodium cation [63, 64]. Similarly, the salt HFE of a cationic molecule is taken to be the sum of the HFE of the molecule and the chlorine anion [63, 64]. The correlation coefficient between salt HFE obtained from experimental and Poltype/AMOEBA is 0.95, and the unsigned mean relative error is less than 3%. Phosphor and sulfur containing cationic molecules produced salt HFE that agreed well with experiment. Some of the largest errors come from the oxonium cations. As mentioned previously, a simple scaling has been applied to all atoms with a formal negative charge or hydrogen atoms bonded to atoms with a positive charge. Further investigation of accurate ion parameters is required.

4 Conclusions

A protocol to develop AMOEBA models for small molecules has been established. In this work, we have described a standard approach to generate the AMOEBA force field for small and drug-like molecules. Although the parameterization process for the AMOEBA polarizable force field requires more sophistication compared to the process for fixed-charge force fields, a straightforward procedure is described here. Additionally, the Poltype utility allows one to generate the AMOEBA polarizable force field for a small molecule in a fully automated manner. We have shown good agreement with quantum mechanics measurements in gas phase for monomers and dimers for neutral as well as charged molecules. Furthermore, parameters obtained in gas phase are transferred directly to liquid-phase systems without modification. The hydration free energy (HFE) of neutral and charged molecules have been calculated with the Bennett acceptance ratio and compared with experimental values. The RMS error for the HFE of neutral molecules is less than 1 kcal/mol, while the unsigned mean relative error is less than 3% and a correlation coefficient is 0.95 for the HFE of salts containing charged molecules. Although some assignments such as the van der Waals diameters of atoms with formal charge of ionic molecules need further investigation, Poltype readily facilitates the systematic study of these chemical functional groups. Since the requirement to perform quantum mechanical may be computationally demanding, future work of parameterization for the AMOEBA force field would be to develop a semi-empirical method such as AM1-BCC [8, 59] to assign atomic multipoles. Although the HFE RMS error of the neutral molecules in this study demonstrate improved solvation energies than fixed-charge force fields, confirmation with a more extensive dataset is still necessary. We believe the advantage of polarizable force fields such as AMOEBA will be further illustrated in processes where environmental changes are involved.

Acknowledgments This research was supported by grants from the National Institute of General Medical Sciences (R01GM079686) and the Robert A. Welch Foundation (F-1691) to P.R.

References

1. Buckingham AD (1967) Permanent and induced molecular moments and long-range intermolecular forces. Adv Chem Phys 12:107–142
2. Jensen L, Astrand PO, Osted A, Kongsted J, Mikkelsen KV (2002) Polarizability of molecular clusters as calculated by a dipole interaction model. J Chem Phys 116(10):4001–4010
3. Gresh N, Cisneros GA, Darden TA, Piquemal JP (2007) Anisotropic, polarizable molecular mechanics studies of inter- and intramoecular interactions and ligand-macromolecule complexes. A bottom-up strategy. J Chem Theory Comput 3(6):1960–1986
4. Piquemal JP, Williams-Hubbard B, Fey N, Deeth RJ, Gresh N, Giessner-Prettre C (2003) Inclusion of the ligand field contribution in a polarizable molecular mechanics: SIBFA-LF. J Comput Chem 24(16):1963–1970

5. Rick SW, Stuart SJ, Berne BJ (1994) Dynamical fluctuating charge force-fields: application to liquid water. J Chem Phys 101(7):6141–6156

6. Xie W, Orozco M, Truhlar DG, Gao J (2009) X-Pol potential: an electronic structure-based force field for molecular dynamics simulation of a solvated protein in water. J Chem Theory Comput 5(3):459–467

7. Senn HM, Thiel W (2009) QM/MM methods for biomolecular systems. Angewandte Chemie-Int Ed 48(7):1198–1229

8. Jakalian A, Jack DB, Bayly CI (2002) Fast, efficient generation of high-quality atomic charges. AM1-BCC model: II. Parameterization and validation. J Comput Chem 23(16):1623–1641

9. Cieplak P, Dupradeau FY, Duan Y, Wang JM (2009) Polarization effects in molecular mechanical force fields. J Phys Condens Matter 21(33):333102

10. Lopes PEM, Roux B, MacKerell AD (2009) Molecular modeling and dynamics studies with explicit inclusion of electronic polarizability: theory and applications. Theor Chem Acc 124(1–2):11–28

11. Klimovich PV, Mobley DL (2010) Predicting hydration free energies using all-atom molecular dynamics simulations and multiple starting conformations. J Comput-Aided Mol Des 24(4):307–316

12. Bauer BA, Warren GL, Patel S (2009) Incorporating phase-dependent polarizability in nonadditive electrostatic models for molecular dynamics simulations of the aqueous liquid-vapor interface. J Chem Theory Comput 5(2):359–373

13. Cezard C, Bouvier B, Brenner V, Defranceschi M, Millie P, Soudan JM, Dognon JP (2004) Theoretical investigation of small alkali cation-molecule: a model potential approach. J Phys Chem B 108(4):1497–1506

14. Jiao D, King C, Grossfield A, Darden TA, Ren PY (2006) Simulation of Ca2+ and Mg2+ solvation using polarizable atomic multipole potential. J Phys Chem B 110(37):18553–18559

15. Yu HB, Whitfield TW, Harder E, Lamoureux G, Vorobyov I, Anisimov VM, MacKerell AD, Roux B (2010) Simulating monovalent and divalent ions in aqueous solution using a drude polarizable force field. J Chem Theory Comput 6(3):774–786

16. Wu JC, Piquemal JP, Chaudret R, Reinhardt P, Ren PY (2010) Polarizable molecular dynamics simulation of Zn(II) in water using the AMOEBA force field. J Chem Theory Comput 6(7):2059–2070

17. Jiao D, Golubkov PA, Darden TA, Ren P (2008) Calculation of protein-ligand binding free energy by using a polarizable potential. Proc Natl Acad Sci USA 105(17):6290–6295

18. Jiao D, Zhang JJ, Duke RE, Li GH, Schnieders MJ, Ren PY (2009) Trypsin-ligand binding free energies from explicit and implicit solvent simulations with polarizable potential. J Comput Chem 30(11):1701–1711

19. Shi Y, Jiao DA, Schnieders MJ, Ren PY (2009) Trypsin-ligand binding free energy calculation with AMOEBA. EMBC 2009 Ann Int Conf IEEE Eng Med Biol Soc 1–20:2328–2331. http://dx.crossref.org/10.1109%2FIEMBS.2009.5335108

20. Xie WS, Pu JZ, MacKerell AD, Gao JL (2007) Development of a polarizable intermolecular potential function (PIPF) for liquid amides and alkanes. J Chem Theory Comput 3(6):1878–1889

21. Wang JM, Wang W, Kollman PA, Case DA (2006) Automatic atom type and bond type perception in molecular mechanical calculations. J Mol Graphics Model 25(2):247–260

22. Moriarty NW, Grosse-Kunstleve RW, Adams PD (2009) Electronic ligand builder and optimization workbench (eLBOW): a tool for ligand coordinate and restraint generation. Acta Crystallogr Sect D Biol Crystallogr 65:1074–1080

23. Ren PY, Ponder JW (2002) Consistent treatment of inter- and intramolecular polarization in molecular mechanics calculations. J Comput Chem 23(16):1497–1506

24. Ren PY, Ponder JW (2003) Polarizable atomic multipole water model for molecular mechanics simulation. J Phys Chem B 107(24):5933–5947

25. Ponder JW, Wu CJ, Ren PY, Pande VS, Chodera JD, Schnieders MJ, Haque I, Mobley DL, Lambrecht DS, DiStasio RA, Head-Gordon M, Clark GNI, Johnson ME, Head-Gordon T (2010) Current Status of the AMOEBA polarizable force field. J Phys Chem B 114(8):2549–2564

26. Allinger NL, Yuh YH, Lii JH (1989) Molecular mechanics: the MM3 force-field for hydrocarbons.1. J Am Chem Soc 111(23):8551–8566

27. Wilson EB, Decius JC, Cross PC (1955) Molecular vibrations: the theory of infrared and raman vibrational spectra. McGraw-Hill, New York

28. Halgren TA (1992) Representation of vanderwaals (Vdw) interactions in molecular mechanics force-fields: potential form, combination rules, and Vdw parameters. J Am Chem Soc 114(20):7827–7843

29. Thole BT (1981) Molecular polarizabilities calculated with a modified dipole interaction. Chem Phys 59(3):341–350

30. Burnham CJ, Li JC, Xantheas SS, Leslie M (1999) The parametrization of a Thole-type all-atom polarizable water model from first principles and its application to the study of water clusters (n = 2–21) and the phonon spectrum of ice Ih. J Chem Phys 110(9):4566–4581

31. Weininger D (1988) Smiles, a chemical language and information-system.1. Introduction to methodology and encoding rules. J Chem Inf Comput Sci 28(1):31–36

32. Morgan HL (1965) The generation of a unique machine description for chemical structures-a technique developed at chemical abstracts service. J Chem Document 5(2):107–113

33. Stone AJ, Alderton M (1985) Distributed multipole analysis: methods and applications. Mol Phys 56(5):1047–1064

34. Frisch MJ, Trucks GW, Schlegel HB, Scuseria GE, Robb MA, Cheeseman JR, Scalmani G, Barone V, Mennucci B, Petersson GA, Nakatsuji H, Caricato M, Li X, Hratchian HP, Izmaylov AF, Bloino J, Zheng G, Sonnenberg JL, Hada M, Ehara M, Toyota K, Fukuda R, Hasegawa J, M. Ishida TN, Y. Honda, O. Kitao HN, Vreven T, J. A. Montgomery J, Peralta JE, Ogliaro F, Bearpark M, Heyd JJ, Brothers E, Kudin KN, Staroverov VN, Kobayashi R, Normand J, Raghavachari K, Rendell A, Burant JC, Iyengar SS, Tomasi J, Cossi M, Rega N, Millam JM, Klene M, Knox JE, Cross JB, Bakken V, Adamo C, Jaramillo J, Gomperts R, Stratmann RE, Yazyev O, Austin AJ, Cammi R, Pomelli C, Ochterski JW, Martin RL, Morokuma K, Zakrzewski VG, Voth GA, Salvador P, Dannenberg JJ, Dapprich S, A. D. Daniels, Farkas Ö, Foresman JB, Ortiz JV, Cioslowski J, Fox DJ (2009) Gaussian 09. Revision D.1 edn. Gaussian, Inc., Wallingford CT

35. Stone AJ (2005) Distributed multipole analysis: stability for large basis sets. J Chem Theory Comput 1(6):1128–1132

36. Shi Y, Wu CJ, Ponder JW, Ren PY (2011) Multipole electrostatics in hydration free energy calculations. J Comput Chem 32(5):967–977

37. Singh J, Thornton JM (1992) Atlas of protein side-chain interactions, vol I & II. IRL press, Oxford

38. Thornton JM (2009) http://www.ebi.ac.uk/thornton-srv/databases/sidechains/

39. Abraham MH, Whiting GS, Fuchs R, Chambers EJ (1990) Thermodynamics of solute transfer from water to hexadecane. J Chem Soc Perkin Trans 2(2):291–300

40. Chambers CC, Hawkins GD, Cramer CJ, Truhlar DG (1996) Model for aqueous solvation based on class IV atomic charges and first solvation shell effects. J Phys Chem 100(40):16385–16398

41. Berka K, Laskowski R, Riley KE, Hobza P, Vondrasek J (2009) Representative amino acid side chain interactions in proteins. A

comparison of highly accurate correlated ab initio quantum chemical and empirical potential procedures. J Chem Theory Comput 5(4):982–992

42. Tsuzuki S, Honda K, Uchimaru T, Mikami M (2005) Ab initio calculations of structures and interaction energies of toluene dimers including CCSD(T) level electron correlation correction. J Chem Phys 122(14):144323

43. Sinnokrot MO, Sherrill CD (2004) Highly accurate coupled cluster potential energy curves for the benzene dimer: Sandwich, T-shaped, and parallel-displaced configurations. J Phys Chem A 108(46):10200–10207

44. Hobza P, Sponer J (1999) Structure, energetics, and dynamics of the nucleic Acid base pairs: nonempirical ab initio calculations. Chem Rev 99(11):3247–3276

45. Kaminski GA, Friesner RA, Tirado-Rives J, Jorgensen WL (2001) Evaluation and reparametrization of the OPLS-AA force field for proteins via comparison with accurate quantum chemical calculations on peptides. J Phys Chem B 105(28):6474–6487

46. Duan Y, Wu C, Chowdhury S, Lee MC, Xiong GM, Zhang W, Yang R, Cieplak P, Luo R, Lee T, Caldwell J, Wang JM, Kollman P (2003) A point-charge force field for molecular mechanics simulations of proteins based on condensed-phase quantum mechanical calculations. J Comput Chem 24(16):1999–2012

47. Rapcewicz K, Ashcroft NW (1991) Fluctuation attraction in condensed matter: a nonlocal functional-approach. Phys Rev B 44(8):4032–4035

48. Andersson Y, Hult E, Apell P, Langreth DC, Lundqvist BI (1998) Density-functional account of van der Waals forces between parallel surfaces. Solid State Commun 106(5):235–238

49. Ahmed HU, Blakeley MP, Cianci M, Cruickshank DWJ, Hubbard JA, Helliwell JR (2007) The determination of protonation states in proteins. Acta Crystallog Sect D Biol Crystallogr 63:906–922

50. Fenn TD, Schnieders MJ, Brunger AT, Pande VS (2010) Polarizable atomic multipole X-Ray refinement: hydration geometry and application to macromolecules. Biophys J 98(12):2984–2992

51. Faver JC, Benson ML, He XA, Roberts BP, Wang B, Marshall MS, Kennedy MR, Sherrill CD, Merz KM (2011) Formal estimation of errors in computed absolute interaction energies of protein-ligand complexes. J Chem Theory Comput 7(3):790–797

52. Bennett CH (1976) Efficient estimation of free-energy differences from monte-carlo data. J Comput Phys 22(2):245–268

53. Berendsen HJC, Postma JPM, Vangunsteren WF, Dinola A, Haak JR (1984) Molecular-dynamics with coupling to an external bath. J Chem Phys 81(8):3684–3690

54. Essmann U, Perera L, Berkowitz ML, Darden T, Lee H, Pedersen LG (1995) A smooth particle mesh ewald method. J Chem Phys 103(19):8577–8593

55. Darden T, York D, Pedersen L (1993) Particle mesh ewald: an N.Log(N) method for ewald sums in large systems. J Chem Phys 98(12):10089–10092

56. Sagui C, Darden TA (1999) Molecular dynamics simulations of biomolecules: Long-range electrostatic effects. Annu Rev Biophys Biomol Struct 28:155–179

57. Mobley DL, Bayly CI, Cooper MD, Shirts MR, Dill KA (2009) Small molecule hydration free energies in explicit solvent: an extensive test of fixed-charge atomistic simulations. J Chem Theory Comput 5(2):350–358

58. Wang JM, Wolf RM, Caldwell JW, Kollman PA, Case DA (2004) Development and testing of a general amber force field. J Comput Chem 25(9):1157–1174

59. Jakalian A, Bush BL, Jack DB, Bayly CI (2000) Fast, efficient generation of high-quality atomic Charges. AM1-BCC model: I. Method. J Comput Chem 21(2):132–146

60. Schmid R, Miah AM, Sapunov VN (2000) A new table of the thermodynamic quantities of ionic hydration: values and some applications (enthalpy-entropy compensation and Born radii). Phys Chem Chem Phys 2(1):97–102

61. Krishnan CV, Friedman HL (1970) Solvation enthalpies of various ions in water and heavy water. J Phys Chem 74(11):2356

62. Tissandier MD, Cowen KA, Feng WY, Gundlach E, Cohen MH, Earhart AD, Tuttle TR, Coe JV (1998) The proton's absolute aqueous enthalpy and Gibbs free energy of solvation from cluster ion solvation data (vol 102A, pg 7791, 1998). J Phys Chem A 102(46):9308

63. Grossfield A, Ren PY, Ponder JW (2003) Ion solvation thermodynamics from simulation with a polarizable force field. J Am Chem Soc 125(50):15671–15682

64. Kelly CP, Cramer CJ, Truhlar DG (2006) Aqueous solvation free energies of ions and ion-water clusters based on an accurate value for the absolute aqueous solvation free energy of the proton. J Phys Chem B 110(32):16066–16081

Theor Chem Acc (2012) 131:1152
DOI 10.1007/s00214-012-1152-8

REGULAR ARTICLE

How polarization damping affects ion solvation dynamics

**Elvira Guàrdia · Ausias March Calvo ·
Marco Masia**

Received: 29 April 2011 / Accepted: 23 September 2011 / Published online: 19 February 2012
© Springer-Verlag 2012

Abstract In a recent paper (Sala et al. in J Chem Phys 133:234101, 2010), static properties of chloride in water have been addressed using a polarizable force field and by adding screening functions to damp short-range electrostatic interactions. In this contribution, we further explore the impact of damping polarizable interactions on system dynamics. To this end, results from Car–Parrinello molecular dynamics simulations have been used as benchmark for assessing the impact of damping schemes on the ion solvation dynamics of chloride in water. The results are of general validity, and the methodology could be easily implemented in *all methods* used to include polarization.

Keywords Polarization · Force field · Damping · Car–Parrinello · Ion solvation dynamics

Published as part of the special collection of articles: From quantum mechanics to force fields: new methodologies for the classical simulation of complex systems.

E. Guàrdia · A. M. Calvo
Departament de Física i Enginyeria Nuclear, Universitat Politècnica de Catalunya, Campus Nord B4-B5,
08034 Barcelona, Spain
e-mail: elvira.guardia@upc.edu

A. M. Calvo
e-mail: ausias.march@upc.edu

M. Masia (✉)
Dipartimento di Chimica, Istituto Officina dei Materiali del CNR, UOS SLACS, Università degli Studi di Sassari,
Via Vienna 2, 07100 Sassari, Italy
e-mail: marco.masia@uniss.it

1 Introduction

It is widely accepted that the inclusion of polarization effects in force field molecular dynamics (MD) simulations is of high importance for studying both homogeneous and inhomogeneous systems [1–6]. The contribution of many body terms to the total interaction potential (neglected in non-polarizable empirical potentials) is known to vary with the systems, and it is not straightforward to assess it a priori [7]. It is certain, though, that a proper description of polarizable interactions represents the next milestone in force field development. In fact, even though dipolar interactions decay faster than Coulomb ones, they are responsible of interesting and non-negligible physical chemical properties; as an example, the anion surface propensity in water/air interfaces could be explained only by using polarizable force fields [8–11]. The same applies to solvation of highly charged cations: for lanthanides in water, it has been found that EXAFS and XANES spectra could be reproduced only by adding polarizability [12]. Besides these applications, strictly related to ions in water, the study of many different systems might be faced with polarizable force fields, for example, molten salts [13, 14], ionic liquids [15] and biosystems [16].

The most widespread methods to deal with polarization in MD simulations are the Drude oscillator model (or charge on spring), the fluctuating charge model and the Polarizable Point Dipoles method [5, 17, 18]. The latter has been applied in the present contribution; it should be stressed, though, that the conclusions of our study are of general validity and that they could be easily ported to any of the above-mentioned methods. The main focus of our research is to ascertain which features a polarizable force field should have in order to accurately describe the intermolecular potential. We believe that these aspects are important for developing next generation of force fields

that might be used to study, *inter alia*, heterogeneous systems in the presence of high electric fields.

Recently, many authors [19–24] have pointed out that one of the largest limitations of standard molecular-mechanics polarizable force fields lies in that the transferability of gas phase derived potentials to condensed phase is hindered by the absence of polarization exchange-coupling in classical models (since it is due to short-range electron cloud repulsion, it is also known as *Pauli effect* [25]). In 1981 Thole [26] published a seminal paper on the use of damping functions for hindering the so-called *intramolecular polarization catastrophe*, that is, the divergence of dipole moments causing the dynamics to breakdown. Although his results could be extended also to *intermolecular interactions*, only recently this approach has been used in MD simulations [27, 28]. This perspective guided one of us in pursuing a method to develop polarizable models, which could faithfully reproduce the electrostatic properties of simple systems such as ion-molecule (either water or carbon tetrachloride) dimers [17–22, 23]. The method was then extended to bulk systems [28], where we assessed the impact of damping short-range electrostatic interactions on static properties. A model system, namely a chloride ion in water, was studied by comparing force field and Car–Parrinello MD. We found that, while the structure (radial distribution function) is already well reproduced by using a low value for the polarizability (α_{Cl} = 3.2 Å^3) without damping, electrostatic properties are not. In particular we proved that, with such systems, the ion dipole moment is overestimated (broad dipole moment distribution, shifted to high values) and that the ion is characterized by a high polarization anisotropy. Finally we showed that all these quantities are well reproduced if gas-phase polarizability is used (α_{Cl} = 5.48 Å^3), together with damping functions. Recently, it has been pointed out that including short-range damping allows to reproduce thermodynamical properties of halide ions at interfaces [11]. It must be noticed that many authors suggest the use of low polarizabilities for halide in water, thus introducing implicitly the interaction with the electron density of surrounding molecules. With their approach, the polarizability at condensed phase would be smaller than that at gas phase. One weak point of this approach is that the polarization at intermediate distances is lower than expected, as highlighted by high-level quantum chemical calculations on a simple ion–water dimer [22]. On the other hand, our approach is based on the assumption that the ionic polarizability is the same as that at gas phase and that, using suitable damping functions, we account for the dynamical response of the anion to the electron density of the environment.

In this article we investigate the impact of using short-range damping functions on the dynamical properties of the systems. This is of particular interest mainly for people studying ion solvation dynamics. Indeed, these features are the focus of many experimental and theoretical studies such as in references [8, 9, 29–32, 33], to cite just the most outstanding ones. In particular the properties which are mostly observed/simulated are related to the ion (diffusion coefficients) and to its environment (rotational relaxation, exchange times and mechanism, etc). Since in the bulk, ionic and molecular polarization influences solvation shell dynamics [32–36], we focus on ion diffusion, characteristic exchange times of first shell water molecules and rotational dynamics. The manuscript is organized as follows: In Sect. 2 the computational approach used is explained, in Sect. 3 the results are discussed, and then, follow our conclusions in Sect. 4.

2 Computational details

2.1 Electrostatic damping

The total electrostatic energy of a system of charges and dipoles can be partitioned into different contributions arising from charge–charge, charge–dipole and dipole–dipole interactions plus the dipole polarization term:

$$U_{\mathrm{el}} = U_{qq} + U_{q\mu} + U_{\mu\mu} + U_{\mathrm{dip}}^{\mathrm{pol}}$$

$$= \sum_i \sum_{j > i} \left(q_i \hat{T}_{ij} q_j + \mu_i^\alpha \hat{T}_{ij}^\alpha q_j - q_i \hat{T}_{ij}^\alpha \mu_j^\alpha - \mu_i^\alpha \hat{T}_{ij}^{\alpha\beta} \mu_j^\beta \right)$$

$$+ \frac{1}{2} \sum_{i=1}^{N_\mu} \boldsymbol{\mu}_i \cdot \hat{\alpha}_i^{-1} \cdot \boldsymbol{\mu}_i, \tag{1}$$

where $\hat{\alpha}_i$ is the ith atom polarizability tensor; in the following discussion, we consider it to be isotropic, thus substituting the tensor with a scalar. In the above equation we have introduced the *electrostatic interaction tensors* $\hat{T}_{ij}$, $\hat{T}_{ij}^\alpha$ and $\hat{T}_{ij}^{\alpha\beta}$, which are useful for calculating, energies, forces and electric fields. Their functional form [28, 37] is given by:

$$\hat{T}_{ij} = [s_0(r)] \frac{1}{r} \tag{2}$$

$$\hat{T}_{ij}^\alpha = \nabla_\alpha \hat{T}_{ij} = -[s_1(r)] \frac{r_\alpha}{r^3} \tag{3}$$

$$\hat{T}_{ij}^{\alpha\beta} = \nabla_\alpha \hat{T}_{ij}^\beta = [s_2(r)] \frac{3 r_\alpha r_\beta}{r^5} - [s_1(r)] \frac{\delta_{\alpha\beta}}{r^3} \tag{4}$$

$$\hat{T}_{ij}^{\alpha\beta\gamma} = \nabla_\alpha \hat{T}_{ij}^{\beta\gamma} = -[s_3(r)] \frac{15}{r^7} r_\alpha r_\beta r_\gamma + [s_2(r)] \frac{3}{r^5} \left(r_\alpha \delta_{\beta\gamma} + r_\beta \delta_{\alpha\gamma} \right.$$

$$\left. + r_\gamma \delta_{\alpha\beta} \right), \tag{5}$$

where r_α, r_β and r_γ are the cartesian components of the vector $\mathbf{r} = \mathbf{r}_i - \mathbf{r}_j$ defining the distance (which norm is

$r = |\mathbf{r}|$) between particles i and j, and the Kronecker delta function $\delta_{\alpha\beta}$ returns 1 if $\alpha = \beta$ and 0 otherwise. The appropriate screening functions $s_n(r)$ describe the kind of interacting charges' distributions. In traditional point charge schemes the charge distribution is a delta function centered in $\mathbf{r}_i$ ($s_n(r) = 1$); some applications deal with smeared charges, and the screening functions are not unit, but rather a nonlinear function of the distance. It can be easily shown that, knowing $s_0(r)$, higher-order screening functions are recursively obtained applying

$$s_k(r) = s_{k-1}(r) - \frac{r}{2k-1}\frac{\partial}{\partial r}s_{k-1}(r). \tag{6}$$

The use of screening functions is well established both for the Ewald summation method (accounting for the periodic boundary conditions) and for electrostatic damping schemes. The latter arise naturally if one considers that the charges or dipoles are not points, but rather distributed according to same a priori assumed distribution. This is, indeed, a realistic situation when the molecules come "close enough". In the case of halide ions, it has been shown that "close enough" means ca. 4 Å [22, 23]. In this paper we studied both the exponential and the gaussian charge distributions, which have been shown to be the most promising; furthermore, they are easy to implement and they imply a negligible computational overhead (see [28] for further details).

Here we would like to stress that, although the polarizable point dipole scheme allows for more flexibility, the method could be easily extended to other schemes accounting for polarizability. In fact, the overall effect of damping functions is to screen the electric field for short-range interactions. Therefore, given that both in Drude oscillators and in fluctuating charges methods there are not dipoles, but only charge, it suffices to consider the damping of the tensors accounting for charge–charge interactions. Recently, the method has been implemented in CHARMM force field [38, 39], together with Drude oscillators.

2.2 Classical MD

Classical MD simulations were performed with an in-house program which will be let freely under request. The system is composed of 96 water molecules and one chloride anion. The size of system was chosen in order to compare with Car–Parrinello MD simulations; in fact, although the Ewald summation technique allows to deal with charged systems, according to our calculations, static and dynamical properties are slightly affected by the size of the simulation box.

The polarizable point dipole method has been used for accounting for polarization. For water we implemented the

Table 1 List of names and main features of the studied models

Model	Charge distribution	$\hat{\alpha}$ (Å^3)
A3.2-none	None	3.25
A3.2-gau	Gaussian	3.25
A4.0-gau	Gaussian	4.00
A5.5-gau	Gaussian	5.48
A3.2-exp	Exponential	3.25
A4.0-exp	Exponential	4.00
A5.5-exp	Exponential	5.48

See [28] for further details

RPOL model [40]: The charges associated with each atom reproduce the water dipole moment at gas phase. On top of that, site polarizabilities are associated with each atom, which allow to obtain a dipole moment distribution at condensed phase peaked at ca. 2.6 D. The force field parameters are divided into seven models, differing among them for the value of chloride polarizability and for the type of the assumed charge distribution; the main features of these models are resumed in Table 1. The entire set of parameters for the force fields used are given in Tables I and II of [28].

In order to accelerate the computational time, the ASPC scheme [41] has been implemented. After having equilibrated the system at 298 K for 500 ps, we have run six NVE 1 ns simulations, starting from different initial configurations.

2.3 Car–Parrinello MD

Ab initio simulations have been performed, using the Car–Parrinello (CP) [42] scheme for propagating the wavefunction and the ionic configurations as implemented in the CPMD package [43]. The BLYP density functional [44, 45] was used for the electronic structure calculations. The

Table 2 Diffusion coefficients for the chloride ion obtained from the mean square displacement (D_{msd}) and from the velocity autocorrelation function (D_{vacf})

Model	D_{msd} (10^{-9} m^2 s^{-1})	D_{vacf} (10^{-9} m^2 s^{-1})	T (K)
A3.2-none	0.67 (0.09)	0.72 (0.12)	299.1 (2.4)
A3.2-exp	0.78 (0.17)	0.86 (0.15)	297.0 (5.3)
A3.2-gau	0.83 (0.12)	0.74 (0.11)	307.0 (5.1)
A4.0-exp	1.11 (0.12)	1.10 (0.12)	305.4 (2.5)
A4.0-gau	0.87 (0.14)	0.93 (0.10)	298.6 (2.7)
A5.5-exp	1.12 (0.10)	1.22 (0.09)	295.2 (3.9)
A5.5-gau	0.78 (0.17)	0.85 (0.16)	291.5 (4.9)
CPMD	1.07	–	300.0

Values for classical MD are averaged over six NVE simulations (mean standard deviations in parenthesis). The average temperatures are also reported cutoff for the wavefunction was set to 80 Ry, the time step was set to 4 a.u., and the fictitious mass for the orbital was chosen to be 400 a.m.u. In the present study we have used dispersion-corrected atom-centered pseudopotentials (DCACPs) [46] in the Troullier–Martins [47] format for oxygen and hydrogen. Norm-conserving Goedecker pseudopotentials [48–50] have been used for chloride. Production runs of 15 ps in the microcanonical ensemble followed an NVT equilibration run of 3 ps. The initial configuration was taken from classical molecular dynamics simulations. Ionic and molecular dipole moments were computed using the Wannier center [51, 52] analysis as explained in [23, 53].

2.4 Computed properties

The trajectories from both classical and Car–Parrinello MD have been post-processed in order to compute the following dynamical properties:

1. *ion diffusion*: chloride diffusion was studied computing the ion mean square displacement $\Delta_r(t)$ and velocity autocorrelation function $C_v(t)$ as

$$\Delta_r(t) = \langle [\mathbf{r}(t) - \mathbf{r}(0)]^2 \rangle, \tag{7}$$

$$C_v(t) = \frac{\langle \mathbf{v}(t) \cdot \mathbf{v}(0) \rangle}{\langle v^2(0) \rangle}. \tag{8}$$

the diffusion coefficient was obtained by calculating the slope of the $\Delta_r(t)$ and the integral of $C_v(t)$ as

$$D_{\mathrm{msd}} = \frac{1}{6} \lim_{t \to \infty} \frac{\mathrm{d}}{\mathrm{d}t} \Delta_r(t), \tag{9}$$

$$D_{\mathrm{vacf}} = \frac{k_B T}{m} \int_0^\infty C_v(t)\mathrm{d}t. \tag{10}$$

2. *residence time of first shell molecules*: water molecules on the first shell could escape from there and then be substituted by outer molecules; this exchange process is quite fast for chloride, and the mean residence time of water molecules in the first shell could be evaluated from

$$n(t) = \left\langle \frac{1}{N_{1\mathrm{st}}} \sum_{i=1}^{N_{1\mathrm{st}}} \theta_i(t)\theta_i(0) \right\rangle, \tag{11}$$

where the sum runs over the $N_{1\mathrm{st}}$ hydration molecules present in the first shell at $t = 0$, θ_i is unity if the ith molecule is in the first shell, and it is zero otherwise. Given the high lability of the first hydration shell, $n(t)$ is evaluated by allowing first shell molecules to leave the first shell for a maximum period of time t^*. As in previous works [54], we used $t^* = \infty$. The resulting function could be fitted with a double exponential

function $\tilde{n}(t) = A \exp(-k_1 t) + (1 - A) \exp(-k_2 t)$. The integral of this function yields the characteristic residence time:

$$\tau_{1\mathrm{st}} = \int_0^\infty n(t)\mathrm{d}t, \tag{12}$$

where the integral was evaluated numerically up to $t = 5$ ps using $n(t)$ and analytically up to $t = \infty$ using $\tilde{n}(t)$.

3. *rotation of molecules in the first shell*: the study of reorientational motions has been carried out by means of the time correlation function $C_2^{\mathrm{OH}}(t)$ defined as follows:

$$C_2^{\mathrm{OH}}(t) = \langle P_2(\mathbf{u}_{\mathrm{OH}}(t) \cdot \mathbf{u}_{\mathrm{OH}}(0)) \rangle \tag{13}$$

where $\mathbf{u}_{\mathrm{OH}}$ is the unitary vector along the O–H bond of water molecules, and P_2 is the second Legendre polynomial, that is, $P_2(\cos\theta) = \frac{1}{2}(3\cos^2\theta - 1)$. This correlation function shows a "backscattering-like" minimum, known as the free rotor frequency, and then decays exponentially. The long-time decay can be fitted with a single exponential function $\tilde{C}_2^{\mathrm{OH}}(t) = A \exp(-kt)$. In order to interpret the $C_2^{\mathrm{OH}}(t)$ function, it is convenient to calculate its time integral, yielding the so-called reorientational correlation time

$$\tau_2^{\mathrm{OH}} = \int_0^\infty C_2^{\mathrm{OH}}(t)\mathrm{d}t, \tag{14}$$

which basically indicates the mean time employed by a water molecule to rotate around the O–H direction. The above integral was calculated numerically up to $t = 5$ ps using $C_2^{\mathrm{OH}}(t)$ and then analytically up to $t = \infty$ using $\tilde{C}_2^{\mathrm{OH}}(t)$.

4. *hydration shell rotation*: analogously to the previous one, the solvation shell rotation around the ion can be characterized by computing the time correlation function $C_2^{\mathrm{OCl}}(t)$ where in Eq. 13 the unitary vector along O–H bond is substituted by the unitary vector joining the chloride ion to oxygen atoms $\mathbf{u}_{\mathrm{OCl}}$. The rotational time of the solvation shell τ_2^{OCl} was evaluated as in Eq. 14.

Comparing the above dynamical properties with experimental data would require the description of zero point energy contributions with path integral techniques [55, 56]. Nevertheless, since the focus of the present contribution is on the capabilities of classical force field to reproduce dynamical properties of density functional based simulations, neither our classical nor Car–Parrinello MD simulations account for nuclear quantum effects.

3 Results and discussion

The diffusion of a solute is connected to its interaction with the solvent, particularly with the first-shell molecules. For ions in water, it has been shown that there exists a tight coupling between equilibrium and non-equilibrium effects of hydration shell exchange and ion diffusion [57]. Furthermore, in case of halide anions, a proper modeling of the interaction potential is of high importance for reproducing the hydrogen bonds, as the first hydration shell seems to be critically dependent on them. In Fig. 1 the mean square

displacement (MSD) of the ion from CP and classical MD simulations are shown. For the sake of clarity, we show the results in three panels, each containing the MSD of potential models with a fixed value of anion polarizability (the same approach is used also for the following figures). It can be noticed that the MSD obtained in CPMD simulations is not as straight as the one computed in classical simulations. This is due to the shortness of CPMD trajectories which do not allow to gather enough statistics for the MSD. A similar effect was found for the velocity auto-correlation functions (not shown here). Nevertheless, a general trend could be appreciated in the graphs: For low values of the anion polarizability, the diffusion is slower than in the case of high polarizability. In Table 2 the values for the diffusion coefficients are shown. It can be noticed that the difference between exponential and gaussian damping functions correlates well with the difference in temperatures. Nonetheless, the potentials with exponential damping seem to perform slightly better than the ones with gaussian damping.

The lower ion mobility can be explained by considering that the interaction of ion with water dipole moments is badly described when a low polarizability value is taken as reference. In the case of undamped potentials, the induced dipole moment on the ion is much higher than in ab initio calculations. Therefore, high dipolar interactions render the system more viscous. In the case of damped potentials, the dipole moment is lower than the ab initio one at all distances. Hence the Coulomb interaction prevails upon the dipolar one, causing the hydrogen bond to be stronger and to slower the system dynamics. This is confirmed by the inspection of hydration shell exchange and rotational dynamics in Figs. 2, 3 and 4, and in Table 3. The residence time correlation function of first shell molecules decays much faster for high polarizability rather than for low polarizability. In the case of the exponential damping with high anion polarizability, the correlations function overlaps almost completely with CPMD results. A similar trend is observed in the rotational times of first shell molecules around the O–H bond. This measure is intimately connected to the hydrogen bond dynamics of first shell molecules with the anion. It can be seen that the correlation functions (Fig. 3) resemble the CPMD results better for high values of the polarizability.

As mentioned in the introduction, in a previous article, we showed that a higher polarization anisotropy of the first solvation shell is associated with a low anion polarizability. This has a direct consequence on the collective motion of first shell molecules. The rotational correlation function of the Cl–O bond (Fig. 4) conveys a picture of how fast the whole solvation shell rotates around the ion. Again, only at high values of the anion polarizability, we recover the dynamical behavior of ab initio simulations.

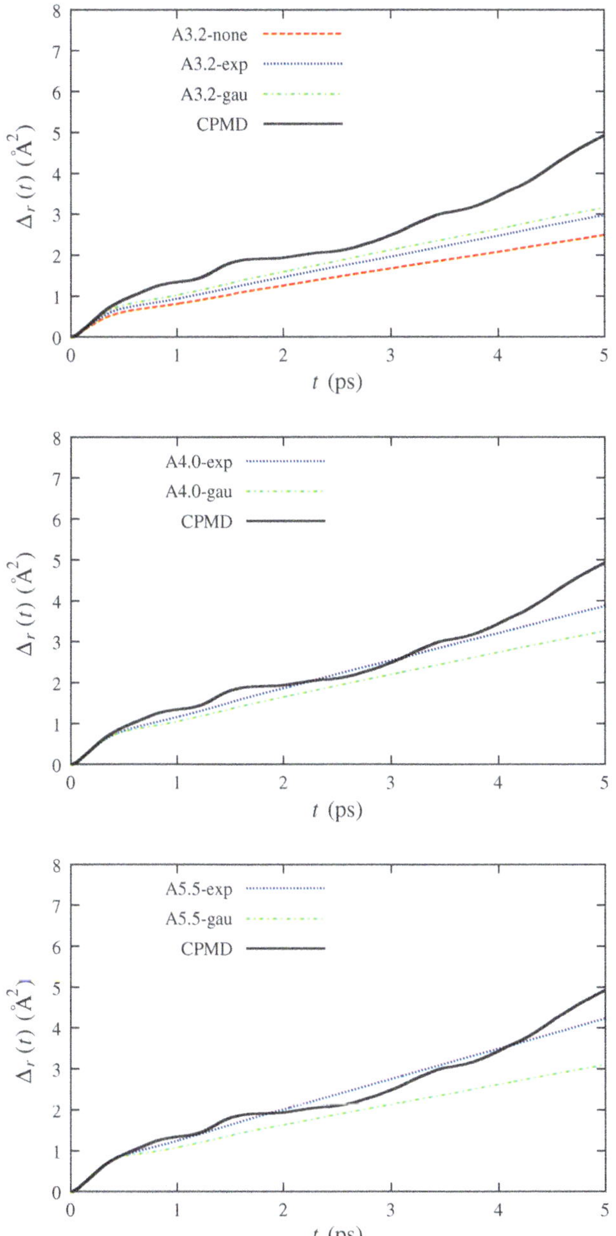

Fig. 1 Chloride mean square displacement of all models compared to Car–Parrinello results (key in the legend of each *panel*)

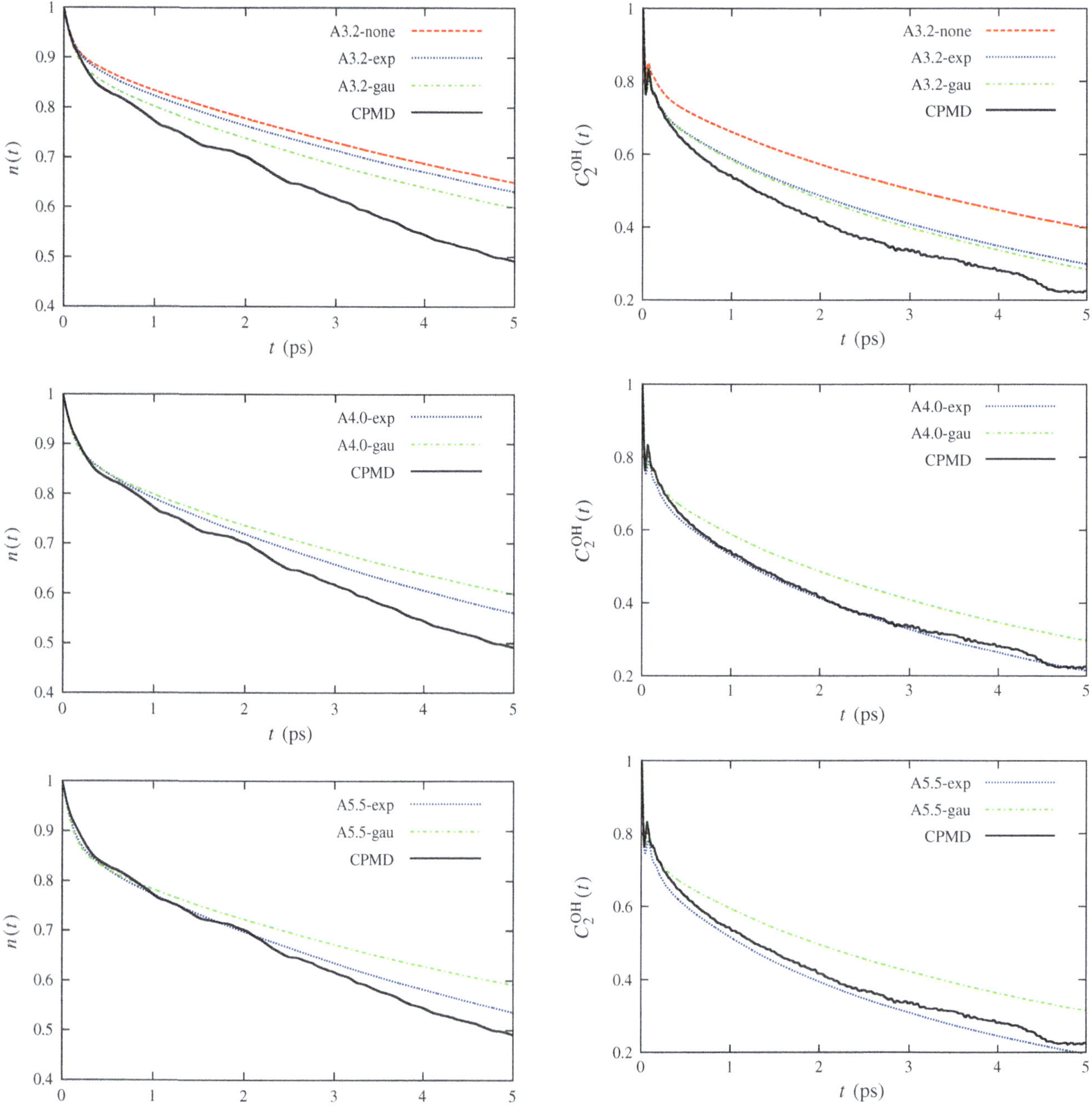

Fig. 2 First shell residence correlation function of all models compared to Car–Parrinello results (key in the legend of each *panel*)

Fig. 3 O–H reorientational correlation functions for the first shell water molecules. Results for second Legendre polynomial are shown for all models, compared to Car–Parrinello results (key in the legend of each *panel*)

From the above results we learn that the ion solvation dynamics is well reproduced by a system with high anion polarizability. Using a low polarizability seems to dampen the dynamics and to render it slower, at least for the motions studied here. Moreover it seems that the exponential damping performs slightly better than the gaussian one. Nonetheless we would like to remark that the damping parameters were not optimized for condensed-phase simulations. Thus, it is not correct to conclude that, *always*, the exponential damping performs better than the gaussian

one; in fact it is only in the case of our parameter set. Finally it can be seen that the results obtained for polarizability values equal to 4 and 5.48 $\AA^3$ do not differ very much, the main difference lying in the decay of the exchange correlation function. From this result it would be tempting to use the anion polarizability equal to 4 $\AA^3$; it must be highlighted, though, that statical properties are not as well reproduced for A4.0-gau and A4.0-exp model potentials (see [28]). Therefore, since it is highly important

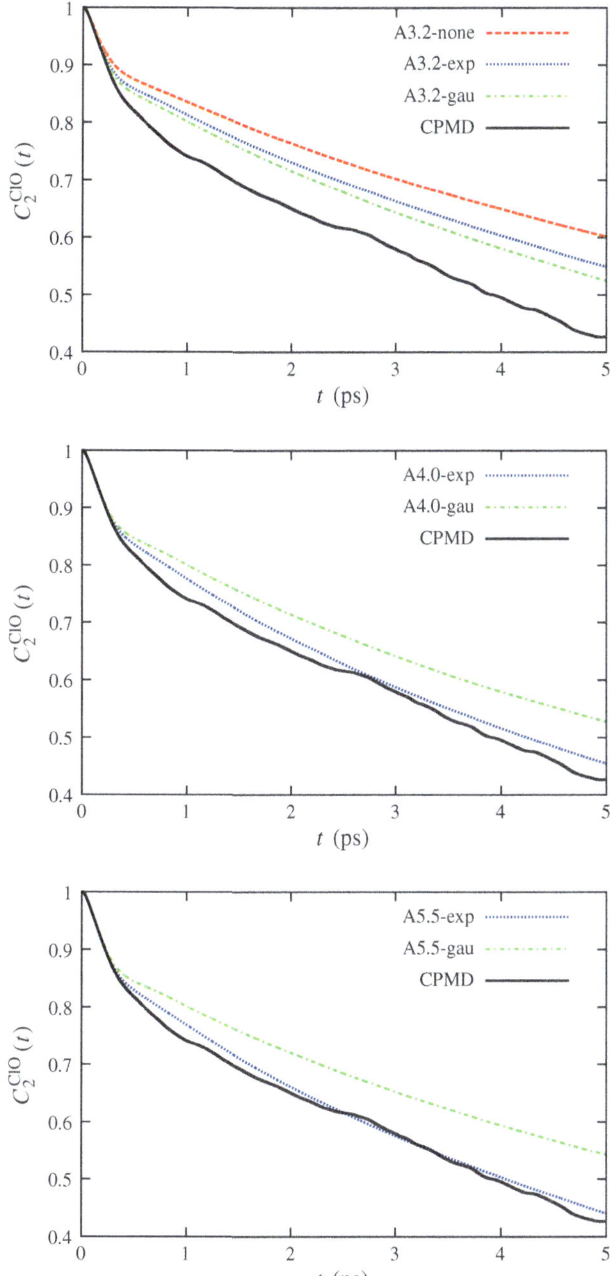

Fig. 4 O–Cl reorientational correlation functions for the first shell water molecules. Results for second Legendre polynomial are shown for all models, compared to Car–Parrinello results (key in the legend of each *panel*)

to reproduce faithfully both static and dynamical properties, we suggest the use of A5.5-gau and A5.5-exp force fields, or similar (optimized) ones with the value of the anion polarizability equal to 5.48 Å^3.

In the above discussion we have never mentioned the role played by water polarizability. The water model is certainly important when considering the solvation shell dynamics; nevertheless it should be observed that the anion shows higher fluctuations in the dipole moment (from almost 0 to 2 Debye), which are relevant in driving the system dynamics. We acknowledge, though, that a better force field for water is needed; we are currently working on this topic.

4 Conclusions and perspectives

The use of damping functions for the simulation of polarizable systems has been introduced since 1981 [26]. The main reason to include *intra-molecular* damping has been the need to hinder the polarization catastrophe for simulations with the polarizable point dipoles method. Though, until few years ago, electrostatic damping was not used to treat also *inter-molecular* interactions; probably this delay was due to the fact that the electric fields in most of the studied systems were not high enough to cause any appreciable divergence of the dipoles. In 2005, by studying a simple ion–water dimer, it was found that the damping of inter-molecular electrostatic interactions was needed to reproduce short-range effects of interacting electron densities [21]. In the case of highly polarizable ions, such as halides, it was remarked that, to reproduce both long- and short-range polarization, the damping functions should be used on top of force field where the halide had its gas-phase polarizability [22].

Recently we showed that the latter conclusion holds also for the case of chloride in bulk water; ab initio results on the static properties of this system were nicely reproduced using the same polarizability and damping functions optimized for gas-phase calculations [28].

In this contribution we showed that the same force field allows to reproduce better also dynamical properties of the ion and of its solvation shell. It seems that, in the absence of damping functions, the dynamics of chloride and of first shell molecules is slower. Therefore, the introduction of damping functions is highly important when studying dynamical properties. Both exponential and gaussian charge distributions yield good results, the former performing slightly better than the latter. It must be stressed that the force field was not parameterized using the above properties as target in any optimization procedure. Thus, it seems that the results have a broad range of validity and that the force field parameters used are portable from gas to condensed phase.

The calculation of damping functions implies a negligible computational overhead, and it can be easily ported also to other methods to include polarizability. In fact, it suffices to use the appropriate screening functions in the electrostatic interaction tensors (Eq. 1). The application to shell models at gas and condensed phase could be found in [21, 38, 39].

Table 3 Residence time, and O–H and O–Cl reorientational correlation times of the first shell molecules

Model	τ_{1st} (ps)	τ_2^{OH} (ps)	τ_2^{ClO} (ps)
A3.2-none	14.0 (1.3)	6.2 (0.5)	11.3 (0.9)
A3.2-exp	13.2 (2.0)	4.4 (0.9)	9.5 (1.7)
A3.2-gau	11.6 (0.7)	4.1 (0.5)	8.6 (0.9)
A4.0-exp	9.9 (1.3)	3.0 (0.3)	6.8 (0.9)
A4.0-gau	11.8 (1.4)	4.3 (0.5)	8.8 (1.2)
A5.5-exp	9.1 (0.5)	2.8 (0.2)	6.4 (0.4)
A5.5-gau	11.9 (1.8)	4.6 (0.7)	9.4 (1.5)
CPMD	7.5	3.0	6.0

Values for classical MD are averaged over six NVE simulations (mean standard deviations in parenthesis)

Finally we would like to address a subtle issue arising from the above conclusions. On the one hand, we showed that, using gas-phase polarizability with the appropriate damping functions allows to reproduce many static and dynamical properties compared to ab initio results. On the other hand, the use of gas-phase polarizability at condensed phase could seem nonsense; in fact, it is usually assumed that, in the bulk, the polarizability is lower than at gas phase. Whether the polarizability is an intrinsic property has still to be answered. What we have shown here is that, merely from the operative point of view, considering it an intrinsic property allows for a better description of structural and dynamical properties.

Acknowledgments The authors thankfully acknowledge the computer resources, technical expertise and assistance provided by the Barcelona Supercomputing Center - Centro Nacional de Supercomputación for the projects QCM-2009-1-0014, QCM-2008-3-0012 and QCM-2008-2-0010. The research institution INSTM is also acknowledged by M.M., who is also thankful for the resources given by the *Cybersar Project* managed by the "Consorzio COSMOLAB". E.G. acknowledges financial support from the Direcció General de Recerca de la Generalitat de Catalunya (Grant 2009SGR-1003) and from the Ministerio de Ciencia e Innovación (MICINN) of Spain (Grant FIS2009-13641-C02-01).

References

1. Jorgensen WL (2007) J Chem Theory Comput 3:1877
2. Iuchi S, Izvekov S, Voth GA (2007) J Chem Phys 126:124505
3. Piquemal JP, Perera L, Cisneros GA, Ren P, Pedersen LG, Darden TA (2006) J Chem Phys 125:054511
4. Jiao D, King C, Grossfield A, Darden TA, Ren P (2006) J Phys Chem B 110:18553
5. Rick SW, Stuart SJ (2002) Rev Comp Chem 18:89
6. Burnham CJ, Li J, Xantheas SS, Leslie M (1999) J Chem Phys 110:4566
7. Stone AJ (2008) Science 321:787
8. Jungwirth P, Tobias DJ (2006) Chem Rev 106:1259
9. Kuo IFW, Mundy CJ (2004) Science 303:658
10. Wick CD, Xantheas SS (2009) J Phys Chem B 113:4141
11. Wick CD (2009) J Chem Phys 131:084715
12. Duvail M, Vitorge P, Spezia R (2009) J Chem Phys 130:104501
13. Hull S, Keen DA, Madden PA, Wilson M (2007) J Phys Cond Matter 19:406214
14. Bitrian V, Trullas J (2008) J Phys Chem B 112:1718
15. Krekeler C, Dommert F, Schmidt J, Zhao YY, Holm C, Berger R, Delle Site L (2010) Phys Chem Chem Phys 12:1817
16. Harder E, Mackerell AD, Roux B (2009) J Am Chem Soc 131:2760
17. Masia M, Probst M, Rey R (2004) J Chem Phys 121:7362
18. Lopes PEM, Roux B, MacKerell AD (2009) Theor Chem Acc 124:11
19. Kaminski GA, Stern HA, Berne BJ, Friesner RA (2004) J Phys Chem A 108:621
20. Giese TJ, York DM (2005) J Chem Phys 123:164108
21. Masia M, Probst M, Rey R (2005) J Chem Phys 123:164505
22. Masia M, Probst M, Rey R (2006) Chem Phys Lett 420:267
23. Masia M (2008) J Chem Phys 128:184107
24. Piquemal JP, Chelli R, Procacci P, Gresh N (2007) J Phys Chem A 111:8170
25. Söderhjelm P, Öhrn A, Ryde U, Karlström G (2008) J Chem Phys 128:014102
26. Thole BT (1981) Chem Phys 59:341
27. Souaille M, Loirat H, Borgis D, Gaigeot MP (2009) Comp Phys Commun 180:276
28. Sala J, Guàrdia E, Masia M (2010) J Chem Phys 133:234101
29. Smith JD, Saykally RJ, Geissler PL (2007) J Am Chem Soc 129:13847
30. Bakker HJ (2008) Chem Rev 108:1456
31. Mallik BS, Semparithi A, Chandra A (2008) J Chem Phys 129:194512
32. Omta AW, Kropman MF, Woutersen S, Bakker HJ (2003) Science 301:347
33. Laage D, Hynes JT (2007) Proc Natl Acad Sci USA 104:11167
34. Masia M, Rey R (2005) J Chem Phys 122:094502
35. Kropman MF, Nyenhuys HK, Bakker HJ (2002) Phys Rev Lett 88:77601
36. Kropman MF, Bakker HJ (2001) Science 291:2118
37. Stone AJ (1996) The theory if intermolecular forces. Clarendon Press, Oxford
38. Lamoureux G, Roux B (2006) J Phys Chem B 110:3308
39. Harder E, Anisimov VM, Whiteld T, MacKerrell AD, Roux B (2008) J Phys Chem B 112:3509
40. Dang LX (1992) J Chem Phys 97:2659
41. Kolafa J (2004) J Comp Chem 25:335
42. Car R, Parrinello M (1985) Phys Rev Lett 55:2471
43. Copyright IBM Corp. 1990–2006, computer code CPMD version 3.11, (MPI für Festköorperforschung Stuttgart 1997–2001)
44. Becke AD (1988) Phys Rev A 38:3098
45. Lee C, Yang W, Parr RG (1988) Phys Rev B 37:785
46. Lin IC, Seitsonen AP, Coutinho-Neto MD, Tavernelli I, Rothlisberger U (2009) J Phys Chem B 13:1127
47. Troullier N, Martins JL (1991) Phys Rev B 43:1993
48. Goedecker S, Teter M, Hutter J (1996) Phys Rev B 54:1703
49. Hartwigsen C, Goedecker S, Hutter J (1998) Phys Rev B 58:3641
50. Krack M (2005) Theor Chem Acc 114:145
51. Marzari N, Vanderbilt D (1997) Phys Rev B 56:12847
52. Silvestrelli PL, Parrinello M (1999) Phys Rev Lett 82:3308; erratum: (1999) Phys Rev Lett 82:5415
53. Guàrdia E, Skarmoutsos I, Masia M (2009) J Chem Theory Comput 5:1449
54. Guàrdia E, Laria D, Martí J (2006) J Phys Chem B 110:6332
55. Paesani F, Iuchi S, Voth GA (2007) J Chem Phys 127:074506
56. Paesani F, Yoo S, Bakker HJ, Xantheas SS (2010) J Phys Chem Lett 1:2316
57. Møller KB, Rey R, Masia M, Hynes JT (2005) J Chem Phys 122:114508

Theor Chem Acc (2012) 131:1146
DOI 10.1007/s00214-012-1146-6

REGULAR ARTICLE

Achieving fast convergence of ab initio free energy perturbation calculations with the adaptive force-matching method

Eric R. Pinnick · Camilo E. Calderon ·
Andrew J. Rusnak · Feng Wang

Received: 19 June 2011 / Accepted: 9 September 2011 / Published online: 19 February 2012
© Springer-Verlag 2012

Abstract This paper studies the possibility of improving
the convergence of ab initio free energy perturbation (FEP)
calculations by developing customized force fields with the
adaptive force-matching (AFM) method. The ab initio FEP
method relies on a molecular mechanics (MM) potential to
sample configuration space. If the Boltzmann weight of the
MM sampling is close to that of the ab initio method, the
efficiency of ab initio FEP will be optimal. The difference
in the Boltzmann weights can be quantified by the relative
energy difference distribution (REDD). The force field
developed through AFM significantly improves the REDD
when compared with standard MM models, thus improving
the convergence of the ab initio FEP calculation. The static
dielectric constant ε_s of ice-Ih was studied with PW-91
through ab initio FEP. With a customized force field
developed through AFM, we were able to converge ε_s to
80 ± 4 with 3,600 configurations. A similar ab initio FEP
calculation with the TIP4P model would require 220 times
more configurations to achieve the same accuracy. Our
study indicates that the PW-91 functional underestimates
ice-Ih ε_s by about 20%.

Keywords Ab initio free energy perturbation ·
Adaptive force-matching · Static dielectric constant · Ice

Published as part of the special collection of articles: From quantum
mechanics to force fields: new methodologies for the classical
simulation of complex systems.

E. R. Pinnick · C. E. Calderon · A. J. Rusnak · F. Wang (✉)
Boston University, Boston, MA, USA
e-mail: fengwang@bu.edu

1 Introduction

In a sense, accurate determination of potential energy
surfaces (PESs) of molecular systems can be considered a
solved problem. With a sufficiently high level of correlation and a large enough basis set, post-Hartree–Fock (post-
HF) methods such as many-body perturbation theory [1–5],
configuration interaction [6–8], and coupled cluster (CC)
[9, 10] can, in principle, provide arbitrarily high accuracy
within the Born-Oppenheimer approximation. In practice,
however, most post-HF methods scale as high-order poly-
nomials of system size, and accurate PES determination
quickly becomes computationally intractable.

Recent advances in computational chemistry have gone
a long way toward reducing the computational cost of
post-HF methods. For example, variants of the MP2
method [2, 4, 11] such as RI-MP2 [12, 13] and LMP2
[14–16] have been made so efficient that systems with a
few hundred atoms can be described with reasonable
computational cost. However, for even larger systems or
for systems that require higher order treatment of correlation, one would have to resort to density functional
theory (DFT).

DFT provides a good trade-off between cost and accuracy [17–19]. Relying on a mean field approximation
similar to Hartree–Fock, DFT produces surprisingly high
accuracy for a wide range of problems and has become the
de facto standard to obtain PESs for systems that are
computationally intractable with more rigorous post-HF
treatments. Due to the simplicity of the DFT formalism,
near-linear scaling computational cost is a real possibility
[20–25]. The popular Kohn-Sham DFT implementations
can model systems with around one thousand atoms;
methods such as orbital-free DFT [17, 26–29] and other
variants have been applied to even larger systems.

In addition, DFT has been used to drive molecular dynamics sampling, although the time scale that can be achieved by DFT-based molecular dynamics is typically limited to tens of picoseconds. Sampling any non-trivial configuration space at nano-second time scale or longer with an electronic-structure method is likely to remain a significant challenge for years to come.

In this paper, we explore the possibility of using the recently proposed adaptive force-matching (AFM) method to parameterize a customized potential to best reproduce the PES of an electronic-structure method [30–34]. The AFM potential will be used to perform sampling on the electronic-structure-quality PES. This approach significantly extends the time scale that can be sampled with electronic-structure PES. The configurations sampled with the AFM force field will be used to calculate ensemble properties using the ab initio free energy perturbation (FEP) approach [35–38]. We demonstrate that such an approach is much more efficient than performing ab initio FEP with a non-customized molecular mechanics (MM) potential.

Philosophically, our AFM method bears some resemblance to the work by Ischtwan and Collins [39], where a potential energy surface sampled from a MM potential is successively improved by additional electronic-structure calculations. However, the Collins approach never reparametrizes the MM potential used for sampling.

In Sect. 2, we provide a brief introduction of the AFM method; in Sect. 3, we briefly review the ab initio FEP method; in Sect. 4, we demonstrate the use of an AFM force fields with ab initio FEP to calculate the static dielectric constant of ice-Ih; a summary will be provided in Sect. 5.

2 The adaptive force-matching method

The AFM method is designed to parameterize an otherwise general-purpose force field to a set of thermodynamic conditions related to the investigation of one specific problem. A typical MM force field is composed of rather simple energy expressions. These force fields are often required to be transferable so that they can model a range of molecules under different thermodynamic conditions. Due to the simplicity of the energy expressions, requiring a force field to be transferable will lead to compromises in accuracy. When the force field parameters are required to be balanced for all conditions, they cannot provide optimal accuracy for any one condition.

Because molecular interactions are dictated by complicated quantum mechanical forces between the electrons and between the electrons and nuclei, modeling these quantum interactions accurately with a transferable potential is extremely challenging. In the context of force field development, these quantum interactions are typically classified into more intuitive physical concepts. A very short and incomplete list includes the various types of molecular orbital interactions, polarization, dispersion, charge transfer, exchange-repulsion, and charge penetration [40–50]. Capturing all these quantum mechanical interactions through MM is a grand challenge that is extremely difficult to solve. It is likely that once all quantum mechanical interactions are represented appropriately, a truly transferable and accurate MM force field may be as computationally demanding as an electronic-structure method.

Fortunately, modeling any one specific problem requires only a small variety of molecules to be simulated under a limited number of thermodynamic states. The philosophy of AFM is to generate problem-specific force fields. AFM is designed to adapt an otherwise general-purpose force field to answer one specific question with the best possible accuracy. Within the limitation of simple force field expressions, AFM achieves better accuracy by sacrificing transferability. For example, if the melting temperature of ice-Ih is to be determined, the AFM force field will be required to describe water interactions around the melting point in both the liquid and the solid phases. Being able to describe water interactions in a hydrophobic environment or at higher temperatures is irrelevant for ice melting and therefore not necessary for that force field to model.

When designing a problem-specific force field, the answer to the question being addressed should not be fitted. In fact, no experimental properties are fitted in the AFM approach. Our published works only fit atomic forces and molecular forces and torques from electronic-structure calculations. AFM uses an iterative mechanism to create a training set that is most representative of the target configuration space. AFM also provides a mechanism to adjust the cost of the electronic-structure calculations allowing the least expensive electronic-structure method for a desired accuracy to be used [30]. Previous realizations of AFM utilize the singular value decomposition (SVD) method to obtain parameters [51, 52]. SVD guarantees the global minimum of the objective function to be identified but is restricted to the use of linear parameters in the force field. Fortunately, most standard force field terms can be fit using just linear parameters.

Starting from an initial guess force field, a typical AFM study is composed of three steps:

The sampling step: In the sampling step, the guess force field is used to sample the configuration space most relevant to the question under investigation. If the configuration space is difficult to transverse, an enhanced sampling method can be applied to ensure proper sampling of all important basins in free energy. If the force field is expected to model several state points, an independent sampling can be performed for each state point. Configurations from all the state points will be fed to the next step.

The ab initio step: Taking configurations from the sampling step, quantum mechanics (QM) or QM/MM calculations will be performed to obtain reference forces for subsequent fitting. Published work from our group has mostly focused on using QM/MM methods in the ab initio step [30–32, 34]. The use of QM/MM allows post-HF methods to be used for the QM region. The accuracy of the QM forces can be judged by performing convergence tests with regard to the size of the QM region, the level of theory, and the basis set [31]. Convergence tests can provide information for minimizing the computational cost of QM/MM calculations while maintaining the desired accuracy. The maximum accuracy achievable by the force field can be established by calculating the root mean square error (RMSE) of the fit in the fitting step. Since the maximum accuracy is frequently bottlenecked by the quality of the energy expressions, performing QM/MM calculations with a quality much higher than that provided by the energy expressions is not necessary.

The fitting step: After QM forces are obtained, the force field will be parameterized to best reproduce the QM forces. This can be accomplished with a two-step process by minimizing the molecular force RMSE,

$$\chi^2_{\mathrm{mole}} = \sum_i w_i \left(F_i^{\mathrm{ref}} - F_i^{\mathrm{fit}}\right)^2 + \sum_i u_i \left(\tau_i^{\mathrm{ref}} - \tau_i^{\mathrm{fit}}\right)^2$$
$$+ \lambda \sum_\mu \left(\sum_v n_v q_{\mu v}\right), \tag{1}$$

in the first step and the atomic force RMSE,

$$\chi^2_{\mathrm{atom}} = \sum_m s_m \left[f_{\mathrm{intra},m}^{\mathrm{fit}} + f_{\mathrm{mole},m}^{\mathrm{fit}} - f_i^{\mathrm{ref}}\right]^2, \tag{2}$$

in the second step.

In the first step, only intermolecular parameters are being determined. In the last term of Eq. 1, q_{uv} is the product of charges on atoms u and v; this term is included to ensure charge neutrality when SVD is used to fit Eq. 1 [31, 53]. Since SVD can only fit linear parameters, only products of charges can be fit. The partial charges are solved using a system of nonlinear equations from the knowledge of charge products. w_i and u_i in Eq. 1 are relative weights, and we typically use

$$w_i = \frac{s_i}{\sum_i s_i |F_i^{\mathrm{ref}}|^2}, \tag{3}$$

and

$$u_i = \frac{s_i}{\sum_i s_i |\tau_i^{\mathrm{ref}}|^2}, \tag{4}$$

where s_i is the solvation parameter chosen to be one for molecules buried in the QM region and zero for molecules at the interface [32]. The solvation parameters are used to remove QM/MM interface molecules from the fit; the forces on QM molecules at interface are influenced by the presence of nearby MM particles and are not a good reflection of the true QM PES. The sum of square forces and square torques is used in Eqs. 3 and 4 to make the molecular force and torque terms in Eq. 1 dimensionless.

In the second step, the intramolecular parameters are optimized to make the sum of the intramolecular forces, $f_{\mathrm{intra},m}^{\mathrm{fit}}$, and the intermolecular forces, $f_{\mathrm{mole},m}^{\mathrm{fit}}$, best reproduce the reference QM forces f_i^{ref}. $f_{\mathrm{mole},m}^{\mathrm{fit}}$ is calculated using the intermolecular parameters obtained in the first step, and s_{m} is the solvation parameter.

We prefer a two-step fitting procedure, because it is more important for a force field to provide a good description of intermolecular interactions when compared with intramolecular interactions. A two-step fitting procedure should not be considered to be set in stone. One could imagine all parameters to be fit in one step with appropriate weights applied to intermolecular and intramolecular terms. For protein molecules, one could consider fitting inter-residue forces and intra-residue forces separately.

If possible, we prefer to use the SVD method to fit the force field parameters. As mentioned previously, the SVD method guarantees the global minimum of the objective function to be obtained. On the other hand, SVD requires the forces to be linear functions of the parameters. To date, we have been able to fit harmonic, quartic, and Morse bond terms, harmonic and quartic angle terms, harmonic cosine dihedral terms, Lennard-Jones interactions, and a hydrogen-bond terms with SVD [30–34]. The only term we fit nonlinearly is the exponential parameter in a Buckingham potential.

After the fitting step, the forces produced by the new force field will better represent the QM forces. With the improved accuracy, the new force field will lead to better sampling and will improve the quality of the QM/MM calculations in the ab initio step. When QM/MM calculations are performed in the ab initio step, the improved MM parameters will provide a better representation of the true electrostatic environment for the QM region through Coulombic embedding.

Each iteration of these three steps will be referred to as a generation of AFM. Several generations of AFM will be performed until convergence is reached. Typically, additional generations are run even after convergence to allow all the QM forces from all the converged generations to be fit together in a global fit. The objective of the global fit is to reduce the error bar on the final parameters.

3 Ab initio free energy perturbation method

The free energy perturbation (FEP) method was introduced by Zwanzig [54] in 1954. Assuming an ensemble of configurations governed by a Hamiltonian H_1 is sampled, the FEP approach allows thermodynamic properties of a different ensemble governed by Hamiltonian H_2 to be determined by reweighing the configurations sampled with H_1. Although the FEP approach is most frequently used to calculate the free energy of a different Hamiltonian [55, 56], as outlined in the original Zwanzig paper, it is applicable to any thermodynamic property that can be determined from the partition function. One popular application of FEP is the ab initio FEP approach [35–38]. In ab initio FEP, the H_1 for sampling is an MM force field. Thermodynamic properties associated with an electronic-structure Hamiltonian H_2 can be calculated by reweighting each configuration in the MM partition function by $\exp[(U_F - U_A)/k_B T]$, where U_A is the configuration energy according to the electronic-structure method and U_F is the configuration energy of the force field [57].

Due to the exponential factor in the weight, FEP will not converge with any practical sample size unless the difference between the two Hamiltonians is sufficiently small. The difference between the two Hamiltonians can be judged by studying the distribution of $U_F - U_A$. When the MM force field exactly reproduces the QM PES, the distribution of $U_F - U_A$ will be a δ function. Thus, all the exponential weights will be identical to $\exp[(\Delta U_0)/k_B T]$, where ΔU_0 is the difference between the zeros of energies for the MM and electronic-structure PES. It can be easily shown that for the calculation of any property using ab initio FEP, the result does not depend on ΔU_0. When the distribution of $U_F - U_A$ is calculated, we choose the medians of the distributions of both the QM and MM energies to be zero, thus making $\Delta U_0 = 0$. We will refer to the distribution of $U_F - U_A$ as the FEP relative energy difference distribution (REDD) in the subsequent discussions.

When ab initio FEP is applied, an off-the-shelf MM force field is frequently used without checking the REDD. In this work, we propose to use AFM to fit an MM potential to ab initio forces. We demonstrate that the force field fit by AFM provides a much better description of the QM PES. The standard deviation of the REDD is much narrower when an AFM force field is used when compared with off-the-shelf potentials. The significantly narrower distribution allows FEP to converge exponentially faster. In addition, other methods designed to reduce the error of ab initio FEP can be used together with a customized AFM force field [58].

4 Calculation of the static dielectric constant of ice-Ih with ab initio FEP and AFM

In order to demonstrate the effectiveness of combining an AFM force field with ab initio FEP, we decided to calculate the static dielectric constant ε_s of ice-Ih. The calculation of ε_s for ice is notoriously challenging even with an MM potential [59–63]. Although the dielectric constant associated with electronic polarization in ice-Ih has been determined using DFT [64], to the best of our knowledge, ε_s for ice-Ih has not been calculated with an electronic-structure method.

The ε_s for ice-Ih is larger than the ε_s of liquid water. The large ε_s for ice-Ih can be attributed to hydrogen-bond rearrangements under an external electric field. The structure of ice-Ih satisfies the ice rules: each water molecule in ice will form four hydrogen bonds with its first-solvation-shell water molecules. The water will serve as a hydrogen acceptor in two of the hydrogen bonds and a hydrogen donor in the other two. For the infinite bulk phase of ice-Ih, there are infinite hydrogen-bond arrangements that satisfy the ice rules. The many possible hydrogen-bond arrangements give rise to the Pauling entropy. While Pauling assumed that all the hydrogen-bond arrangements are degenerate, the configuration energies of various proton arrangements are not identical. In order to determine ε_s for ice-Ih, these hydrogen-bond arrangements have to be sampled.

Below the melting temperature, the relaxation time of the hydrogen-bond arrangements is much larger than typical time scale accessible to molecular dynamics simulations. While ε_s for ice-Ih has been determined with Monte-Carlo, it has only recently been calculated in the framework of molecular dynamics with the electrostatic switching method [60]. However, direct application of the electrostatic switching method in electronic-structure-based MD is not feasible. In this work, we calculate DFT-level ε_s of ice-Ih through the ab initio FEP approach by developing a force field for FEP with the AFM method. We choose the Perdew–Wang 91 (PW-91) [65] exchange-correlation functional to provide the electronic-structure energy and forces under Kohn–Sham DFT [66].

Before the ab initio FEP calculation was performed, a customized force field was developed to achieve the best sampling for the ab initio FEP procedure. In order to develop such a problem-specific force field, the sampling step in AFM was performed at simulation conditions identical to those needed for the ab initio FEP study. In both cases, we used an orthorhombic simulation box with the lattice vectors $a = b = 9.06$ Å, $c = 14.72$ Å. The angle between lattice vectors a and b is $120°$. The supercell contains 32 molecules with a volume of $1{,}040.96$ Å^3 (Fig. 1). The sampling was performed with the electrostatic switching method. All other parameters for the electrostatic

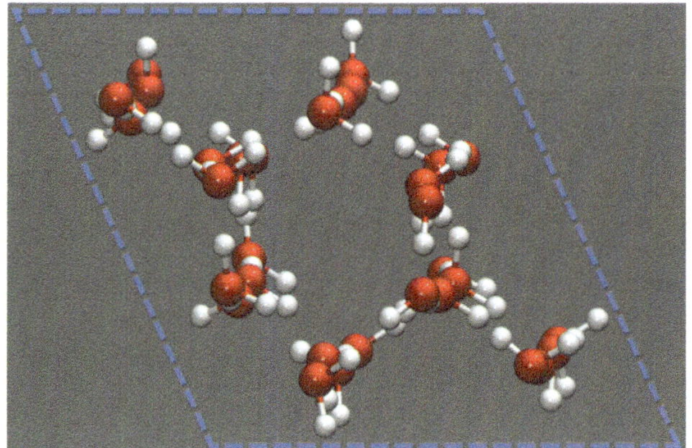

Fig. 1 An ice configuration with 32 water molecules in orthorhombic periodic *box* viewed from the *top*. The *dotted line* indicates simulation *box* boundaries

switching calculation were taken from our previous work [30]. The sampling was performed at 253.15 K.

A total of 300 configurations were generated for each generation of AFM. As mentioned previously, the PW-91 exchange-correlation functional was used to calculate reference forces. Since DFT is fairly inexpensive, we are able to model the whole box with DFT instead of relying on QM/MM.

The reference forces were obtained using the Vienna Ab-initio Simulation Package (VASP) [67–70], along with the generalized gradient approximation PW-91 exchange-correlation functional [65]. The core electrons are approximated with the Vanderbilt ultrasoft pseudo-potentials [71]. A plane wave basis set with a kinetic energy cutoff of 500 eV was used. The electronic energies are required to converge to 10^{-5} eV before the self-consistent iterations were allowed to terminate. Atomic forces were calculated with the Hellman-Feynman theorem [72]. Integration of the electronic distribution at the Fermi level was facilitated by the use of first order Methfessel–Paxton smearing [73] with a smearing width of 0.005 eV. Sampling of the first Brillouin zone was performed using a discrete $3 \times 3 \times 2$ Monkhorst–Pack k-point mesh [74].

The energy expression for the AFM force field is the 4-point flexible model published previously [30], with charges only on hydrogen atoms and the off-atom virtual site. The importance of each term in this force field has been established with AFM. Only the terms that most significantly reduce the force RMSE were included [30, 32]. This force field contains 14 parameters; only 12 of the 14 parameters were optimized in this work. The r_c and M-site a parameters were taken from our previous work and not reoptimized.

Due to the use of periodic boundary conditions, our current force-matching program does not support fitting Columbic terms with SVD when the Ewald summation

method [75, 76] is used to model long-range electrostatics. Therefore, in this work, we used the simplex method in the fitting step [77]. Since simplex is a nonlinear optimization algorithm, we do not need to fit charge products as done previously. The atomic partial charges were fit directly. Since charge products were not being fit, the last term of Eq. 1 was omitted.

In order to avoid being trapped in a local minimum, three random trial moves are performed after simplex converges[1] by perturbing each of the converged force field parameters. This is accomplished by multiplying each parameter by a random number from a uniform distribution between 0.9 and 1.1, inclusive. Additional simplex optimizations are performed starting from these perturbed initial guesses. If the additional simplex optimizations find a better minimum, the above process is repeated from the best minimum. The simplex iterations will terminate if no new minimum can be found after the random perturbations. We will refer to this variant of simplex method as simplex with random perturbations. The initial guess for the very first simplex optimization is always the parameters from the previous generation of AFM. In our studies, the perturbed guesses occasionally lead to a lower minimum. The optimal RMSE obtained by the simplex with random perturbation is in the same range as the RMSEs produced previously with SVD. Thus, we have good reasons to believe the true global minimum for the objective functions can be obtained with our method.

The ice-Ih ε_s was calculated by the fluctuation-dissipation theorem with the formula

$$\varepsilon_s = \varepsilon_0 + \frac{4\pi}{3Vk_\mathrm{B}T}\left(\langle M^2 \rangle - \langle M \rangle^2\right) \tag{5}$$

where V is the volume and M is the dipole moment of the box. The electronic contribution to the dielectric constant of 1.78 was used as ε_0 [64]. For determination of the PW91 ε_s, the dipole fluctuation and average dipole moments were determined following the standard FEP method using

$$\langle M \rangle_\mathrm{A} = \frac{\left\langle M_i \cdot e^{-(U_\mathrm{A}-U_\mathrm{F})/k_\mathrm{B}T} \right\rangle_\mathrm{F}}{\left\langle e^{-(U_\mathrm{A}-U_\mathrm{F})/k_\mathrm{B}T} \right\rangle_\mathrm{F}}, \tag{6}$$

and

$$\langle M^2 \rangle_\mathrm{A} = \frac{\left\langle M_i^2 \cdot e^{-(U_\mathrm{A}-U_\mathrm{F})/k_\mathrm{B}T} \right\rangle_\mathrm{F}}{\left\langle e^{-(U_\mathrm{A}-U_\mathrm{F})/k_\mathrm{B}T} \right\rangle_\mathrm{F}}, \tag{7}$$

---

[1] It is well known that the simplex algorithm occasionally suffers from false termination and should always be restarted from the converged parameters to make sure no further optimization is possible. In this work, we will not consider a simplex algorithm converged unless such a restart has been performed to verify that the final parameters are indeed a minimum.

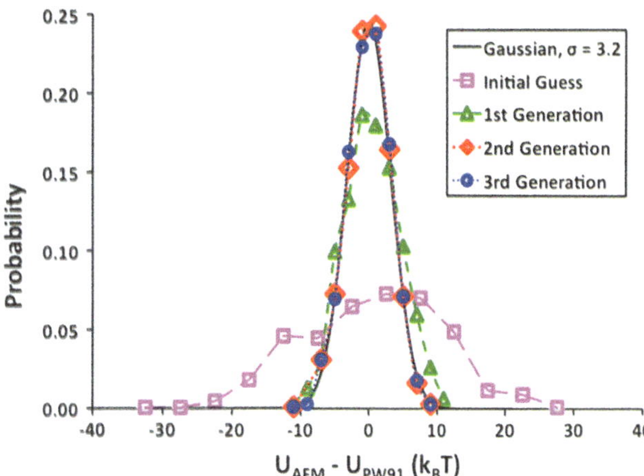

Fig. 2 Relative energy difference distribution for each generation of AFM. The medians of PW-91 and force field energies are set to be zero. A number of 300 configurations from each generation were used to produce this figure. The final generation was fit to a Gaussian distribution (*black curve*) with a standard deviation of 3.2 $k_B T$. The area *below* each curve was normalized to one

Table 1 Parameters for the PW91-Ice$_f$ force field

Parameter	Value	Parameter	Value
Q_M (e)	−1.484	r_e (Å)	0.964
q_H (e)	0.742	k_2 (kcal/(mol Å^2))	1,170
A_{OO} (kcal/mol)	239,000	k_3 (kcal/(mol Å^3))	−4,540
α (Å^{-1})	4.06	k_4 (kcal/(mol Å^4))	7,610
C_{OO} (kcal Å^6/mol)	1,020	θ_e (°)	106.73
A_4 (kcal Å^4/mol)	82.5	k_θ (kcal/(mol rad^2))	81.06

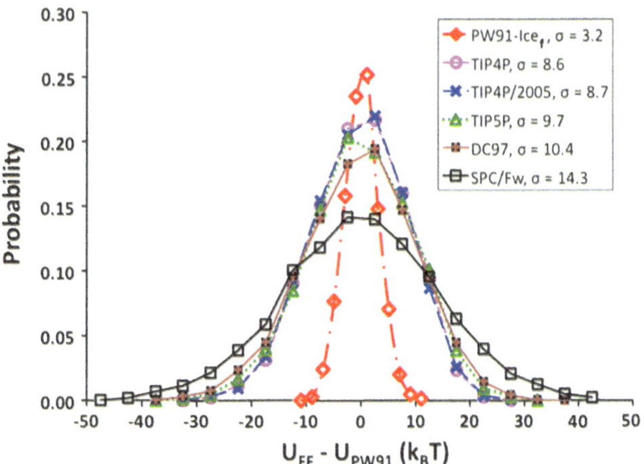

Fig. 3 Relative energy difference distribution of the PW91-ice$_f$ force field developed with AFM and several off-the-shelf models. The medians of PW-91 energies (U_{PW91}) and force field energies (U_{FF}) are set to be zero. A number of 3,600 configurations sampled with PW91-ice$_f$ are used for each force field. The standard deviation, σ, in unit of $k_B T$ is reported with the legend. The area *below* each curve was normalized to one

such that the ab initio dipole moment M of configuration i is reweighed according to the potential energy difference $U_A - U_F$, where U_F is the force field energy and U_A is the DFT energy.

The DFT dipole moment was determined using the Bader's atoms-in-molecules approach [78] via a modified version of the *Bader* module [79–81]. Both the partial charge contribution and the first moment contribution to the dipole moment were calculated [82, 83]. The molecular dipole moment determined by this method does not depend on the partitioning between different atoms of the same molecule and is rigorously correct as long as the Bader surfaces between molecules are properly determined [82, 84]. Our calculation with a single water in a periodic box produced a dipole moment of 1.84 D following this approach, in excellent agreement with the experimental dipole moment of water [85].

The initial guess for AFM was the BLYP-SP$_f$ model published by Wang et al [34]. Figure 2 plots the REDD for each generation of AFM. We note that only the PW-91 forces were fitted in AFM; the relative energies were only used to calculate the REDD and not fitted. As mentioned previously, the medians of the energies were set to zero. If the force field created by AFM correctly reproduces the PW-91 relative energies, the distribution in Fig. 2 will be a δ function centered at zero. It is clear from Fig. 2 that standard deviation of the REDD, the relative energy RMSE, quickly reduces for each generation of AFM and converges to 1.6 kcal/mol after generation 2. The fast convergence of AFM was observed in all previous works [30–34]. After convergence, the relative energy RMSE of 1.6 kcal/mol is 3.2 $k_B T$ at the sampling temperature of

253.15 K. This is a significant improvement over the initial guess, which gives a relative energy RMSE of 4.9 kcal/mol (9.7 $k_B T$) for the REDD.

The converged AFM force field will be referred to as PW91-ice$_f$. The parameters for this potential are summarized in Table 1. A number of 3,600 configurations are obtained with the PW91-ice$_f$ potential. The REDD for the 3,600 configurations is plotted in Fig. 3; the relative energy RMSE for these configurations is 1.59 kcal/mol (3.2 $k_B T$), in perfect agreement with that estimated with the 300 configurations. The REDDs for selected off-the-shelf models, Simple Point Charge/Flexible Wu (SPC/FW) [86], Transferable Intermolecular Potential 4 Point (TIP4P) [87], TIP4P/2005 [88], Dang-Chang 97 (DC97) [89], and Transferable Intermolecular Potential 5 Point (TIP5P) [90] are also reported in this figure. SPC/FW is a flexible 3-site model. TIP4P and TIP4P/2005 are rigid 4-site models with the TIP4P/2005 model specifically optimized to reproduce the melting temperature of ice-Ih and heat of vaporization of liquid water. DC97 is a polarizable but rigid 4-site model. TIP5P is a rigid 5-site model fit to reproduce the density maximum of water. Compared with the customized force field PW91-ice$_f$, the off-the-shelf models perform significantly worse in reproducing the PW91 relative energies. The TIP4P and TIP4P/2005 models give relative energy RMSEs of 4.3 and 4.4 kcal/mol, respectively. The DC97 model produced a relative energy RMSE of 5.2 kcal/mol. The fact that the polarizable DC97 model is slightly worse than the non-polarizable TIP4P is somewhat surprising. TIP5P produces a relative energy RMSE of 4.9 kcal/mol. The TIP5P model being slightly worse than TIP4P is not surprising. Although the ice-Ih melting temperature of TIP5P water is close to the experimental value of 273 K, it has been established that TIP5P predicts ice-Ih to be metastable. The most stable phase for the TIP5P model at 273 K is ice II [91, 92]. The three-point SPC/FW model performs significantly worse, giving a relative energy RMSE of 7.2 kcal/mol. This is in agreement with the common understanding that a 4-site model is generally better than a 3-site model at describing the solid phases of water.

As mentioned previously, the weight of each micro-state in ab initio FEP depends exponentially on the relative energy difference. The AFM model produces a standard deviation of 3.2 k_BT for the REDD, which is significantly narrower than the 8.6 k_BT standard deviation of the best off-the-shelf model tested. The convergence of the FEP calculation is thereby improved by approximately a factor of $e^{5.4} = 221$ with the AFM force field. We note the development of the AFM potential only incurred a fraction of the cost of the ab initio FEP.

The ε_s of PW-91 ice-Ih was calculated to be 80 ± 4, slightly lower than the experimental value of 99.4 [85]. The conclusion that PW-91 underestimates ice-Ih ε_s by about 20% supports the general belief that common DFT functions are insufficient for modeling the strongly hydrogen-bonded ice and water [93–95]. DFT tends to predict a liquid water structure that is too ordered under constant volume conditions [96–98]. At the same time, popular DFT functionals overestimate the ice-Ih melting temperature [99–101], underestimate liquid water density [93, 102], and overestimate ice density [103, 104].

Figure 4 reports the dipole moments of individual water molecules predicted by PW-91, the PW91-ice$_f$ force field, and the DC97 potential. The average water dipole moment is 2.55 D according to PW-91, which is also about 20% smaller than an estimate of 3.09 D based on a recent self-consistent induction model fit to experimental multiple moments and polarizability [105]. Since the ε_s, scales linearly with the square of dipole moments, it is possible that the 3.09 D estimate is too high. Given that ice-Ih is a crystal, the large variance of the water dipole moment distribution is noteworthy. As seen in Fig. 4, the most polarized water has a dipole moment of about 3.1 D, and the least polarized water only has a dipole moment close to

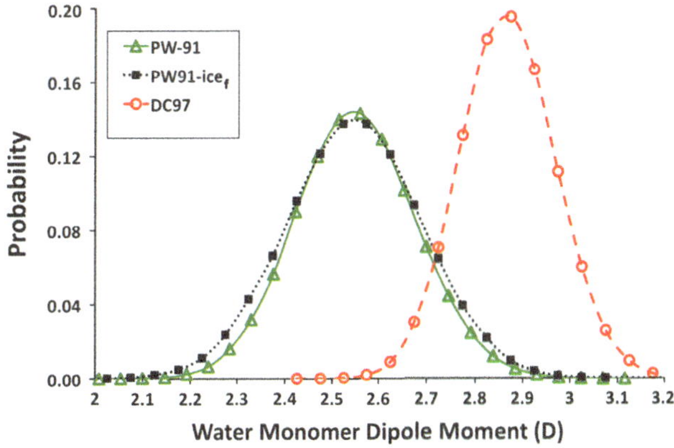

Fig. 4 Distribution of water monomer dipole moments. The area *below* the curve was normalized to one

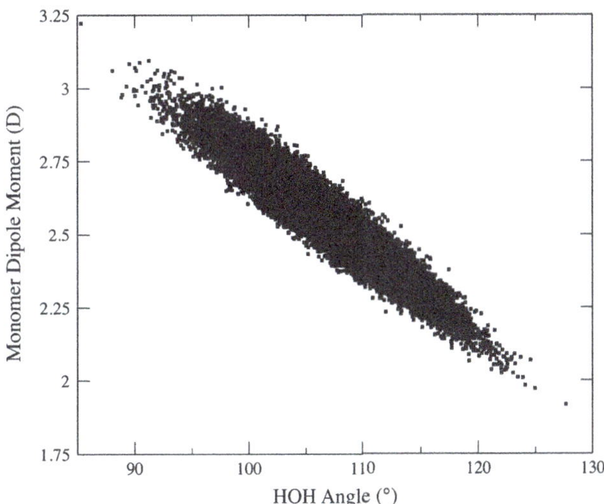

Fig. 5 *Scatter plot* showing the correlation of the dipole moment and the HOH angle for each water molecule. Each water molecule from the 3,600 ice configurations is represented as a point in this scatter plot

2.0 D. The PW91-ice$_f$ force field provides a very similar distribution of dipole moments centered around 2.55 D. This is very surprising considering PW91-ice$_f$ is not a polarizable model. Figure 5 shows a scatter plot of each water's dipole moment versus its HOH angle. Each water molecule in the 3,600 configurations is represented as a point in this figure. It is clear from Fig. 5 that the HOH angle is strongly anti-correlated with the dipole moment, allowing a flexible model to reproduce a dipole moment distribution similar to that of PW-91 waters by varying its angle.

The DC-97 model predicts a larger average dipole moment at 2.87 D but produced a smaller ε_s of 75 ± 1 at 253.15 K. This is caused by DC-97 not producing the correct ensemble weight for the various micro-states. The DC97 water model has a monomer polarizability of

1.44 $\mathring{A}^3$, which is identical to the experimental polarizability of water in gas phase. Since the PW-91 polarizability of water in gas phase was calculated to be 1.55 $\mathring{A}^3$[104], it is somewhat surprising to see the induced dipole moment for water in ice-Ih is actually lower for PW-91.

We also calculated the ice-Ih ε_s using the AFM-ice model without FEP. With the same 3,600 configurations, the ε_s was determined to be 56.4 ± 0.4. The error bar is noticeably smaller. This is due to the finite width of the REDD. The FEP weight makes the statistics poorer with the reweighting. The ε_s estimate with the PW91-ice$_f$ force field is significantly lower than the PW-91 functional.

Two contributions cause the ε_s estimated with the force field to be different from that obtained using FEP. The first contribution is due to slightly different Boltzmann weight in the equilibrium ensemble. The second contribution is from the difference in box dipole moments predicted by the PW-91 and the unpolarizable force field. We will refer to the two contributions as the Boltzmann contribution and the polarization contribution.

In order to disentangle the two contributions, the reweighted PW91-ice$_f$ ε_s is reported in Table 2. This quantity was obtained by using the force field dipole moment but reweight each configuration to have the same Boltzmann weight as the DFT method. The reweighted PW91-ice$_f$ model predicted an ε_s of 76 ± 7, which is in good agreement with the PW-91 estimate. The contribution due to polarization is thus only 4. Figure 6 reports a scatter plot where each point is one of the x, y, and z components of the dipole moment vector. The x-axis of each point is the PW-91 dipole moment component of the box and y-axis is the corresponding component of the PW91-ice$_f$ dipole moment. All of the points are very close to the diagonal line, indicating the PW91-ice$_f$ model is doing a very good job at reproducing the PW-91 box dipole moment. The RMSE of the dipole moment component is about 1.1 D, which is significantly smaller than the 13 D root mean square dipole moment for each component of the simulation box. Not surprisingly, with the DFT dipole moments and the Boltzmann weight of the force field, an ε_s of 58.3 is produced. Compared with a value of 56.4 obtained with the force field dipole moments, the polarization contribution is again negligibly small.

It is worth commenting on the small 32-water box size used in this study. Potentially a finite size effect could skew the ε_s measured with the DFT method. It has been established previously that a 128-water box predicts the same ε_s as a 300-water box with a non-polarizable model [60]. We reinvestigated the finite size effect with the DC97 polarizable model. The ε_s calculated with the 32-water box is 75 ± 1, while 300 configurations using a 300-water box give a ice-Ih ε_s of 74 ± 1 at the same temperature. This indicates that a 32-water box is indeed sufficient.

Table 2 Static dielectric constants of ice-Ih for PW-91 and MD force fields at 253.15 K

	PW-91	PW91-ice$_f$	Reweighted PW91-ice$_f$	DC97
ε_s	80 ± 4	56.4 ± 0.4	76 ± 7	75 ± 1

The experimental ε_s is 99.4

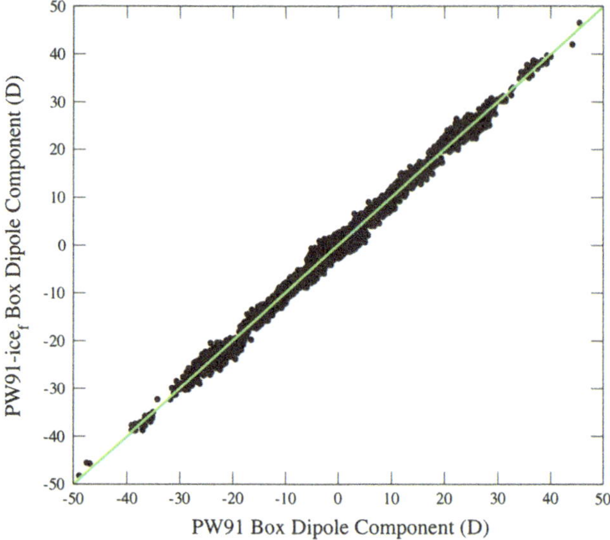

Fig. 6 *Box* dipole moment component scatter plot. Each component of the dipole moment vector is plotted separately. The y axis is dipole moment components produced by the PW91-ice$_f$ force field, the x axis is the dipole moment components calculated with PW-91. The root mean square error of the dipole moment components is 1.1 D

In our previous work, bifurcated hydrogen-bond defects were reported in ice-Ih at temperature as low as 200 K. It has been argued whether such a hydrogen-bond defect could be an artifact of the force field [60]. With the more accurate PW91-ice$_f$ model, one hydrogen-bond defect was detected from the 3,600 configurations sampled at 253.15 K, indicating such hydrogen-bond defects are likely to be real.

5 Summary

Ab initio FEP is a powerful method for obtaining electronic-structure-quality thermodynamic properties with sampling performed by inexpensive MM force fields. It is demonstrated in this work that creating a customized force field for ab initio FEP with AFM significantly accelerates the convergence of ab initio FEP, thus making ab initio FEP an even more powerful method. The PW-91 ε_s of ice-Ih was calculated through ab initio FEP. The customized force field created with AFM produced a REDD with a standard deviation of 1.6 kcal/mol. This compares favorably with a standard deviation of 4.3 kcal/mol at 253.15 K produced by the best off-the-shelf model tested, TIP4P. Such an improvement of the REDD makes sampling 220 times more efficient with the AFM force field than with TIP4P for ab initio FEP.

In this work, the number of electronic-structure calculations used for the development of the force field is a small fraction of the number of electronic-structure calculations used for converging the ab initio FEP. The savings from improved sampling decisively outweighs the cost associated with developing the AFM force field.

The PW-91 method predicts a gas phase water dipole moment of 1.84 D, in good agreement with experimental measurement. The PW-91 average water dipole moment in ice-Ih is predicted to be 2.55 D. The PW-91 ε_s for ice-Ih is determined to be 80 ± 4, underestimating the experimental estimate by about 20%. It is a pleasant surprise that the non-polarizable force field developed through AFM is able to reproduce the dipole moment distribution for the PW-91 reference method for the ice-Ih configurations. This is achieved through the flexible HOH bond.

Acknowledgments This work is supported by NSF CAREER award CHE0748628. The computer resource for this study is provided by the National Center for Supercomputing Applications under grant TG-CHE070060 and by the Boston University Center for Scientific Computing.

References

1. Head-Gordon M, Pople JA (1988) A method for two-electron Gaussian integral and integral derivative evaluation using recurrence relations. J Chem Phys 89(9):5777–5786
2. Moller C, Plesset MS (1934) Note on an approximation treatment for many-electron systems. Phys Rev 46(7):618
3. Saebo S, Almlof J (1989) Avoiding the integral storage bottleneck in LCAO calculations of electron correlation. Chem Phys Lett 154(1):83–89
4. Frisch MJ, Head-Gordon M, Pople JA (1990) A direct MP2 gradient method. Chem Phys Lett 166(3):275–280
5. Krishnan R, Pople JA (1978) Approximate fourth-order perturbation theory of the electron correlation energy. Int J Quantum Chem 14(1):91–100. doi:10.1002/qua.560140109
6. Raghavachari K, Pople JA (1981) Calculation of one-electron properties using limited configuration interaction techniques. Int J Quantum Chem 20(5):1067–1071. doi:10.1002/qua.560200503
7. Pople JA, Seeger R, Krishnan R (1977) Variational configuration interaction methods and comparison with perturbation theory. Int J Quantum Chem 12(S11):149–163. doi:10.1002/qua.560120820
8. Krishnan R, Schlegel HB, Pople JA (1980) Derivative studies in configuration—interaction theory. J Chem Phys 72(8):4654–4655
9. Cizek J (1966) On the correlation problem in atomic and molecular systems. calculation of wavefunction components in Ursell-type expansion using quantum-field theoretical methods. J Chem Phys 45(11):4256–4266
10. Kummel HG (2003) A biography of the coupled cluster method. Int J Modern Phys B 17(28):14
11. Head-Gordon M, Head-Gordon T (1994) Analytic MP2 frequencies without fifth-order storage. Theory and application to bifurcated hydrogen bonds in the water hexamer. Chem Phys Lett 220(1–2):122–128
12. Bernholdt DE, Harrison RJ (1996) Large-scale correlated electronic structure calculations: the RI-MP2 method on parallel computers. Chem Phys Lett 250(5–6):477–484
13. Feyereisen M, Fitzgerald G, Komornicki A (1993) Use of approximate integrals in ab initio theory. An application in MP2 energy calculations. Chem Phys Lett 208(5–6):359–363
14. Schutz M, Hetzer G, Werner H-J (1999) Low-order scaling local electron correlation methods. I. Linear scaling local MP2. J Chem Phys 111(13):5691–5705
15. Hetzer G, Schutz M, Stoll H, Werner H-J (2000) Low-order scaling local correlation methods II: splitting the Coulomb operator in linear scaling local second-order Moller–Plesset perturbation theory. J Chem Phys 113(21):9443–9455
16. Lee MS, Maslen PE, Head-Gordon M (2000) Closely approximating second-order M[o-slash]ller–Plesset perturbation theory with a local triatomics in molecules model. J Chem Phys 112(8):3592–3601
17. Parr RG, Yang W (1994) Density-functional theory of atoms and molecules. Oxford University Press, New York
18. Kohn W, Sham LJ (1965) Self-consistent equations including exchange and correlation effects. Phys Rev 140(4A):A1133
19. Parr RG, Yang WT (1995) Density-functional theory of the electronic-structure of molecules. Annu Rev Phys Chem 46:701–728
20. Bowler DR, Miyazaki T (2010) Calculations for millions of atoms with density functional theory: linear scaling shows its potential. J Phys Condens Matter 22(7):074207
21. Shimojo F, Kalia RK, Nakano A, Vashishta P (2005) Embedded divide-and-conquer algorithm on hierarchical real-space grids: parallel molecular dynamics simulation based on linear-scaling density functional theory. Comput Phys Commun 167(3):151–164
22. Nakano A, Kalia RK, Nomura K-I, Sharma A, Vashishta P, Shimojo F, van Duin ACT, Goddard WA, Biswas R, Srivastava D (2007) A divide-and-conquer/cellular-decomposition framework for million-to-billion atom simulations of chemical reactions. Comput Mater Sci 38(4):642–652
23. Bowler DR et al (2008) Introductory remarks: linear scaling methods. J Phys Condens Matter 20(29):290301
24. Goedecker S (1999) Linear scaling electronic structure methods. Rev Modern Phys 71(4):1085
25. Carter EA (2008) Challenges in modeling materials properties without experimental input. Science 321(5890):800–803. doi:10.1126/science.1158009
26. Smargiassi E, Madden PA (1994) Orbital-free kinetic-energy functionals for first-principles molecular dynamics. Phys Rev B 49(8):5220
27. Wang YA, Govind N, Carter EA (1999) Orbital-free kinetic-energy density functionals with a density-dependent kernel. Phys Rev B 60(24):16350
28. Gavini V, Bhattacharya K, Ortiz M (2007) Quasi-continuum orbital-free density-functional theory: a route to multi-million atom non-periodic DFT calculation. J Mech Phys Solids 55(4):697–718
29. Wang YA, Carter EA (2002) Theoretical methods in condensed phase chemistry, vol 5. Kluwer, Dordrecht
30. Akin-Ojo O, Wang F (2011) The quest for the best nonpolarizable water model from the adaptive force matching method. J Comput Chem 32(3):453–462. doi:10.1002/jcc.21634
31. Akin-Ojo O, Song Y, Wang F (2008) Developing ab initio quality force fields from condensed phase quantum-mechanics/ molecular-mechanics calculations through the adaptive force matching method. J Chem Phys 129(6):064108
32. Akin-Ojo O, Wang F (2009) Improving the point-charge description of hydrogen bonds by adaptive force matching. J Phys Chem B 113(5):1237–1240. doi:10.1021/jp809324x

33. Wei D, Song Y, Wang F (2011) A simple molecular mechanics potential for mum scale graphene simulations from the adaptive force matching method. J Chem Phys 134(18):184704

34. Wang F, Akin-Ojo O, Pinnick ER, Song Y (2011) Approaching Post-Hartree–Fock quality potential energy surfaces with simple pair-wise expressions: parameterizing point-charge based force fields for liquid water using the adaptive force matching method. Mol Simul 37:591

35. Sakane S, Yezdimer EM, Liu W, Barriocanal JA, Doren DJ, Wood RH (2000) Exploring the ab initio/classical free energy perturbation method: the hydration free energy of water. J Chem Phys 113(7):2583–2593

36. Wesolowski T, Warshel A (1994) Ab initio free energy perturbation calculations of solvation free energy using the frozen density functional approach. J Phys Chem 98(20):5183–5187. doi:10.1021/j100071a003

37. Muller RP, Warshel A (1995) Ab initio calculations of free energy barriers for chemical reactions in solution. J Phys Chem 99(49):17516–17524. doi:10.1021/j100049a009

38. Wood RH, Yezdimer EM, Sakane S, Barriocanal JA, Doren DJ (1999) Free energies of solvation with quantum mechanical interaction energies from classical mechanical simulations. J Chem Phys 110(3):1329–1337

39. Ischtwan J, Collins MA (1994) Molecular potential energy surfaces by interpolation. J Chem Phys 100(11):8080–8088

40. Piquemal J-P, Marquez A, Parisel O, Giessner-Prettre C (2005) A CSOV study of the difference between HF and DFT inter-molecular interaction energy values: the importance of the charge transfer contribution. J Comput Chem 26(10):1052–1062. doi:10.1002/jcc.20242

41. Kristyán S, Pulay P (1994) Can (semi)local density functional theory account for the London dispersion forces? Chem Phys Lett 229(3):175–180

42. Burnham CJ, Xantheas SS (2002) Development of transferable interaction models for water. III. Reparametrization of an all-atom polarizable rigid model (TTM2-R) from first principles. J Chem Phys 116(4):1500–1510

43. Basch H, Stevens WJ (1995) Hydrogen bonding between aromatics and cationic amino groups. J Mol Struct THEOCHEM 338(1–3):303–315

44. Li H, Gordon MS, Jensen JH (2006) Charge transfer interaction in the effective fragment potential method. J Chem Phys 124(21):214108

45. Reed AE, Curtiss LA, Weinhold F (1988) Intermolecular interactions from a natural bond orbital, donor-acceptor viewpoint. Chem Rev 88(6):899–926. doi:10.1021/cr00088a005

46. Murdachaew G, Mundy CJ, Schenter GK (2010) Improving the density functional theory description of water with self-consistent polarization. J Chem Phys 132(16):164102

47. Chang DT, Schenter GK, Garrett BC (2008) Self-consistent polarization neglect of diatomic differential overlap: application to water clusters. J Chem Phys 128(16):164111

48. Tsuzuki S, Luthi HP (2001) Interaction energies of van der Waals and hydrogen bonded systems calculated using density functional theory: assessing the PW91 model. J Chem Phys 114(9):3949–3957

49. Gresh N, Cisneros GA, Darden TA, Piquemal J-P (2007) Anisotropic, polarizable molecular mechanics studies of inter- and intramolecular interactions and ligand-macromolecule complexes. A bottom–up strategy. J Chem Theory Comput 3(6):1960–1986. doi:10.1021/ct700134r

50. Piquemal J-P, Chelli R, Procacci P, Gresh N (2007) Key role of the polarization anisotropy of water in modeling classical polarizable force fields. J Phys Chem A 111(33):8170–8176. doi:10.1021/jp072687g

51. Golub G, Kahan W (1965) Calculating the singular values and pseudo-inverse of a matrix. J Soc Ind Appl Math Se B Numer Anal 2(2):205–224

52. Trefethen LN, Bau D (1997) Numerical linear algebra. Society for Industrial and Applied Mathematics, Philadelphia

53. Izvekov S, Parrinello M, Burnham CJ, Voth GA (2004) Effective force fields for condensed phase systems from ab initio molecular dynamics simulation: a new method for force-matching. J Chem Phys 120(23):10896–10913

54. Zwanzig RW (1954) High-temperature equation of state by a perturbation method. I. Nonpolar gases. J Chem Phys 22(8):1420–1426

55. Dang LX, Pearlman DA, Kollman PA (1990) Why do A.T base pairs inhibit Z-DNA formation? Proc Natl Acad Sci 87(12): 4630–4634

56. Dang LX, Merz KM, Kollman PA (1989) Free energy calculations on protein stability: Thr-157 Val-157 mutation of T4 lysozyme. J Am Chem Soc 111(22):8505–8508. doi:10.1021/ja00204a027

57. Allen MP, Tildesley DJ (1999) Computer simulation of liquids. Clarendon Press, Oxford

58. Wood RH, Dong H (2011) Communication: combining non-Boltzmann sampling with free energy perturbation to calculate free energies of hydration of quantum models from a simulation of an approximate model. J Chem Phys 134(10):101101

59. Rick SW, Haymet ADJ (2003) Dielectric constant and proton order and disorder in ice Ih: Monte Carlo computer simulations. J Chem Phys 118(20):9291–9296

60. Lindberg GE, Wang F (2008) Efficient sampling of ice structures by electrostatic switching. J Phys Chem B 112(20):6436–6441. doi:10.1021/jp800736t

61. Leach AR (1996) Molecular modelling: principles and applications. Longman, Harlow

62. Rahman A, Stillinger FH (1972) Proton distribution in ice and the Kirkwood correlation factor. J Chem Phys 57(9):4009–4017

63. Aragones JL, MacDowell LG, Vega C (2010) Dielectric constant of ices and water: a lesson about water interactions. J Phys Chem A 115(23):5745–5758. doi:10.1021/jp105975c

64. Lu D, Gygi F, Galli G (2008) Dielectric properties of ice and liquid water from first-principles calculations. Phys Rev Lett 100(14):147601

65. Perdew JP, Wang Y (1992) Accurate and simple analytic representation of the electron-gas correlation energy. Phys Rev B 45(23):13244

66. Perdew JP, Chevary JA, Vosko SH, Jackson KA, Pederson MR, Singh DJ, Fiolhais C (1993) Erratum: Atoms, molecules, solids, and surfaces: applications of the generalized gradient approximation for exchange and correlation. Phys Rev B 48(7):4978

67. Kresse G, Hafner J (1993) Ab initio molecular dynamics for liquid metals. Phys Rev B 47(1):558

68. Kresse G, Furthmüller J (1996) Efficient iterative schemes for ab initio total-energy calculations using a plane-wave basis set. Phys Rev B 54(16):11169

69. Kresse G, Hafner J (1994) Ab initio molecular-dynamics simulation of the liquid-metal–amorphous-semiconductor transition in germanium. Phys Rev B 49(20):14251

70. Kresse G, Furthmüller J (1996) Efficiency of ab-initio total energy calculations for metals and semiconductors using a plane-wave basis set. Comput Mater Sci 6(1):15–50

71. Vanderbilt D (1985) Optimally smooth norm-conserving pseudopotentials. Phys Rev B 32(12):8412

72. Feynman RP (1939) Forces in molecules. Phys Rev 56(4):340

73. Methfessel M, Paxton AT (1989) High-precision sampling for Brillouin-zone integration in metals. Phys Rev B 40(6):3616

74. Monkhorst HJ, Pack JD (1976) Special points for Brillouin-zone integrations. Phys Rev B 13(12):5188

75. Darden T, York D, Pedersen L (1993) Particle mesh Ewald: an N·log(N) method for Ewald sums in large systems. J Chem Phys 98(12):10089–10092. doi:10.1063/1.464397

76. Essmann U, Perera L, Berkowitz ML, Darden T, Lee H, Pedersen LG (1995) A smooth particle mesh Ewald method. J Chem Phys 103(19):8577–8593. doi:10.1063/1.470117

77. Press WH (2007) Numerical recipes: the art of scientific computing. Cambridge University Press, New York

78. Bader RFW (1990) Atoms in molecules: a quantum theory. Clarendon Press, Oxford

79. Tang W et al (2009) A grid-based Bader analysis algorithm without lattice bias. J Phys Condens Matter 21(8):084204

80. Henkelman G, Arnaldsson A, Jûnsson H (2006) A fast and robust algorithm for Bader decomposition of charge density. Comput Mater Sci 36(3):354–360

81. Sanville E, Kenny SD, Smith R, Henkelman G (2007) Improved grid-based algorithm for Bader charge allocation. J Comput Chem 28(5):899–908. doi:10.1002/jcc.20575

82. Bader RFW, Larouche A, Gatti C, Carroll MT, MacDougall PJ, Wiberg KB (1987) Properties of atoms in molecules: dipole moments and transferability of properties. J Chem Phys 87(2):1142–1152. doi:10.1063/1.453294

83. Gatti C, Silvi B, Colonna F (1995) Dipole moment of the water molecule in the condensed phase: a periodic Hartree–Fock estimate. Chem Phys Lett 247(1–2):135–141. doi:10.1016/0009-2614(95)01190-0

84. Bader RFW, Matta CF (2001) Properties of atoms in crystals: dielectric polarization. Int J Quantum Chem 85(4–5):592–607. doi:10.1002/qua.1540

85. Haynes WM (2011) CRC handbook of chemistry and physics. Taylor and Francis, London

86. Wu Y, Tepper HL, Voth GA (2006) Flexible simple point-charge water model with improved liquid-state properties. J Chem Phys 124(2):024503. doi:10.1063/1.2136877

87. Jorgensen WL, Chandrasekhar J, Madura JD, Impey RW, Klein ML (1983) Comparison of simple potential functions for simulating liquid water. J Chem Phys 79(2):926–935

88. Abascal JLF, Vega C (2005) A general purpose model for the condensed phases of water: TIP4P/2005. J Chem Phys 123(23):234505

89. Dang LX, Chang T-M (1997) Molecular dynamics study of water clusters, liquid, and liquid–vapor interface of water with many-body potentials. J Chem Phys 106(19):8149–8159

90. Mahoney MW, Jorgensen WL (2000) A five-site model for liquid water and the reproduction of the density anomaly by rigid, nonpolarizable potential functions. J Chem Phys 112(20):8910–8922

91. Aragones JL, Noya EG, Abascal JLF, Vega C (2007) Properties of ices at 0 K: a test of water models. J Chem Phys 127(15):154518

92. Vega C, Abascal JLF, Conde MM, Aragones JL (2009) What ice can teach us about water interactions: a critical comparison of the performance of different water models. Faraday Discuss 141:251–276

93. Schmidt J, VandeVondele J, Kuo IFW, Sebastiani D, Siepmann JI, Hutter J, Mundy CJ (2009) Isobaric, isothermal molecular dynamics simulations utilizing density functional theory: an assessment of the structure and density of water at near-ambient conditions. J Phys Chem B 113(35):11959–11964. doi:10.1021/jp901990u

94. McGrath MJ, Siepmann JI, Kuo IFW, Mundy CJ, VandeVondele J, Hutter J Jr, Mohamed F, Krack M (2005) Simulating fluid-phase equilibria of water from first principles. J Phys Chem A 110(2):640–646. doi:10.1021/jp0535947

95. Sprik M, Hutter J, Parrinello M (1996) Ab initio molecular dynamics simulation of liquid water: comparison of three gradient-corrected density functionals. J Chem Phys 105(3):1142–1152

96. Lee H-S, Tuckerman ME (2006) Structure of liquid water at ambient temperature from ab initio molecular dynamics performed in the complete basis set limit. J Chem Phys 125(15):154507

97. Grossman JC, Schwegler E, Draeger EW, Gygi F, Galli G (2004) Towards an assessment of the accuracy of density functional theory for first principles simulations of water. J Chem Phys 120(1):300–311

98. Schwegler E, Grossman JC, Gygi F, Galli G (2004) Towards an assessment of the accuracy of density functional theory for first principles simulations of water. II. J Chem Phys 121(11):5400–5409

99. Yoo S, Zeng XC, Xantheas SS (2009) On the phase diagram of water with density functional theory potentials: the melting temperature of ice I_h with the Perdew–Burke–Ernzerhof and Becke–Lee–Yang–Parr functionals. J Chem Phys 130(22):221102

100. Yoo S, Xantheas SS (2011) Communication: the effect of dispersion corrections on the melting temperature of liquid water. J Chem Phys 134(12):121105

101. Schwegler E, Sharma M, Gygi F, Galli G (2008) Melting of ice under pressure. Proc Natl Acad Sci 105(39):14779–14783. doi:10.1073/pnas.0808137105

102. McGrath MJ, Siepmann JI, Kuo I-FW, Mundy CJ (2006) Vapor–liquid equilibria of water from first principles: comparison of density functionals and basis sets. Mol Phys Int J Interface Between Chem Phys 104(22):3619–3626

103. Fortes AD, Wood IG, Brodholt JP, Vocadlo L (2003) Ab initio simulation of the ice II structure. J Chem Phys 119(8):4567–4572

104. Leung K, Rempe SB (2006) Ab initio rigid water: effect on water structure, ion hydration, and thermodynamics. Phys Chem Chem Phys 8(18):2153–2162

105. Batista ER, Xantheas SS, Jonsson H (1999) Multipole moments of water molecules in clusters and ice Ih. J Chem Phys 111(13):6011–6015

Theor Chem Acc (2012) 131:1162
DOI 10.1007/s00214-012-1162-6

REGULAR ARTICLE

A coarse-grained model for β-D-glucose based on force matching

Sergiy Markutsya · Yana A. Kholod ·
Ajitha Devarajan · Theresa L. Windus ·
Mark S. Gordon · Monica H. Lamm

Received: 20 October 2011 / Accepted: 17 January 2012 / Published online: 2 March 2012
© Springer-Verlag 2012

Abstract Cellulosic ethanol production is a two-stage process that involves the hydrolysis of cellulose to form simple sugars and the fermentation of these sugars to ethanol. Hydrolysis of cellulose is the rate-limiting step, and there is a great need to characterize the process with numerical simulations to better understand the complex mechanisms involved. The ultimate goal is to generate accurate coarse-grained molecular models that are capable of predicting the structure of lignocellulose before and after pretreatment so that subsequent ab initio calculations can be performed to probe the degradation pathways. As a first step toward that goal, the force-matching method is used to derive coarse-grained models for β-D-glucose molecules in aqueous solution. Using the same reference, an all-atom molecular dynamics simulation trajectory, two sets of three- and six-site coarse-grained models of β-D-glucose are developed using two definitions of the coarse-grained center site location: center of mass (CG-CM) and geometric center (CG-GC). The performance of these coarse-grained models is evaluated by comparing the coarse-grained predictions for bond-length distributions and radial distribution functions to those obtained from the all-atom reference simulation. The six-site coarse-grained models retain more structural details than the three-site coarse-grained models. Comparison between center site definitions shows that CG-CM models generally predict local ordering better, while CG-GC models predict long-range structure better.

Keywords Coarse-grain force fields · Glucose · Glucopyranose · Molecular dynamics simulation

1 Introduction

Biomass conversion of cellulosic material, such as grasses and wood chips, is a promising route for producing renewable energy and chemical feedstocks [1]. The most important steps in the biomass-to-energy production process are acidic or enzymatic hydrolysis to break the crystalline cellulose structure into glucose molecules, and subsequent fermentation to convert glucose to ethanol as the final product [2]. The hydrolysis step is a bottleneck in this process due to the long reaction time and low efficiency. To understand how to overcome this bottleneck, accurate and detailed studies of the hydrolysis reaction are required. Numerical simulations are one tool that is able to deliver the desired level of detail for the process and also to satisfy accuracy requirements.

Molecular dynamics simulations for an all-atom crystalline cellulose system solvated with water molecules ($\sim 10^9$ atoms) is a very challenging computational task. An alternative is to design a numerical simulation of the equivalent coarse-grained system to reduce the number of degrees of freedom. Such coarse-grained models should be

Published as part of the special collection of articles: From quantum mechanics to force fields: new methodologies for the classical simulation of complex systems.

S. Markutsya · M. H. Lamm
Department of Chemical and Biological Engineering,
Iowa State University, Ames, IA, USA

S. Markutsya · Y. A. Kholod · A. Devarajan ·
T. L. Windus · M. S. Gordon (✉) · M. H. Lamm
Ames Laboratory, Ames, IA 50011, USA
e-mail: mark@si.fi.ameslab.gov

Y. A. Kholod · A. Devarajan · T. L. Windus · M. S. Gordon
Department of Chemistry, Iowa State University,
Ames, IA, USA

able to accurately represent most important physical and chemical properties of the system while significantly reducing the computational cost. Coarse-grained models are usually created by combining multiple atoms into one group and then representing this group as a single effective coarse-grained site, thereby significantly reducing the complexity of the system. There are generally many alternative choices for the coarse-graining sites, and one must make sensible choices, usually based on physical, chemical, and structural properties of the system. However, it is often difficult to predict whether a particular set of coarse-grained sites are the best equivalent representation of the corresponding all-atom system.

Once the coarse-grained sites have been chosen, the effective forces/potentials between these coarse-grained sites must be derived and several coarse-graining strategies have been applied. Energy-based coarse-graining methods [3, 4] parameterize the coarse-grained potentials to match thermodynamic properties, such as the partitioning of a species between oil and water. In other strategies, such as Boltzmann inversion (BI) [5], iterative Boltzmann inversion (IBI) [6], and inverse Monte Carlo (IMC) [6–9], the coarse-grained potentials are parameterized to reproduce the average structure observed in all-atom simulations. The force-matching [10–12] scheme takes a different approach and bases the coarse-grained model on the reference inter- and intra-atom forces. In force matching, the pairwise effective force field is derived from a trajectory obtained with all-atom molecular dynamics simulations. The original force-matching approach [10] is not suitable for high molecular weight biological molecules because of the large number of coarse-grained sites in the system. To overcome this problem, a new multiscale coarse-graining (MS-CG) approach has been developed [11, 12]. With the MS-CG approach, the force field is assumed to depend linearly on the fitting parameters and the problem is reduced to the solution of an over-determined system. This over-determined system is solved for smaller sets of all-atom configurations (subsets), and the final solution is obtained by averaging over all subsets. The coarse-grained potential obtained in this way is a potential of mean force (PMF), which is the derivative of the free energy in the phase space with reduced degrees of freedom. Thus, the coarse-grained model can only be used for the identical thermodynamic state point at which the all-atom system is simulated. In any coarse-graining approach, configurational entropy is lost as the degrees of freedom in the system are reduced. The loss of configurational entropy was quantified for hydrocarbon chains and shown to increase as the flexibility of the chain increases [13]. The configurational entropy loss is significant when the degrees of freedom are drastically reduced, such as when an implicit solvent coarse-grained model is derived for a polymer or protein. Kim and

Lamm recently demonstrated a method for restoring the lost configurational entropy to coarse-grained models of flexible macromolecules in solution when the solvent degrees of freedom have been removed [14].

A coarse-grained model for malto-oligosaccharides and their aqueous mixtures has been developed by Molinero and Goddard [15, 16] using an IMC method to derive non-bonded interactions and the BI method to extract bonded interactions. In this model, the glucose monomer is represented by three sites, and each water molecule by a single site. In their approach, the non-bonded interactions are described with two-body Morse potentials, and valence interactions between two coarse-grained sites are described as conventional bond, angle, and torsion interactions. However, the Morse function is not necessarily optimal for interactions between coarse-grained sites. Therefore, their model does not always accurately predict such structural properties of the system as the radial distribution function, bond lengths, angles, and dihedrals.

Izvekov and Voth have used their multiscale coarse-graining (MS-CG) method [11, 12] that is based on force matching to build a three-site coarse-grained model for α-D-glucose and α-(1 → 4)-D-glucan with 14 glucose units in aqueous solution [17]. The MS-CG method was used to build non-bonded interactions, whereas the bonded interactions are obtained with a Boltzmann statistical analysis of the all-atom trajectory. With this approach, they were able to reproduce many experimental structural and thermodynamical properties in the constant NPT ensemble.

A hybrid of the force-matching and Boltzmann inversion (BI) methods was used by Hynninen et al. to develop a three-site model to describe β-D-glucose, cellobiose, and cellotetraose [18]. They used a pure force-matching method for β-D-glucose and hybrid methods for cellobiose and cellotetraose in aqueous solution. In their hybrid method, the non-bonded interactions were obtained from the force-matching method, while a BI was used to fit the bond distances, angles, and dihedral parameters. Their coarse-grained model yields a good match with all-atom simulations for structural properties.

Recently, Srinivas et al. [19] have used the Boltzmann inversion method to develop a one-site model for a cellulose crystalline fibril containing 36 chains with 40 cellobiose units in each chain in aqueous solution. In addition, a coarse-grained simulation of fully amorphous cellulose fibrils was performed. The method provides an accurate and constraint-free approach to derive coarse-grained models for cellulose with a wide range of crystallinity.

To develop a coarse-grained model that can accurately represent physical and chemical properties of all-atom systems, it is important to have both an accurate, robust coarse-graining method and representative coarse-grained sites that are able to provide an acceptable level of accuracy. Coarse-grained models have been built for many systems. However, there appears to be no systematic study available on how the choice of the number and locations of coarse-grained sites might influence the predictions of a coarse-grained model. Moreover, it is difficult to predict a priori the most appropriate and effective coarse-graining parameters.

In this work, the force-matching approach [11, 12] is employed to develop a coarse-grained model of β-D-glucose in water solution as a structural unit of cellulose. For this system, two different coarse-grained mapping schemes (three-site and six-site) are investigated. In addition to varying the number of atoms in a coarse-grained site, two definitions of the center site location for the coarse-grained sites, center of mass (CM) and geometric center (GC), were considered and evaluated in detail. This provides a systematic comparative analysis of these coarse-grained mapping alternatives.

The paper is organized as follows: In Sect. 2, the force-matching method is briefly described, the coarse-grained mapping schemes are defined, and the parameters used for the molecular dynamics (MD) simulations are given. In Sect. 3, the results obtained from all-atom MD and coarse-grained MD simulations are presented and discussed. Section 4 contains the conclusions to the work.

2 Methods

2.1 Force-matching method

Coarse-grained potentials for β-D-glucose and water were derived using a multiscale coarse-graining approach based on force matching [11, 12]. A detailed description of this method can be found elsewhere [20–23]. Briefly, in the force-matching (FM) method, a coarse-grained pairwise effective force field is derived from the corresponding all-atom trajectory and force data obtained from a reference all-atom molecular dynamics simulation. For a given configuration from the reference all-atom simulation, the positions of N coarse-grained sites and the net forces $\mathbf{F}_i^{\mathrm{ref}}$ acting along them are computed. The force acting on the ith coarse-grained site due to the jth coarse-grained site is modeled as $\mathbf{f}_{ij}(\mathbf{r}_i, \mathbf{r}_j, p_1, p_2, \ldots, p_m)$ where $p_1, p_2, \ldots p_m$ are the m unknown parameters that need to be determined. Since the functional form for the force between coarse-grained sites is not known a priori, cubic splines are chosen to conveniently and systematically construct $\mathbf{f}_{ij}(\mathbf{r}_i, \mathbf{r}_j, p_1, p_2, \ldots, p_m)$ as a linear function of unknowns. Hence, the following system of N linear equations with m unknowns is obtained.

$$\sum_{j=1}^{N} \mathbf{f}_{ij}(r_i, r_j) = \mathbf{F}_i^{\mathrm{ref}}, \quad i = 1, 2, 3, \ldots, N \tag{1}$$

The system of equations is overdetermined ($N > m$), and the values for the m unknown parameters are found using a

singular value decomposition [24]. The parameters obtained in this way from each all-atom configuration are averaged over the total number of all-atom configurations sampled.

2.2 Coarse-grain mapping

Because a goal of this work is to evaluate coarse-grain mapping strategies, four different coarse-grain models for β-D-glucose were evaluated. The first two coarse-grained models consist of the same three coarse-grain sites as shown in Fig. 1 and two different definitions for the center of each coarse-grained site: center of mass (three-site CG-CM) and geometric center (three-site CG-GC). The 3-site coarse-grain mapping scheme is similar to coarse-grained models for glucose used previously [17, 18]. The advantage of this mapping is that it contains the minimum number of coarse-grain sites that preserve the anisotropic shape of glucose. The disadvantage of this mapping scheme is that it cannot represent the chair conformation that is the preferred conformation of β-D-glucose. The third and fourth coarse-grained models consist of six coarse-grain sites as shown in Fig. 2, again with two definitions for the center of each coarse-grained site: six-site CG-CM and six-site CG-GC. With this mapping scheme, the chair conformation

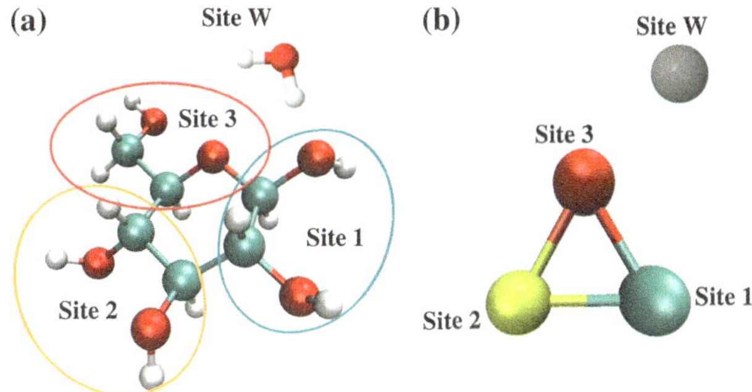

Fig. 1 **a** All-atom structures for β-D-glucose and water molecules; **b** three-site coarse-grained representation of β-D-glucose and single-site water molecules

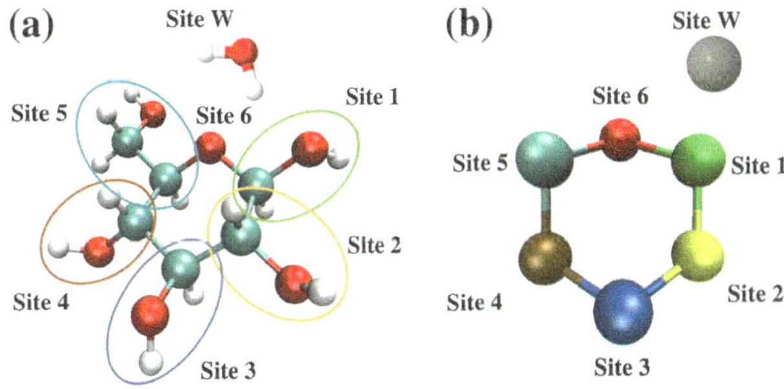

Fig. 2 **a** All-atom structures for β-D-glucose and water molecules; **b** six-site coarse-grained representation of β-D-glucose and single-site water molecules

of glucose can be sampled; however, this mapping doubles the number of coarse-grain sites and requires additional computational time. For all of the coarse-grained simulations, the water molecules were modeled with a single coarse-grained site that has been shown to accurately capture the liquid state structure of water [25]. The center site location of each coarse-grain water molecule was defined as either the center of mass or the geometric center in accordance with the particular coarse-grained model for β-D-glucose being used.

2.3 Molecular dynamics simulations

All MD simulations in this study were run using LAMMPS [26, 27]. The all-atom MD simulations used a CHARMM-style force field for glucose and the TIP3P [28] model for water. The CHARMM force field parameters for glucose were taken from parameters derived for carbohydrates [29]. The long-range electrostatic interactions were treated with the particle mesh Ewald (PME) method [30], and all hydrogen bond lengths were fixed by the SHAKE algorithm [31]. A Nose–Hoover thermostat was implemented to control the temperature at 300 K for all simulations [32]. In

addition, the NVT ensemble is used for all simulations in this study. A mixture of 64 β-D-glucose molecules and 27,224 water molecules was placed in a cubic box with side length 9.6 nm and periodic boundary conditions. The system was equilibrated for 800 ps at $T = 300$ K, followed by 8 ns of simulation. A total of 4,000 configurations with positions, velocities, and forces were collected every 2 ps.

The coarse-grained MD simulations used the tabulated form of the coarse-grained force field as input to LAMMPS. Coarse-grained MD simulations were initiated from the same initial system as was used for the all-atom case. The coarse-grained system of β-D-glucose and water molecules was first equilibrated for 800 ps, followed by 8 ns of simulation.

3 Results and discussion

This section describes the results from the coarse-grained models discussed above for β-D-glucose in aqueous solution, beginning with an evaluation of the three-site coarse-grain model for β-D-glucose. Bond-length distributions and radial distribution functions from MD simulations using the

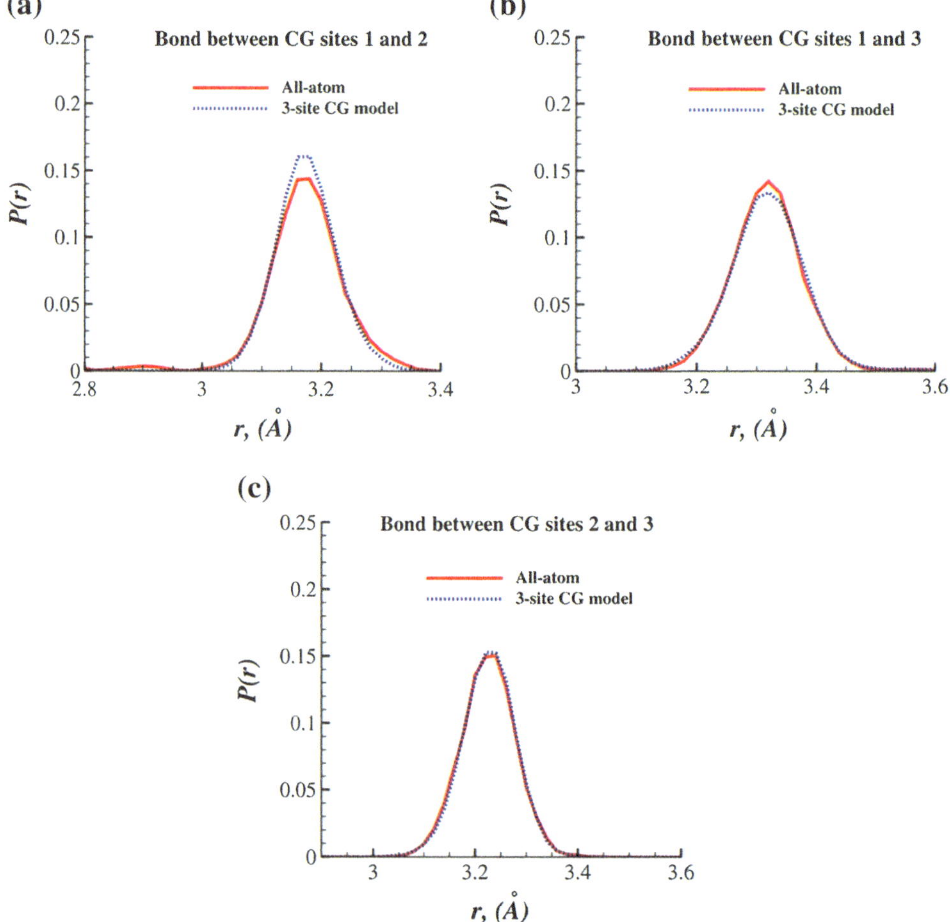

Fig. 3 The bond-length distribution $P(r)$ for the three-site CG-CM model for β-D-glucose: **a** bond between CG sites 1 and 2, **b** bond between CG sites 1 and 3, **c** bond between CG sites 2 and 3. The coarse-grain sites are defined as shown in Fig. 1, and the center of each coarse-grain site is located at the center of mass of the corresponding atoms. For comparison, the bond-length distribution between the coarse-grain sites taken directly from the reference all-atom MD simulation is shown

three-site coarse-grained β-D-glucose model are compared with MD simulation results obtained using an all-atom model. This discussion is followed by an analogous analysis of the six-site coarse-grained β-D-glucose model. When comparing bond-length distributions and radial distribution functions, the data are considered to be in "good agreement" if the peak positions differ by less than 0.1 Å, and the magnitude of peaks differs by not more than 30%. The data are then analyzed to demonstrate how the definition of the center of each coarse-grained site (center of mass vs. geometric center) changes the level of agreement for intramolecular and intermolecular structural properties between the coarse-grained simulation and the all-atom reference simulation. It is also shown how different degrees of coarsening affect the structural predictions derived from simulations with the coarse-grained model.

3.1 Evaluation of the three-site coarse-grained β-D-glucose model

Consider a coarse-grained model where each β-D-glucose molecule is represented by three coarse-grain sites and each water molecule is represented by one coarse-grain site, as shown in Fig. 1. This mapping yields the simplest coarse-grained model consisting of the minimum number of coarse-grain sites that can capture the planar structure and spatial anisotropy of the glucose molecule. With this three-site coarse-grained model for β-D-glucose, the aqueous solution can be fully described with 10 non-bonded pair interactions (six glucose–glucose, three glucose–water, one water–water) and three bonded interactions. When selecting a coarse-grain mapping scheme for an all-atom system, the center of the coarse-grain site must be specified. Two logical choices are the center of mass of all the atoms associated with the coarse-grain site (CG-CM) or the geometric center (CG-GC) of all the atoms associated with the coarse-grain site. Both cases are considered here to investigate how the center site location influences the outcome obtained from simulations using the coarse-grained model.

Figures 3 and 4 show results from MD simulations of aqueous β-D-glucose solutions and compare the three-site CG-CM model with the all-atom reference model. Figure 3 shows a comparison of bond-length distributions, $P(r)$,

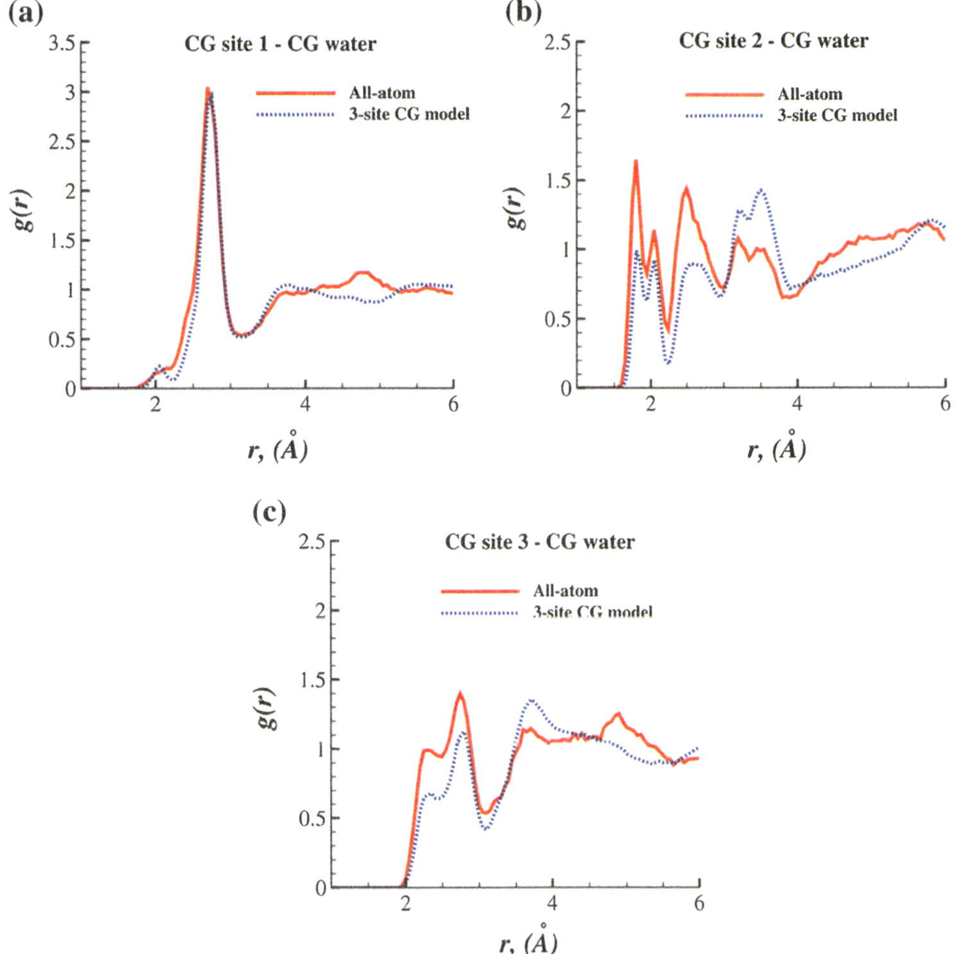

Fig. 4 The radial distribution function $g(r)$ for the three-site CG-CM model for β-D-glucose with the center of each coarse-grain site defined at the center of mass: **a** CG site 1-CG water, **b** CG site 2-CG water, **c** CG site 3-CG water. The coarse-grain sites are defined as shown in Fig. 1. For comparison, the radial distribution function taken directly from the reference all-atom MD simulation is shown

between the three coarse-grain β-D-glucose sites. Here, $P(r) = N(r)/N_{\text{tot}}$ is the probability of finding a bond of length r, $N(r)$ is the number of bonds of length r, and N_{tot} is the total number of bonds. To make direct comparisons between the coarse-grained model and the all-atom model, the configurations from the all-atom MD simulation are first reduced to the three-site CG-CM mapping before calculating the bond-length distributions. Good agreement between the bond-length distributions from the three-site CG-CM MD simulation and the all-atom MD simulation is observed. For the bond between coarse-grain sites 1 and 2, the CG-CM model overestimates the maximum in the bond-length distribution by 15%. This discrepancy can be attributed to the fact that the three-site CG-CM model does not recover the minor peak at $r = 2.9$ Å (Fig. 3a) observed in the bond-length distribution derived from the all-atom simulation.

Figure 4 shows the radial distribution function, $g(r)$, for water around the three CG-CM β-D-glucose sites. The radial distribution function between β-D-glucose sites was not calculated because statistics for intermolecular interactions between β-D-glucose molecules are poor due to its

low concentration in the system. As for the bond-length distribution case, the radial distributions functions for the all-atom system are obtained by reducing the configurations from the all-atom MD simulations to the three-site CG-CM mapping before calculating $g(r)$. Good agreement between the radial distribution functions from the three-site CG-CM simulations and the all-atom simulations is observed in the first neighbor peak for coarse-grain site 1. For coarse-grain sites 2 and 3, the three-site CG-CM model underestimates the magnitude of the peaks in $g(r)$ at short-range (<3 Å) distances and overestimates the midrange (3 Å $< r < 4$ Å). However, the peak positions agree well with those derived from the reference all-atom simulation.

Now consider the three-site CG-GC model with the all-atom reference model. There is good agreement in the bond-length distributions obtained from the three-site CG-GC model and the all-atom model (Fig. 5). Figure 6 shows the radial distribution functions for water around the three CG-GC β-D-glucose sites. The three-site CG-GC model overestimates the magnitude of the first neighbor peak for all three coarse-grained sites. The first neighbor peak for coarse-grain site 1 is sharper and shifted about

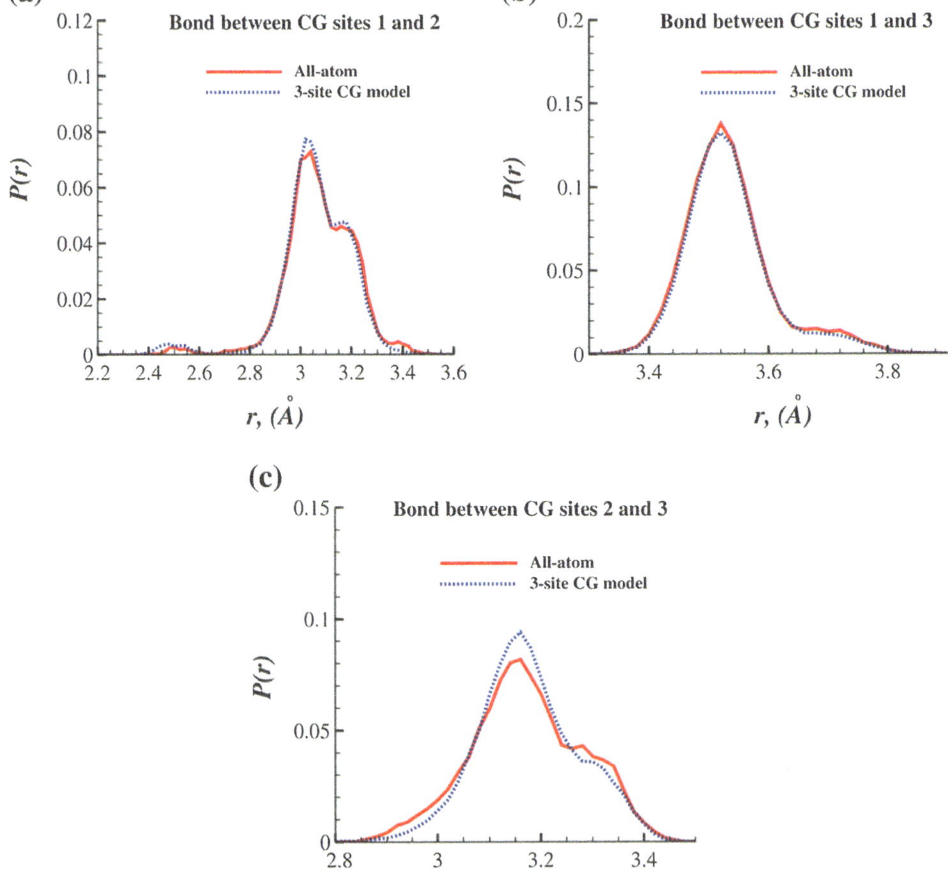

Fig. 5 The bond-length distribution $P(r)$ for the three-site CG-GC model for β-D-glucose: **a** bond between CG sites 1 and 2, **b** bond between CG sites 1 and 3, **c** bond between CG sites 2 and 3. The coarse-grain sites are defined as shown in Fig. 1, and the center of each coarse-grain site is located at the geometric center of the corresponding atoms. For comparison, the bond-length distribution between the coarse-grain sites taken directly from the reference all-atom MD simulation is shown

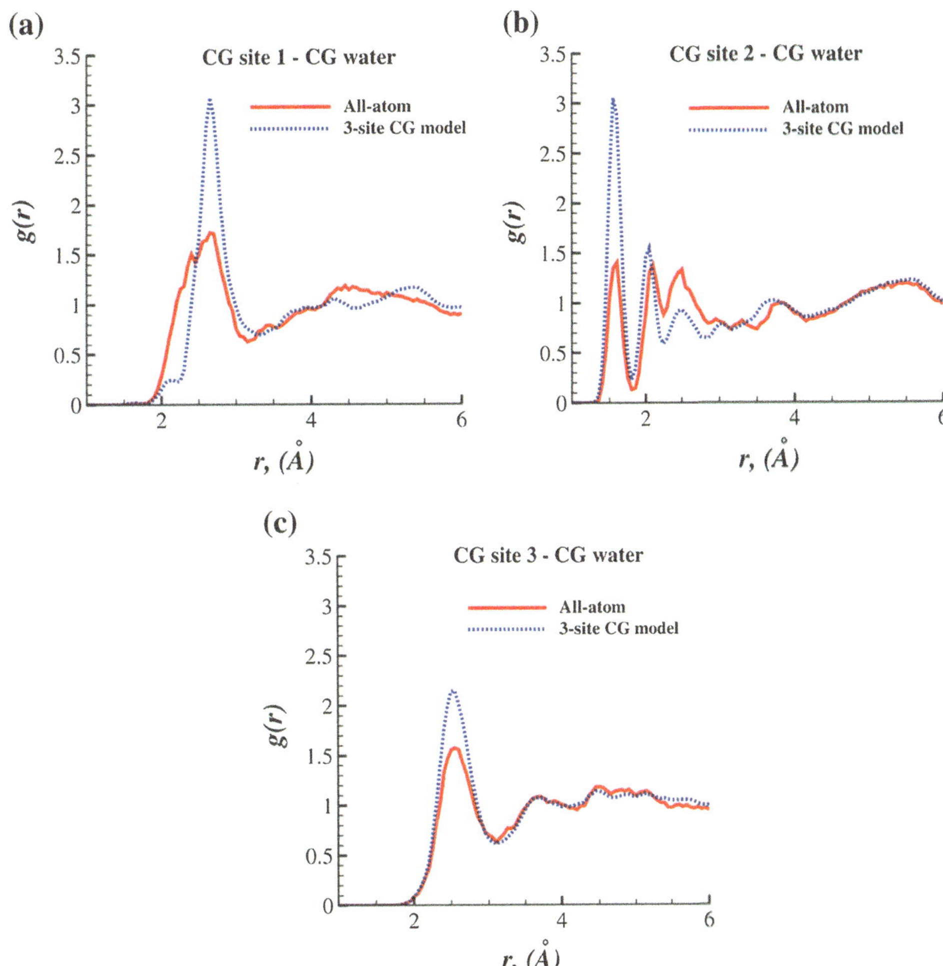

Fig. 6 The radial distribution function $g(r)$ for the three-site CG-GC model for β-D-glucose with the center of each coarse-grain site defined at the geometric center: **a** CG site 1-CG water, **b** CG site 2-CG water, **c** CG site 3-CG water. The coarse-grain sites are defined as shown in Fig. 1. For comparison, the radial distribution function between the coarse-grain sites taken directly from the reference all-atom MD simulation is shown

0.2 Å higher than the all-atom model, while the first neighbor peak positions for coarse-grain sites 2 and 3 do agree well with the all-atom model.

The results for the three-site coarse-grained models represented in the present work can be qualitatively compared with the three-site coarse-grained glucose models developed previously [17, 18]. Direct quantitative comparison is not possible since in these studies, different all-atom force fields are applied, different glucose conformers are used, and different concentrations of glucose are considered. However, qualitatively the three-site coarse-grained models represented in this paper show a good match in the radial distribution functions and bond distribution computed from the all-atom reference simulation and the coarse-grained simulation. In the present study, all coarse-grain interactions (non-bonded, bonds, and angles) are obtained from the coarse-graining procedure. The previously developed models [17, 18] used a hybrid scheme, whereby non-bonded interactions were obtained from force-matching method, and bonds, angles, and dihedrals were derived using Boltzmann inversion methods. Thus, the

three-site CG models in this work are developed with fewer assumptions about the all-atom trajectory.

3.2 Evaluation of the six-site coarse-grained glucose model

Next, consider a coarse-grained model in which each β-D-glucose molecule is represented by six coarse-grain sites and each water molecule is represented by one coarse-grain site, as shown in Fig. 2. With this six-site coarse-grained model for β-D-glucose, the aqueous solution can be fully described with 28 non-bonded pair interactions, six bonded interactions, and three distinct dihedral angles. Figure 7 shows a comparison of bond-length distributions between the six CG-CM β-D-glucose sites. Figure 8 shows the radial distribution functions for water around the six CG-CM β-D-glucose sites. There is good agreement between the radial distribution functions obtained for the six-site CG-CM model and the all-atom model for the first neighbor peaks around coarse-grain sites 1–5. However, the CG-CM model does not recover the sharp secondary

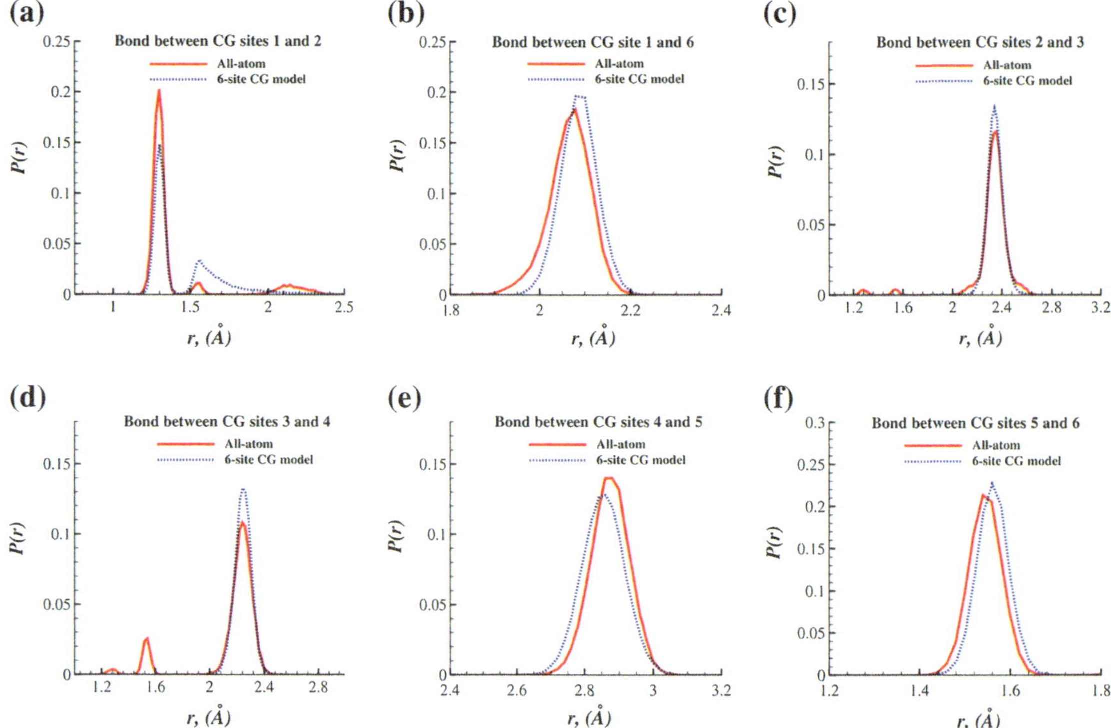

Fig. 7 The bond-length distribution $P(r)$ for the six-site CG-CM model for β-D-glucose: **a** bond between CG sites 1 and 2, **b** bond between CG sites 1 and 6, **c** bond between CG sites 2 and 3, **d** bond between CG sites 3 and 4, **e** bond between CG sites 4 and 5, **f** bond between CG sites 5 and 6. The coarse-grain sites are defined as shown in Fig. 2, and the center of each coarse-grain site is located at the center of mass of the corresponding atoms. For comparison, the bond-length distribution between the coarse-grain sites taken directly from the reference all-atom MD simulation is shown

peaks observed in the all-atom model for coarse-grain sites 1 and 2. Figure 9 shows a comparison of bond-length distributions between the six CG-GC β-D-glucose sites. Good agreement between the bond-length distributions from the six-site CG-GC MD simulation and the all-atom MD simulation is observed for bonds between coarse-grain sites 1–2, 1–6, 2–3, and 4–5. For the bond between coarse-grain sites 1–2, and 3–4, the CG-GC model overestimates the maximum in the bond-length distribution by 40 and 80%, respectively. As discussed above, this discrepancy can be attributed to the fact that the six-site CG-GC model does not recover the secondary peaks observed in the bond-length distribution derived from the all-atom simulation. Figure 10 shows the radial distribution functions for water around the six CG-GC β-D-glucose sites. There is good agreement between the six-site CG-GC model and the all-atom model for coarse-grain sites 2 and 6. The six-site CG-GC model overestimates the magnitude of the first neighbor peaks for coarse-grain sites 1, 3, 4, and 5, but the positions of first and secondary peaks are in close agreement with the all-atom model.

In Fig. 11, the distribution of conformations for the six-site coarse-grain model of β-D-glucose is compared to the all-atom reference data. The conformation distribution is defined by the angles ϕ, calculated between planes formed by coarse-grained sites 6–1–2 and 2–6–5, and ψ, calculated between planes formed by coarse-grained sites 3–4–5 and 3–5–6. The coarse-grained and the all-atom simulation data show occurrences of the chair conformation (where both ϕ, $\psi < 180°$) and the boat conformation (where ϕ or $\psi > 180°$). For both CG-CM and CG-GC systems, the boat conformation is found in 43% of cases. In systems where the all-atom reference data are reduced to the 6-site CG models, about 57% of all conformations are boat conformations. However, when the conformation distributions are calculated for the β-D-glucose rings using all-atom data with no coarse-grain reduction, the boat conformation is found only in 20% cases (not shown). This increase in the presence of boat conformations can be explained by the fact that during the coarse-grain procedure, the positions of the coarse-grained sites in the β-D-glucose rings are significantly different from the carbon centers of the all-atom β-D-glucose rings. Thus, in the coarse-grained models, the occurrence of boat conformations is artificially increased, in this case by a factor of three.

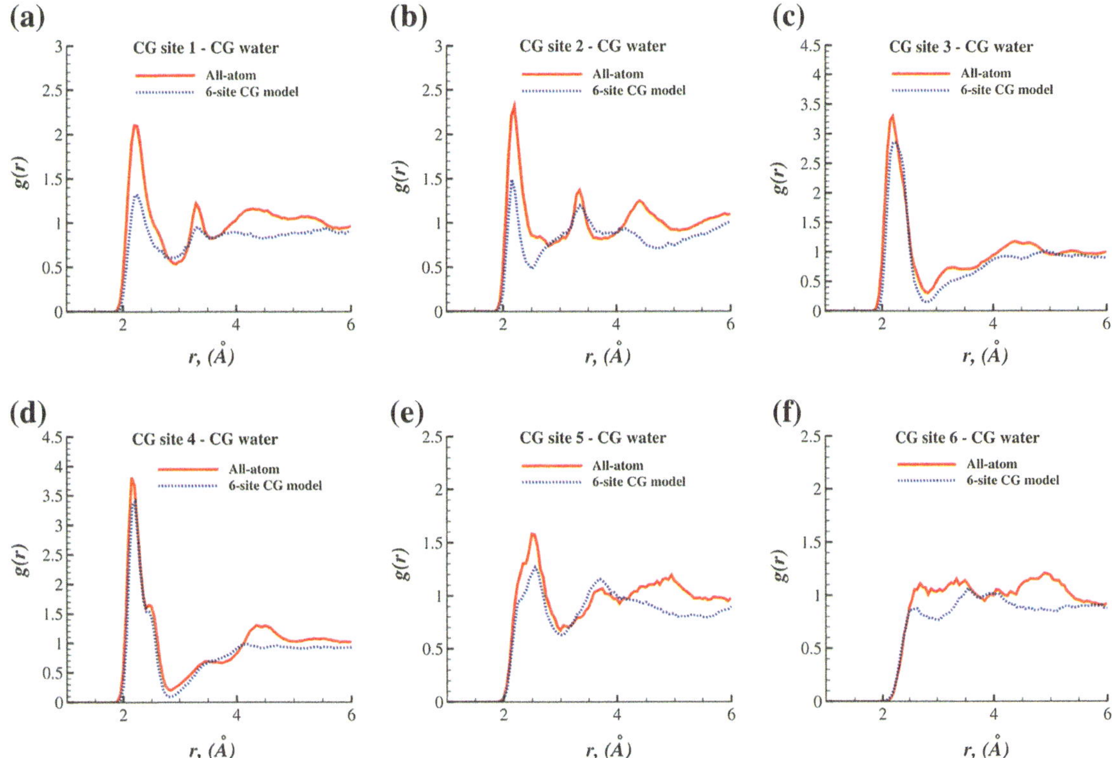

Fig. 8 The radial distribution function $g(r)$ for the six-site CG-CM model for β-D-glucose with the center of each coarse-grain site defined at the center of mass: **a** CG site 1-CG water, **b** CG site 2-CG water, **c** CG site 3-CG water, **d** CG site 4-CG water, **e** CG site 5-CG water, **f** CG site 6-CG water. The coarse-grain sites are defined as shown in Fig. 2. For comparison, the radial distribution function between the coarse-grain sites taken directly from the reference all-atom MD simulation is shown

3.3 Evaluation of the center site location

To facilitate direct comparison between center site locations, difference curves showing $\Delta g(r) = g(r)^{\text{CG}} - g(r)^{\text{AA}}$ for CG-CM and CG-GC are shown for the three-site and six-site models in Figs. 12 and 13, respectively. The CG-CM models reproduce the first neighbor peak in the radial distribution better than the CG-GC models, while the CG-GC models reproduce secondary peaks better than the CG-CM models. Exceptions are that the first neighbor peaks for coarse-grain sites 1 and 2 are better represented by the CG-GC model.

Such differences in the radial distribution functions suggests that defining the center site at different positions results in different effective interactions between coarse-grain sites. To more closely examine the effect that shifting the center site position has on the overall system properties, the difference between the center site for the two definitions (CM and GC) of coarse-grain site centers is calculated for β-D-glucose sites and for water. Table 1 shows the position difference, $R_{\text{CM-GC}}$, between the CM and the GC sites calculated from the all-atom simulations. The distance between the CM and the GC center site position varies

from 0.306 to 0.404 Å for most of the CG sites on glucose; this is comparable to the distance of 0.325 Å for water molecules. It can therefore be concluded that the difference between CG-CM and CG-GC models impacts both β-D-glucose coarse-grain sites and water coarse-grain sites. In the present study, a very dilute solution of β-D-glucose in aqueous solution is considered; therefore, it is difficult to estimate the degree to which the center site location impacts interactions between β-D-glucose molecules. However, the effect of the center site location on water molecules can be investigated in detail. Looking at the water molecule shown in Fig. 14, its center of mass is very close to the center of an oxygen atom (Fig. 14a). On the other hand, the geometric center of the water molecule is relatively far from the center of the oxygen atom and shifted toward the two hydrogen atoms (Fig. 14b). Recall that in this coarse-grain model of water, three atoms are replaced with a single spherical coarse-grain site. Thus, the CG-CM model does not take into account the presence of two hydrogen atoms (in a topological sense) since the center of the coarse-grain site is very close to the center of the oxygen atom (Fig. 14a). With this model, two CG-CM water molecules approaching each other would interact as

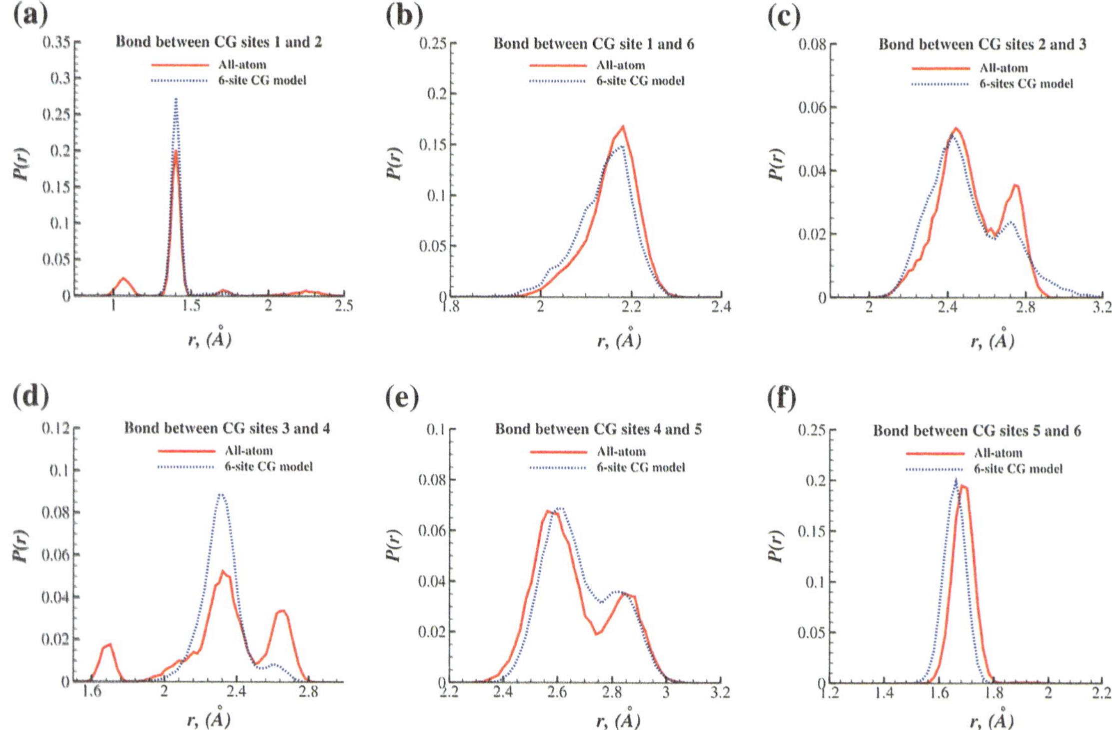

Fig. 9 The bond-length distribution $P(r)$ for the six-site CG-GC model for β-D-glucose: **a** bond between CG sites 1 and 2, **b** bond between CG sites 1 and 6, **c** bond between CG sites 2 and 3, **d** bond between CG sites 3 and 4, **e** bond between CG sites 4 and 5, **f** bond between CG sites 5 and 6. The coarse-grain sites are defined as shown in Fig. 2, and the center of each coarse-grain site is located at the geometric center of the corresponding atoms. For comparison, the bond-length distribution between the coarse-grain sites taken directly from the reference all-atom MD simulation is shown

hard spheres due to the strong repulsion between oxygen atoms. In the CG-GC model (Fig. 14b), the presence of the two hydrogen atoms is taken into account. As two CG-GC water molecules are brought together, they approach from the hydrogen atom sites. As such, this interaction will be more like an elastic sphere interaction.

These arguments are fully supported by the effective force data for the CG water–CG water interaction obtained from the coarse-graining procedure as shown in Fig. 15a. The CG-CM model predicts strong repulsion between CG water molecules for separations that are <2.8 Å—a signature of hard sphere-like behavior. The CG-GC model predicts a shoulder with moderate repulsion at separations between 2.1 and 2.8 Å that can be interpreted as an elastic sphere interaction. Similar behavior is observed for β-D-glucose CG sites–CG water interactions (Fig. 15b). The radial distribution function between the CG water–CG water molecules for the CG-CM and CG-GC water models in comparison to the all-atom reference system are presented in Fig. 16. The positions of the first peak are well-matched for both models and located at 2.8 Å, which is in good agreement with the mean van der Waals diameter of water (2.82 Å) [33]. However, the CG-CM water model is not able to represent secondary peaks, while the CG-GC

water model gives a closer match for the secondary peaks (Fig. 16). This explains why at distances beyond the first neighbor peak, the CG-GC models reproduce the secondary peaks better than the CG-CM models (Figs. 12, 13). Likewise, for systems of glucose and water, the effective elastic sphere interactions that result from coarse-graining to the GC is that the number of nearest neighbor coarse-grained water molecules around a given coarse-grained glucose site increases, thereby overestimating the first neighbor peak in the radial distribution compared to the all-atom reference system (Figs. 12, 13).

Based on this analysis, the CG-GC model for water molecules more accurately represents the overall water–water spatial distribution compared to the CG-CM model. Great attention must be given to the development of an accurate coarse-grained water model. Because water molecules are non-symmetrical, a coarse-grained model that is built based on statistical averaging might not be generally applicable to all systems or even to all state points for a given system. When accurate and reliable coarse-grained structures are required, such as when the intended use of the coarse-grained model is to efficiently generate equilibrium configurations and then restore the all-atom degrees of freedom (e.g., reverse-mapping), it may be

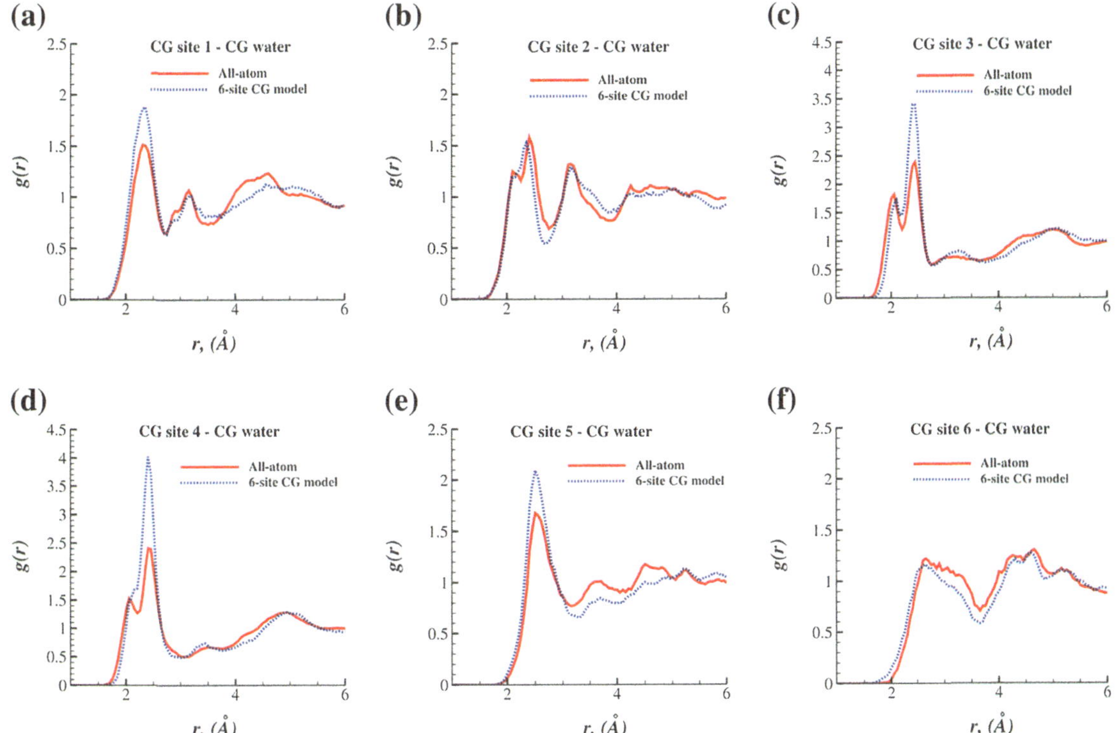

Fig. 10 The radial distribution function $g(r)$ for the six-site CG-GC model for β-D-glucose with the center of each coarse-grain site defined at the geometric center: **a** CG site 1-CG water, **b** CG site 2-CG water, **c** CG site 3-CG water, **d** CG site 4-CG water, **e** CG site 5-CG water, **f** CG site 6-CG water. The coarse-grain sites are defined as shown in Fig. 2. For comparison, the radial distribution function between the coarse-grain sites taken directly from the reference all-atom MD simulation is shown

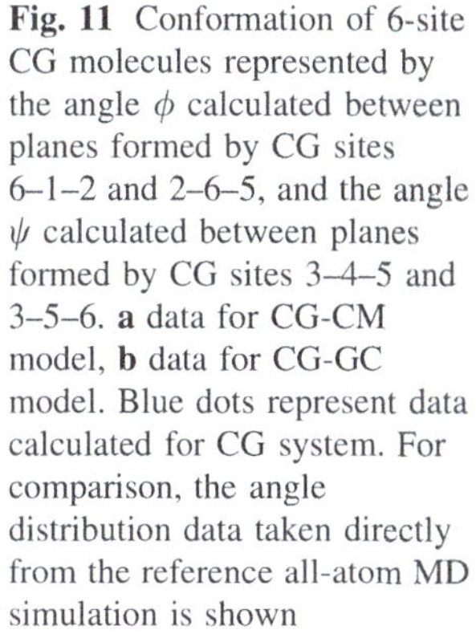

Fig. 11 Conformation of 6-site CG molecules represented by the angle ϕ calculated between planes formed by CG sites 6–1–2 and 2–6–5, and the angle ψ calculated between planes formed by CG sites 3–4–5 and 3–5–6. **a** data for CG-CM model, **b** data for CG-GC model. Blue dots represent data calculated for CG system. For comparison, the angle distribution data taken directly from the reference all-atom MD simulation is shown

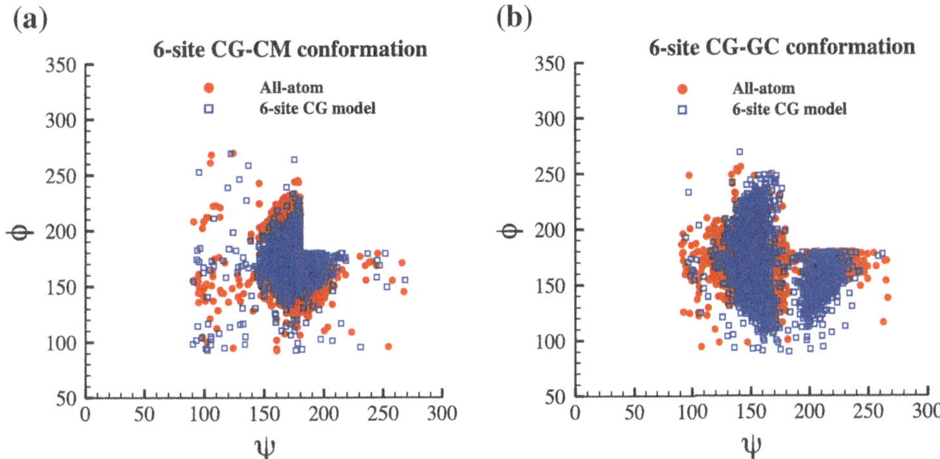

prudent to develop a coarse-grained water molecule model along with the solute molecule, rather than using a generalized coarse-grained water model.

3.4 Evaluation of the coarse-grain mapping

Comparing Fig. 4 to Fig. 8 (CG-CM) and Fig. 6 to Fig. 10 (CG-GC) shows how the mapping scheme affects the accuracy of the coarse-grain model in predicting the structure observed in the all-atom reference simulations. A three-site CG model (Fig. 1) is fully determined when non-bonded interactions, bonds, and angles are defined. A six-site CG model (Fig. 2) is much more complex since the number of unique interaction pairs is significantly larger. Moreover, a dihedral interaction must be included to obtain an accurate coarse-grained model.

Despite its relative simplicity, a three-site CG model is able to represent the structural properties of the system

Fig. 12 The differences of the radial distribution functions calculated between the radial distribution function for the three-site CG-CM and CG-GC models and for the three-site all-atom MD simulation for β-D-glucose. The *positive values* represent overestimation of $g(r)$ by the CG model, and the *negative values* represent underestimation of $g(r)$ by the CG model. **a** CG site 1-CG water, **b** CG site 2-CG water, **c** CG site 3-CG water. The coarse-grain sites are defined as shown in Fig. 1

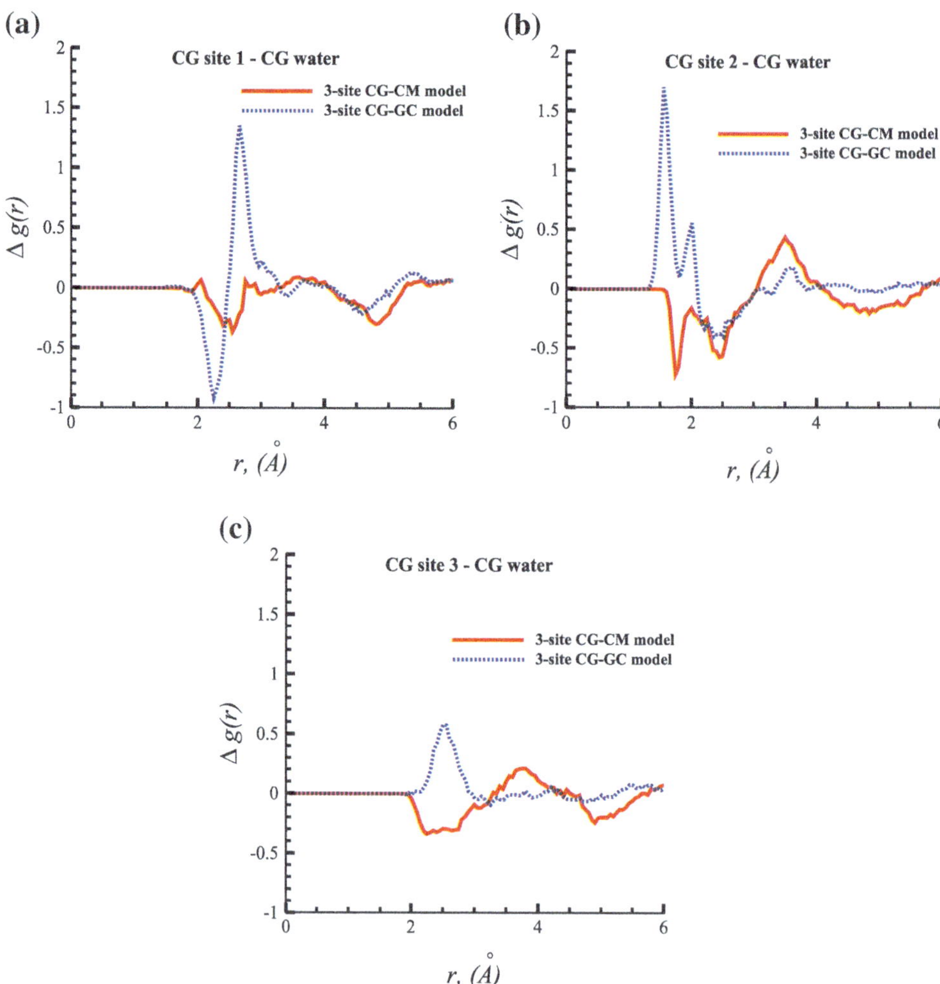

predicted in the all-atom reference simulations well (Figs. 3, 4, 5, 6). The positions of the primary peaks in the radial distribution functions are accurately matched (Figs. 4, 6), while the magnitude of the primary peaks are better predicted by the CG-CM model (Fig. 4). Bond-length distributions show a perfect match with all-atom reference data for both CG-CM and CG-GC models (Figs. 3, 5, respectively). A six-site CG model is able to accurately capture the structural properties of β-D-glucose in a water environment (Figs. 7, 8, 9, 10). The positions and the magnitudes of the primary peaks in the radial distribution function are accurately predicted (Figs. 8, 10). However, the magnitudes for site 1 and site 2 for the CG-CM model are underestimated (Fig. 8), and the magnitudes for site 3 and site 4 for the CG-GC model are overestimated (Fig. 10). The positions and the magnitudes of the secondary peaks are better described by the CG-GC model (Fig. 10) due to the more accurate CG model of water molecules (Fig. 16b). Bond-length distributions for the six-site CG models are in good agreement with the all-atom reference data (Figs. 7, 9). This match is not as

perfect as for the three-site CG models (Figs. 3, 5) especially for the CG-GC model where bond-length distributions have secondary peaks (Fig. 9). However, by taking into account the increased complexity of a six-site CG model, the positions and magnitude of primary and secondary peaks match reasonably well (Figs. 7, 9).

Based on this comparison, it is possible to say that both the three-site CG models and the six-site CG models can be successfully used to accurately predict the structure observed in all-atom reference simulations. If not many details are required for the CG model then a simple three-site CG-CM model can be used. In cases, when more structural details are needed then a more complex six-site CG-GC model should be used.

4 Conclusions

In this work, coarse-grained models of β-D-glucose in water solution were developed using the force-matching approach. For this system, two different coarse-grained

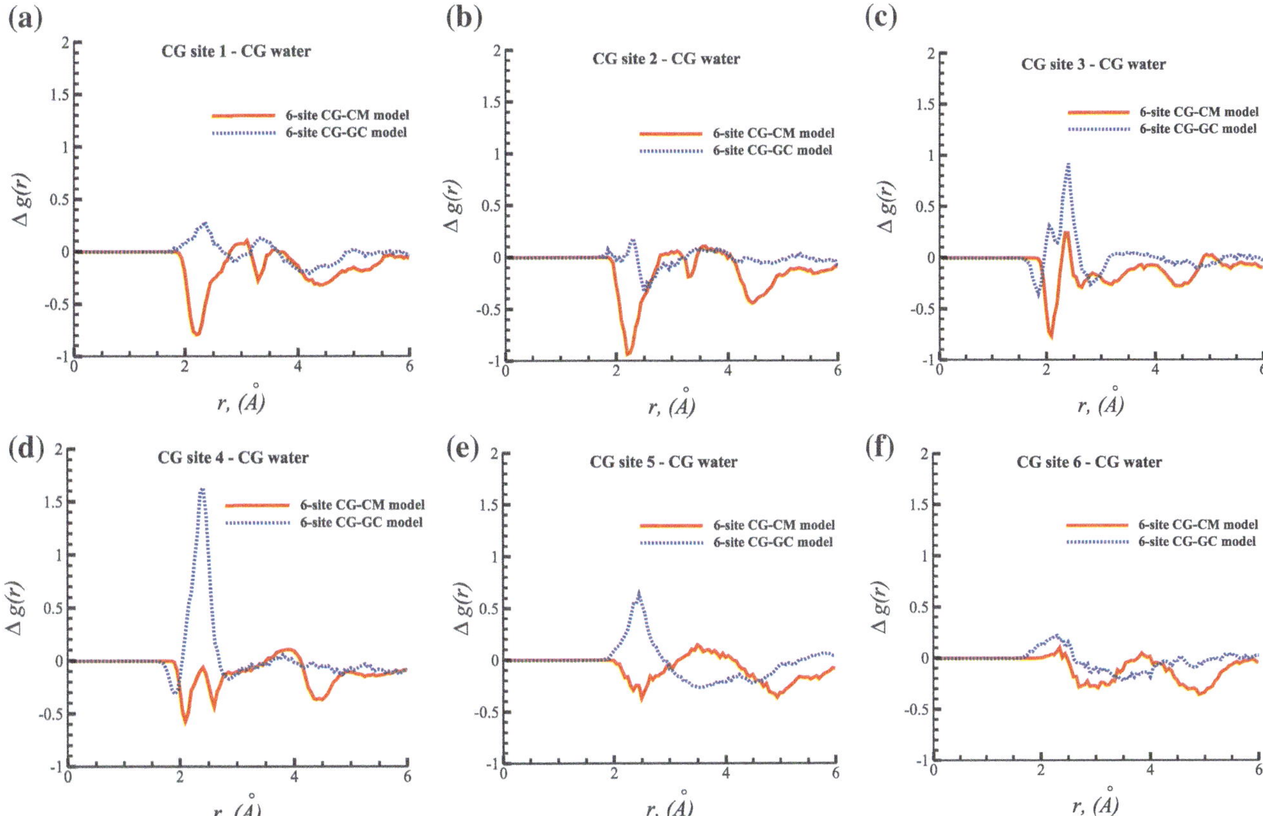

Fig. 13 The differences of the radial distribution functions calculated between the radial distribution function for the six-site CG-CM and CG-GC models and for the six-site all-atom MD simulation for β-D-glucose. The positive values represent overestimation of $g(r)$ by the CG model, and the negative values represent underestimation of $g(r)$ by the CG model. **a** CG site 1-CG water, **b** CG site 2-CG water, **c** CG site 3-CG water, **d** CG site 4-CG water, **e** CG site 5-CG water, **f** CG site 6-CG water. The coarse-grain sites are defined as shown in Fig. 2

Table 1 The distance between the center of mass and geometric center calculated based on all-atom MD simulation results and correspondent standard deviation for all models used in the present study

	$R_{\text{CM-GC}}$, Å	Stdv., Å
3-site model		
1	0.108	0.025
2	0.132	0.052
3	0.404	0.013
6-site model		
1	0.378	0.053
2	0.378	0.060
3	0.365	0.073
4	0.349	0.092
5	0.306	0.020
6	0.000	0.000
1-site water		
1	0.325	0.023

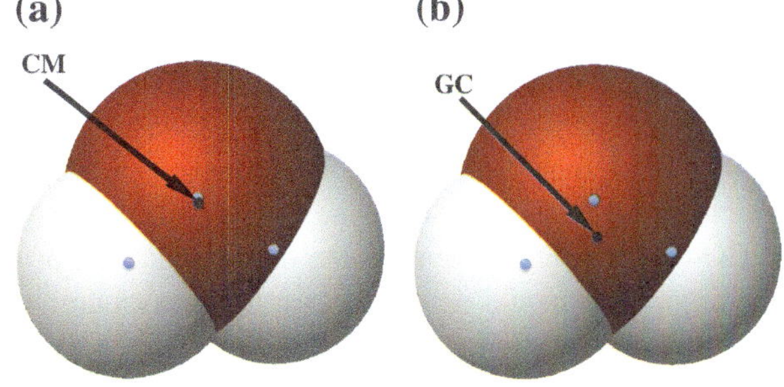

Fig. 14 Water molecule representation. *Red sphere* represents oxygen atom, *white spheres* represent hydrogen atoms. *Blue dots* represent the center of each atom in the molecule. *Black dots* represent **a** center of mass of the water molecule, **b** geometric center of the water molecule

mapping schemes (three-site and six-site) were investigated. In addition to varying the number of atoms in a coarse-grained site, two definitions of the center site location for the coarse-grained sites, center of mass and geometric center, were considered and evaluated in detail.

The CG-CM models generally reproduce the primary peak in the radial distribution function better than the CG-GC models. However, the CG-GC models accurately reproduce secondary peaks in the radial distribution

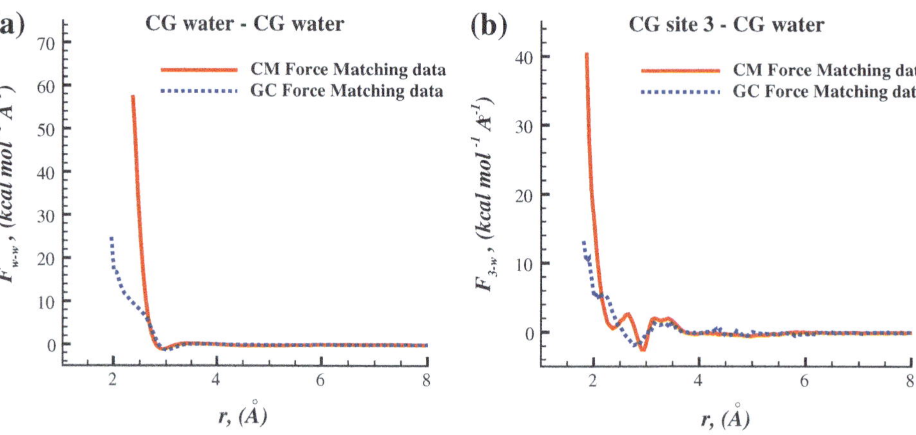

Fig. 15 Effective force–distance curves obtained with coarse-graining method. *Red solid lines* correspond to the CG-CM model and *blue dotted lines* correspond to the CG-GC model. **a** force–distance data for CG water–water interaction, **b** force–distance data for CG site 3–CG water molecules in 3-site CG β-D-glucose model

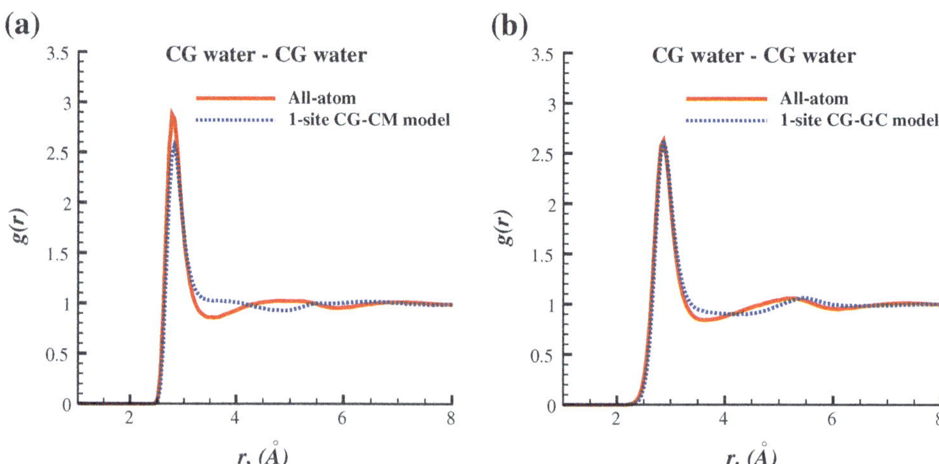

Fig. 16 The radial distribution function $g(r)$ for the water model **a** the center of coarse-grain site defined at the center of mass, **b** the center of coarse-grain site is defined at the geometric center. For comparison, the radial distribution function between the coarse-grain sites taken directly from the reference all-atom MD simulation is shown

functions. This difference in the RDFs for CG-CM and CG-GC models can be explained by the different interactions captured by the two definitions of center site. The coarse-grained sites in the CG-CM water molecules interact like hard spheres, while the CG-GC water molecules interact like soft-spheres. Thus, the choice of using CG-CM or CG-GC depends on whether the desired outcome of the coarse-grained model is to predict local or long-range structure more accurately.

Three-site coarse-grained models reproduce structural properties like radial distribution functions and bond-length distributions observed in all-atom reference simulations reasonably well, although there are some cases where primary peaks are significantly overestimated. For β-D-glucose in water, the three-site CG-CM model gives better predictions than the three-site CG-GC model. The three-site CG-CM model is suitable for situations in which a simple and fast coarse-grained model capable of generating long length scale equilibrium structures is favored over accuracy in local ordering.

The six-site CG models reveal a higher level of structural detail than the three-site CG models. For β-D-glucose

in water, the six-site CG-GC model is a more accurate model than the CG-CM model for most of primary peaks and it yields a nearly perfect match for the secondary peaks in the radial distribution function.

The detailed analysis of coarse-grain modeling shown here underscores the importance of evaluating multiple coarse-grain mapping schemes for a given application. For example, the β-D-glucose molecules considered in the present work are the building blocks for cellulose molecules, and thus, the insight gained about coarse-grained model development for glucose is a critical first step before proceeding to the development of coarse-grain models for the more complicated crystalline cellulose structure.

Acknowledgments This research is sponsored by U.S. Department of Energy's (USDOE) Scientific Discovery through Advanced Computing (SciDAC) program through USDOE's Office of Advanced Scientific Computing Research (ASCR) and Biological and Environmental Research (BER), and performed at the Ames Laboratory, FWP AL-08-330-039. Ames Laboratory is managed by Iowa State University for the U.S. Department of Energy under contract DE-AC02-07CH11358. The authors thank Professor G. A. Voth for providing the multiscale coarse-graining (MS-CG) software developed by his research group.

References

1. Rubin EM (2008) Genomics of cellulosic biofuels. Nature 454:841–845
2. Sanderson K (2011) Lignocellulose: a chewy problem. Nature 474:S12–S14
3. Marrink SJ, Risselada HJ, Yefimov S, Tieleman DP, de Vries AH (2007) The MARTINI force field: coarse grained model for biomolecular simulations. J Phys Chem B 111:7812–7824
4. Monticelli L, Kandasamy SK, Periole X, Larson GR, Tieleman DP, Marrink SJ (2008) The MARTINI coarse-grained force field: extension to proteins. J Chem Theory Comput 4:819–834
5. Ashbaugh HS, Patel HA, Kumar SK, Garde S (2005) Mesoscale model of polymer melt structure: self-consistent mapping of molecular correlations to coarse-grained potentials. J Chem Phys 122:104908
6. Reith D, Pütz M, Müller-Plathe F (2003) Deriving effective mesoscale potentials from atomistic simulations. J Comput Chem 24:1624
7. Muller-Plathe F (2002) Coarse-graining in polymer simulation: from the atomistic to the mesoscopic scale and back. Chem Phys Chem 3:754–769
8. Peter C, Delle Site L, Kremer K (2008) Classical simulations from the atomistic to the mesoscale and back: coarse graining an azobenzene liquid crystal. Soft Matter 4:859–869
9. Lyubartsev AP, Laaksonen A (1995) Calculation of effective interaction potentials from radial distribution functions: a reverse Monte Carlo approach. Phys Rev E 52:3730–3737
10. Ercolessi F, Adams JB (1994) Interatomic potentials from 1st-principles calculations- the force-matching method. Europhys Lett 26:583–588
11. Izvekov S, Voth GA (2005) A multiscale coarse-graining method for biomolecular systems. J Phys Chem B 109:2469–2473
12. Izvekov S, Parrinello M, Burnham CJ, Voth GA (2004) Effective force fields for condensed phase systems from ab initio molecular dynamics simulation: a new method for force-matching. J Chem Phys 120:10896–10913
13. Baron R, de Vries AH, Hunenberger PH, van Gunsteren WF (2006) Comparison of atomic-level and coarse-grained models for liquid hydrocarbons from molecular dynamics configurational entropy estimates. J Phys Chem B 110:8464–8473
14. Kim SH, Lamm MH (2011) Reintroducing explicit solvent to a solvent-free coarse-grained model. Phys Rev E 84:025701
15. Molinero V, Goddard WA (2004) M3B: a coarse grain force field for molecular simulations of malto-oligosaccharides and their water mixtures. J Phys Chem B 108:1414–1427
16. Molinero V, Goddard WA (2005) Microscopic mechanism of water diffusion in glucose glasses. Phys Rev Lett 95:04701
17. Liu P, Izvekov S, Voth GA (2007) Multiscale coarse-graining of monosaccharides. J Phys Chem B 111:11566–11575
18. Hynninen AP, Matthews JF, Beckham GT, Crowley MF, Nimlos MR (2011) Coarse-grain model for glucose, cellobiose, and cellotetraose in water. J Chem Theory Comput 7:2137–2150
19. Srinivas G, Cheng X, Smith JC (2012) A solvent-free coarse grain model for crystalline and amorphous cellulose fibrils. J Chem Theory Comput 7(8):2539–2548
20. Hills Jr R, Lu L, Voth GA (2009) Multiscale coarse-graining of the protein energy landscape. J Phys Chem B 133:4443
21. Thorpe IF, Zhou J, Voth GA (2008) Peptide folding using multiscale coarse-grained model. J Phys Chem B 112:13079–13090
22. Noid WG, Chu JW, Ayton GS, Krishna V, Izvekov S, Voth GA, Das A, Andersen HC (2008) The multiscale coarse-graining method. I. A rigorous bridge between atomistic and coarse-grained models. J Chem Phys 128(24):244114
23. Noid WG, Chu JW, Ayton GS, Krishna V, Izvekov S, Voth GA, Das A, Andersen HC (2008) The multiscale coarse-graining method. II. A rigorous bridge between atomistic and coarse-grained models. J Chem Phys 128(24):244115
24. Lawson CL, Hanson RJ (1974) Solving least squares problems. Prentice-Hall, Englewood Cliffs
25. Izvekov S, Voth GA (2005) Multiscale coarse graining of liquid-state systems. J Chem Phys 123:134105
26. http://lammps.sandia.gov
27. Plimpton SJ (1995) Fast parallel algorithms for short-range molecular dynamics. J Comp Phys 117:1–19
28. Jorgensen WL, Chandrasekhar J, Madura JD (1983) Comparison of simple potential functions for simulating liquid water. J Chem Phys 79:926–935
29. Palma R, Zuccato P, Himmel M, Liang G, Brady JW (2001) In: Himmel ME (ed) Glycosyl hydrolases for biomass conversion, ACS symposium series. American Chemical Society, Washington, DC, pp 112–130
30. Darden T, York D, Pedersen LG (1993) Particle mesh Ewald: an Nlog(N) method for Ewald sums in large systems. J Chem Phys 98:10089–10092
31. Ryckaert JP, Ciccotti G, Berendsen HJC (1977) Numerical integration of the cartesian equations of motion of a system with constraints: molecular dynamics of n-alkanes. J Comput Phys 23:327–341
32. Nose S (1984) A molecular dynamics method for simulations in the canonical ensemble. Mol Phys 52:255
33. Franks F (2000) Water: 2nd edition a matrix of life. Royal Society of Chemistry, Cambridge

Theor Chem Acc (2012) 131:1129
DOI 10.1007/s00214-012-1129-7

REGULAR ARTICLE

CL&P: A generic and systematic force field for ionic liquids modeling

José N. Canongia Lopes · Agílio A. H. Pádua

Received: 2 May 2011 / Accepted: 11 August 2011 / Published online: 17 February 2012
© Springer-Verlag 2012

Abstract In this account, we review the process that led to the development of one of the most widely used force fields in the area of ionic liquids modeling, analyze its subsequent expansions and alternative models, and consider future routes of improvement to overcome present limitations. This includes the description and discussion of (1) the rationale behind the generic and systematic character of the Canongia Lopes & Padua (CL&P) force field, namely its built-in specifications of internal consistency, transferability, and compatibility; (2) the families of ionic liquids that have been (and continue to be) parameterized over the years and those that are the most challenging both in theoretical and applied terms; (3) the steps that lead to a correct parameterization of each type of ion and its homologous family, with special emphasis on the correct modeling of their flexibility and charge distribution; (4) the validation processes of the CL&P and other force fields; and finally (5) the compromises that have to be attained when choosing between generic or specific force fields, coarse-grain or atomistic models, and polarizable or nonpolarizable methods. The application of the CL&P and other force fields to the study of ionic liquids using quantum- and statistical-mechanics methods has led to the discovery and analysis of the unique nature of their liquid phases, that is, the notion that ionic liquids are nano-segregated fluids with structural and dynamic heterogeneities at the nanoscopic scale. This successful contribution of theoretical chemistry to the field of ionic liquids will also serve as a guide throughout the ensuing discussion.

Published as part of the special collection of articles: From quantum mechanics to force fields: new methodologies for the classical simulation of complex systems.

Electronic supplementary material The online version of this article (doi:10.1007/s00214-012-1129-7) contains supplementary material, which is available to authorized users.

J. N. Canongia Lopes (✉)
Centro de Química Estrutural, Instituto Superior Técnico, 1049 001 Lisbon, Portugal
e-mail: jnlopes@ist.utl.pt

J. N. Canongia Lopes
Instituto de Tecnologia Química e Biológica, UNL, Av. República Ap. 127, 2780 901 Oeiras, Portugal

A. A. H. Pádua
Laboratoire Thermodynamique et Interactions Moléculaires, Clermont Université, Université Blaise Pascal, BP 80026, 63171 Aubière, France
e-mail: agilio.padua@univ-bpclermont.fr

A. A. H. Pádua
CNRS, UMR6272, LTIM, 63171 Aubière, France

Keywords Ionic liquids · Force field · Molecular dynamics simulations

1 Introduction

Molecular modeling and simulation include computational techniques developed within the frameworks of quantum or statistical mechanics that are able to analyze the links between the macroscopic behavior of matter and its characteristics at a molecular level. Modeling and simulation studies are traditionally used either as predictive or interpretative research instruments but in the case of ionic liquids—a relatively recent research front—they have also assumed the role of exploratory tools leading to some discoveries that were only later corroborated by experimental evidence. For instance, one of the first works dealing with the recognition of ionic liquids as nano-segregated fluids

originated from molecular dynamics (MD) studies [1], reported in the sequence of the development of a systematic force field for ionic liquids [2, 3]. In other words, the knowledge about the physical chemistry of ionic liquids is rapidly advancing through the interplay between experiments, theory, and modeling, each providing challenges, guidelines, and checks to the others.

Molecular modeling at the level of an atomistic description starts with the definition of a suitable force field capable of describing the intra- and inter-molecular interactions taking place between the different units that constitute the systems to be studied.

At the beginning of the century, the number of force field models capable of describing ionic liquids was very limited and fragmented, with only a few ionic liquids studied in an almost case-by-case basis [4–7]. When Canongia Lopes and Padua developed and introduced their force field [2, 3] (CL&P) their main objective was to provide a systematic and transferable model that could be generalized to describe entire families of ionic liquids.

In order to meet that objective and to take into account the "modular" nature of ionic liquids—substances constituted exclusively of cations and anions where a myriad of feasible ionic combinations are possible—three basic specifications were built into the model: internal consistency, transferability, and compatibility (Fig. 1). In fact these three requirements reflect the intrinsic nature of ionic liquids: internal consistency is needed because unlike molecular substances ionic liquids are composed by anions and cations that can be viewed as discrete units but whose interactions should be modeled in a consistent way; transferability is useful because cations or anions are often members of homologous families of compounds and can also be replaced by other cations or anions—a process called metathesis

that plays a pivotal role in the synthesis of new ionic liquids; and finally, compatibility is necessary because the cations (and sometimes the anions) include organic moieties that can be modeled by other well-established force fields.

Presently, the CL&P force field describes twelve families of ions that compose the most important classes of ionic liquids: four cation families—alkylimidazolia, alkylpyridinia, alkylphosphonia, and alkylammonia—including polyalkylated, cyclic, and/or functionalized derivatives, and eight anion families/individual ions—halogens, triflate, bis(sulfonyl)imide, and its derivatives, alkylsulfates, alkysulfonates, phosphate derivatives, dicyanamide, and nitrate. The different classes of ions are summarized in Table 1, and their parameterization is given as supplementary information to this article.

Moreover, the overall underlying rationale was that, since ionic liquids were a new field of study and these compounds exist in enormous variety, it made more sense to have a more general, though eventually less precise model, instead of a more meticulous model that would represent just one, specific ionic liquid. This somewhat flexible approach proved to be quite successful—nowadays, the CL&P force field is one of the most widely used parameterizations for the molecular simulation of ionic liquids: the five articles that presently describe the force field (2 in 2004, 2006, 2008, and 2010) [2, 3, 8–11] have collected up to date more than 450 citations between them. Most of these studies centered on the prediction of thermophysical and structural properties with various degrees of success (see Sect. 4). Earlier works centered on bulk properties, whereas more recent developments (from 2008 onwards) moved to areas as diverse as surface properties or systems under nonequilibrium conditions. The main conclusion that can be drawn from this wealth of results is that as the problems

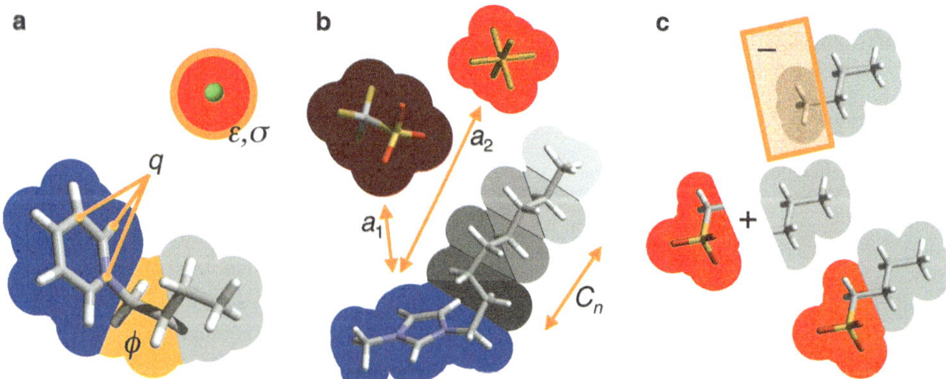

Fig. 1 The three built-in specifications of the CL&P force field: **a** Internal consistency: anions and cations are parameterized with the same force field functional form, with special attention given to the parameterization of atomic partial point charges (q), the flexibility (dihedral angles, ϕ) of the ions, and the adequacy of the repulsive and dispersive forces (ε, σ); **b** transferability: force field parameters are valid within the same homologous family (for instance ions with different alkyl side chain lengths, C_n). They also allow the possibility of ion interchange (a_1, a_2) to yield different ionic liquids; **c** Compatibility: molecular residues and moieties are taken directly from well-established force fields ($-$). Simple rules are established to join seamlessly neutral molecules into an existing ion ($+$)

Table 1 Summary list of ionic liquid ions modeled by the CL&P force field (end of 2010)

Cations		Anions	
Imidazolium		Chloride, Bromide [2,3,9]	Cl^- Br^-
…1,3-dialkyl [2,3]		Triflate [8]	$CF_3SO_3^-$
…1-alkyl [9]		Bis(sulfonyl)imide	
…1,2,3-trialkyl [10]		…bis(trifluoromethyl) [8]	$(CF_3SO_2)_2N^-$
…functionalized [10,11]		…bis(fluoro) [11]	$(FSO_2)_2N^-$
		…bis(perfluoroalkyl) [11]	$(C_nF_{2n+1}SO_2)_2N^-$
N-alkylpyridinium [9]		Alkylsulfates [10]	$C_nH_{2n+1}SO_4^-$
Ammonium		Alkylsulfonates [10]	$C_nH_{2n+1}SO_3^-$
…Tri-, tetra-alkyl [8]		Phosphate	
…N,N-pyrrolidinium [8]		…hexafluoro [2,3,8]	PF_6^-
		…trifluorotrifluoroalkyl [11]	$PF_3(C_nF_{2n+1})_3^-$
Tetralkylphosphonium [9]		Nitrate [2,3,8], Dicyanamide [9]	NO_3^- $N(CN)_2^-$

become more complex (surface vs. bulk conditions, equilibrium vs. transport properties, specific vs. non-specific interactions), the use of the CL&P becomes more difficult. Most of its proposed and undergoing refinements stem from the need of adapting the generic character of the CL&P force field to particular problems/systems without the loss of its inherent internal consistency.

Finally, it must be stressed that ionic liquids are complex fluids that present structural and dynamic heterogeneities at the microscopic scale. These persistent structures of nanometer size [1] mean that ionic liquids can be chemically functionalized for potential applications or to render them more environmentally friendly but without severely modifying their defining physico-chemical properties, such as their low volatility or liquid range. This concept is at the origin of task-specific ionic liquids, [12] or designer solvents. Exploring impact of these modification in chemical structure of the ions on the physico-chemical properties is an important goal of molecular simulation studies; therefore, the possibility of easily representing these task-specific ionic liquids in a force field is an interesting feature, fully within the design spirit based on generality, transferability, and compatibility of the CL&P force field. Several examples have already been published of ionic liquid structures designed for a particular application, such as capture of CO2 [11], enhanced biodegradability [10, 13], or modification of interface properties [14].

2 Development

Ionic liquids share parts of their molecular architectures with molecules that have been parameterized by existing force field frameworks (AMBER [15], OPLS [16, 17], CHARMM [18], etc.), for example, imidazole rings are present in the amino acid histidine. However, ionic liquids have one essential difference: they are composed of ions, and although they may be built by similar structural groups, the distributions of electrostatic charge will be specific. In order to represent correctly the structure and interactions of these new compounds, the sets of parameters that determine charge distributions and conformations (at the least) had to be developed for this class of substances.

The CL&P model was built based on the OPLS-AA functional form [16, 17] (Fig. 2) which means that, technically, it is easy to combine any molecule or residue already defined in the OPLS-AA database with the structures developed for ionic liquids. Compared to other generalist force fields, special attention was devoted in OPLS-AA to the simulation of liquid-state thermodynamic properties, which made this simple and widely used force field an obvious choice for the implementation of an extension to model ionic liquids.

At the time the CL&P force field was proposed, many of the existing ionic liquid models were built by borrowing parameters from different, not always compatible, sources. For instance, it was common to see parameterizations of the cation and of the anion using information from different force fields [4, 6, 7]. In the development of the CL&P force field, in order to respect internal consistency, electronic structure calculations were used extensively to provide essential data for the development of an internally consistent parameter set. This included not only the determination of parameters absent from the OPLS-AA database but, most importantly, the critical re-evaluation of all required terms.

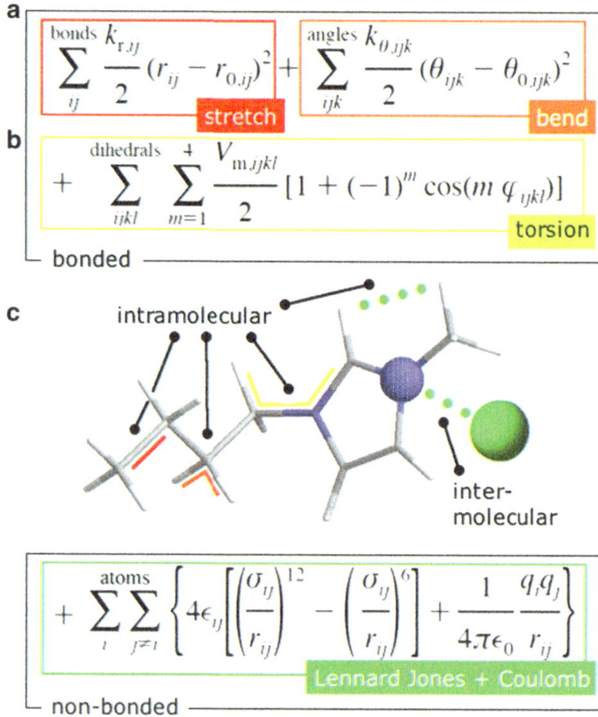

Fig. 2 Schematic representation of the different atom–atom interactions parameterized by the CL&P force field functional for ionic liquids. The three paragraph letters correspond to the sections where each type of interaction is discussed

While developing the CL&P model, two features became the most significant in terms of the (re)-parameterization of the force field: the characterization of the flexibility of the ions (through the torsion energy profiles associated with the different dihedral angles) and the atomic point-charge distributions. Basically, the functional form of the potential energy in the CL&P force field consists of non-bonded and bonded interactions (Fig. 2). The former are repulsive, dispersive, and electrostatic interactions and are represented by Lennard-Jones and Coulomb terms, while the latter are related to the covalent-bond stretching, valence-angle bending, and dihedral angle torsions (internal rotations and flexions), within a given molecular backbone. Because the CL&P is based on the OPLS-AA framework, a large part of the Lennard-Jones, bond, and angle parameters could be transferred without major modifications. This is not surprising (or inconsistent) since many ionic liquids are derived from (neutral) molecules that are contained in the OPLS-AA database (originally developed for organic compounds and residues that are part of biologically relevant molecules like the DNA bases or amino acids). Nevertheless, this sort of compatibility leaves out two of the most important features of many of the ions that compose ionic liquids, viz their asymmetrical charge distribution and their particular conformational flexibility. As it was stressed above, these are parameterized by the atomic point charges,

responsible for modeling the electrostatic field around each ion, and by the parameters associated with the torsions around the dihedral angles, which must reproduce the correct conformational landscape of the ions. These will be discussed in the following Sects. 2.2 and 2.3. First—in Sect. 2.1—we will address the issue of transferring bond and angle parameters.

2.1 Stretching and bending: geometry optimization

Most often than not molecular geometry parameters can be reliably obtained from relatively low-level ab initio or DFT calculations (HF, B3LYP) using simple basis sets (for instance 6-31G) describing a single molecule or ion.

As an example, one can take the ubiquitous cations of the 1-alkyl-3-methylimidazilium family ($[C_n mim]^+$, Fig. 3): several models [2, 3, 5, 6] used similar methods to estimate the parameters describing the imidazolium ring and the published results are comparable with experimental data obtained by diffraction studies [19, 20], as shown in Table 2. Two conclusions can be drawn directly from the table: i) the ring geometry is not strongly affected by the environment of the imidazolium molecule since the ring geometry in two completely different crystals, $[C_2 mim][VOCl_4]$ [19] and $[C_{12} mim][PF_6]$ [20], is comparable between them and similar to the quantum calculation values for the isolated imidazolium cation; and ii) the distortion of the ring caused by different alkyl substituents is so small that the use of a symmetrical ring geometry as in a (symmetrical) $C_1 mim^+$ cation represents a good approximation.

Therefore, the first step in obtaining the CL&P force field parameterization for a new type of ion has been its geometry optimization using HF ab initio calculations and the comparison of the obtained structural data (bond lengths and angles) with relevant published diffraction and other electronic structure results. Most bond and angle parameters have been taken from or based on the OPLS-AA and AMBER force fields [15–17] without any modification. However, whenever significant differences were found between our ab initio geometries and reported OPLS-AA or AMBER parameters, notably in equilibrium distances and angles, we have proposed new values.

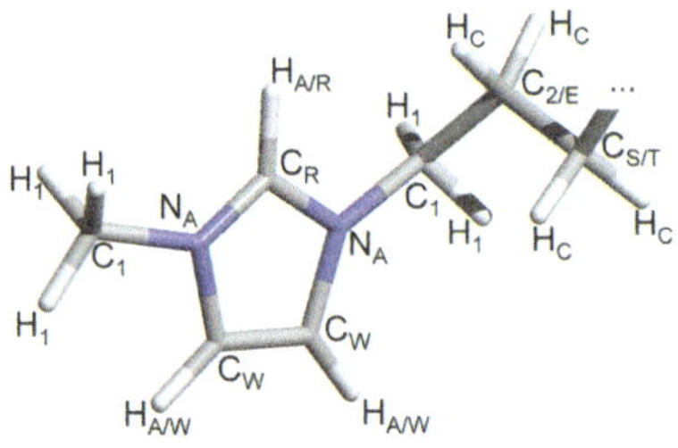

Fig. 3 Nomenclature of 1-alkyl-3-methyl-imidazolium, $[C_n mim]^+$

Table 2 Comparison between experimental x-ray (XR) geometries and single-molecule ab initio (AI) calculations for the imidazolium ring contained in different 1-alky-3-methylimidazolium ions

	$[C_2mim]^+$ XR[a] [19]	$[C_{12}mim]^+$ XR[b] [20]	$[C_2mim]^+$ AI [5]	$[C_2mim]^+$ AI [6]	$[C_2mim]^+$ AI [3]
Bonds (pm)[c]					
N_A–C_R	131.1 (4)	132.2(3)	–	131.4	131.5
	131.1 (4)	132.6(3)		131.5	
N_A–C_W	135.7 (5)	137.3(3)	–	137.8	137.8
	136.0 (6)	137.4(3)		137.8	
C_W–C_W	133.4 (8)	133.4(8)	–	134.2	134.1
N_A–C_1	145.2 (4)	146.8(3)	–	146.6	146.6
	146.8 (4)	147.7(3)		147.8	
C_1–C_E	150.0 (9)	–	–	152.0	–
Angles (°)[c]					
N_A–C_R–N_A	109.6 (3)	–	109.8	109.9	109.8
N_A–C_W–C_W	107.1 (3)	–	108.1	107.0	107.1
	107.6 (4)		108.0	107.2	
C_W–N_A–C_R	108.0 (3)	–	106.9	107.9	108.0
	107.6 (2)		106.8	108.0	
C_W–N_A–C_1	125.9 (3)	–	–	125.6	125.6
	125.2 (3)			125.9	
C_R–N_A–C_1	126.5 (3)	–	–	126.4	126.3
	125.4 (3)			126.1	

[a] $[C_2mim][VOCL_4]$ crystal

[b] $[C_{12}mim][PF_6]$ crystal

[c] Double entries refer to bond lengths and angles on the alkyl and methyl side of the imidazoilum ring, respectively

For instance, and taking again as an example the imidazolium ring, we have established that the use of imidazole bond distances and angles (present in the OPLS-AA database) to model the imidazolium ring of an alkylimidazolium cation can lead to an inaccurate parameterization: unlike the imidazolium ring, the imidazole ring residue is not symmetrical and the distortions of the structure around the carbon connected to the two (non-equivalent) nitrogen atoms are noticeable.

As far as the force constants of the stretching and bending modes are concerned, most of the corresponding parameters could be obtained from the AMBER and OPLS-AA force fields without modification. When these were not available, we have estimated them using normal-mode analyses on the optimized ab initio geometries or (as a last resource) the empirical correlation suggested by Halgren [21]. The main departure from the OPLS-AA force field regards our use of constrained bond lengths for all stretching modes involving hydrogen atoms and constrained 180° angles for ions like dicyanamide [9]—these allow longer simulation time steps (2 fs) to be used in the molecular dynamics simulations. Otherwise, all ions have been modeled as flexible structures (from the stretching and bending points of view). Even the hexafluorophosphate and nitrate anions that were parameterized as rigid ions in the first article describing the CL&P force field [2, 3] have been re-parameterized as flexible ions in subsequent versions [8, 9].

2.2 Torsion: potential energy profiles

The dihedral functions in OPLS-AA, expressed by a cosine series (Fig. 2), must not render the full potential energy related to the torsion around a given covalent bond (the same is true in other force field specifications). This is because non-bonded interactions, that is, Lennard-Jones and Coulombic terms, also act between sites within the same molecule connected by three or more bonds. In OPLS-AA, such non-bonded intramolecular interactions are described by the same parameters used in intermolecular interactions, scaled by a factor of 0.5 in the case of atoms situated exactly 3 bonds apart. These non-bonded (steric/electrostatic) effects account for a part, sometimes predominant, of the torsion energy profile. Thus, the task of fitting force field parameters to torsion energies is particular and specific, in the sense that the non-bonded interactions (Lennard-Jones and coulombic) that contribute to the torsional energy profiles must be accounted for. The significance of these non-bonded contributions is not easily estimated a priori [22]. Therefore, contrary to the cases of bond lengths and valence angles for which parameters can be taken from established force fields without major concerns, the torsion energy profiles depend on the cosine series (cf. Fig. 2) but also on the intermolecular features of the model. Therefore, it is fundamentally inconsistent to take parameters for torsion energies from the literature and then calculate electrostatic charge distributions or modify existing terms of the non-bonded interactions. If this is done, then the model will likely not be able to reproduce conformations in a satisfactory manner. In other words, it is not correct to transpose the parameters of the cosine series between different ions: these should be readjusted on a case-by-case basis to attain agreement with the non-bonded part.

In ionic liquids, only the parts of the ions whose atomic point charges did not suffer substantial re-parameterization (see Sect. 2.3) or whose constrains were dominated by bonded interactions had their torsion parameters taken directly from the AMBER/OPLA-AA force fields [15–17]. These include the parameterization of alkyl chains sufficiently removed from the high charge-density parts ("polar head-groups") of the ions or the proper/improper dihedral angles used to define planar, aromatic rings. All other torsion parameters (specially, those modeling the dihedral angles between polar and non-polar parts, rings and their alkyl side chains, or "conjugate" dihedral angles) were defined with the aid of torsion energy profiles calculated ab initio.

There are several methods to attain internal consistency between the conformational and intermolecular terms. A stepwise approach was chosen [16, 17, 23] in which the model is built up in such a way that the parameters for a given dihedral angle are invariant for any molecular structure within the same homologous series in which that dihedral term may occur, and without loss of accuracy in representing the conformational features of each particular ion. Other strategies exist in which a set of parameters is obtained for each ion individually, but here generality and transferability are lost [24]. The conformational landscapes in the CL&P force field reproduce electronic structure calculations at the post-HF MP2 theoretical level using relatively extended basis sets, cc-pVTZ(-f) [2, 3, 8–11, 23]. These are schematized in Fig. 4.

For a given dihedral, the parameters of the cosine series have been fitted to the difference between the a) total torsion energy profile calculated ab initio, and b) the

torsion energy profile corresponding to the non-bonded contribution (Fig. 4a, b). In the former case (1) the profiles have been obtained using geometry optimizations at the RHF/6-31G(d) level followed by single-point energy calculation at the MP2/cc-pVTZ(-f) level of configurations

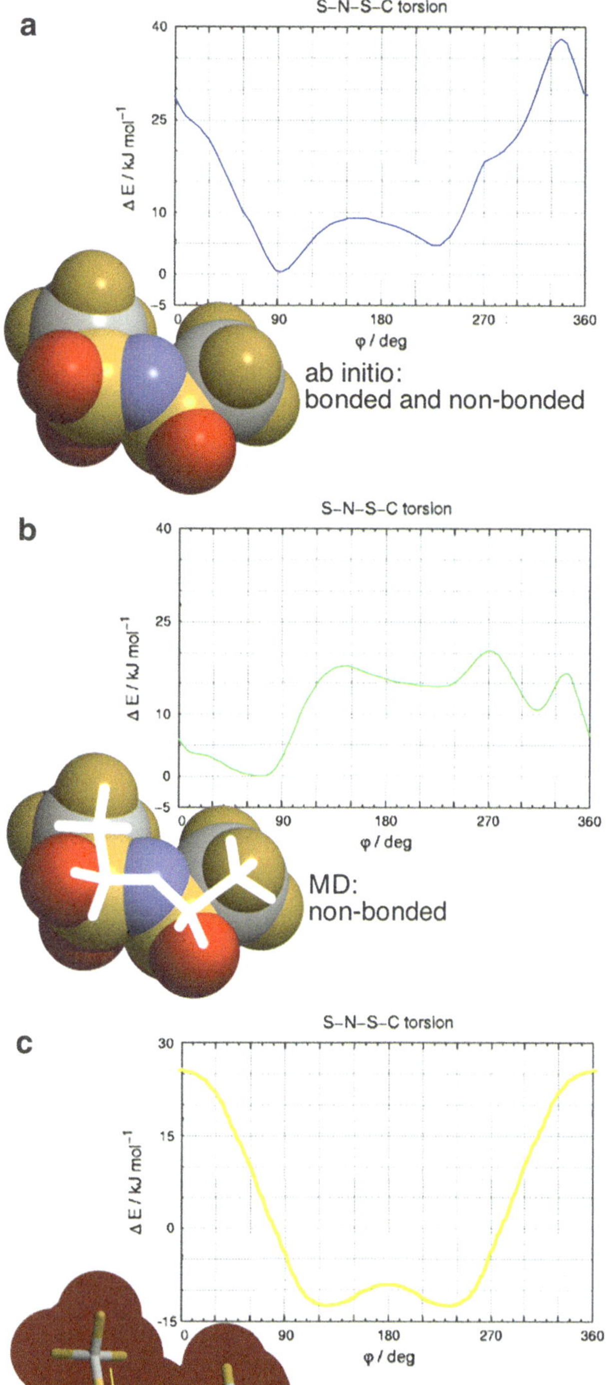

Fig. 4 The parameterization of the S–N–S–C dihedral angle of the bis(trifluoromethylsulfonyl)imide anion, [Ntf₂]⁻, using **a** the total torsion energy profile calculated ab initio, **b** the non-bonded component of the energy profile using molecular dynamics simulations with null dihedral angle parameters, and **c** the fitting of the difference between **a** and **b** to yield the required parameters. It must be stressed that the [Ntf₂]⁻ anion has two interdependent *dihedral angles* around the central imide group that are particularly difficult to parameterize. However, the difference of the two highly asymmetrical profiles given by **a** and **b** yields the symmetrical profile (**c**), proofing that the irregularity of the total torsion profile stems from non-bonded interactions and the soundness of the method used to parameterize this type of dihedral angles. Further validation was also possible using spectroscopic data related to two most stable conformations of this anion [25]

constrained at a fixed value of the dihedral angle under scrutiny. These have been followed by (2) molecular dynamics runs under similar constraint conditions and with the parameters of the selected dihedral cosine series set to zero. Finally, the correct parameterization can be checked by running MD simulations with the full torsion parameterization and evaluating the difference between the obtained results and the total torsion energy profiles calculated ab initio.

2.3 Non-bonded parameters

The OPLS-AA/AMBER models [15–17] both comprise repulsion-dispersion terms represented by a 12–6 Lennard-Jones potential function and an electrostatic term represented by partial charges located at each interaction site of the molecule.

Most of the Lennard-Jones parameters for each type of atom were taken from the OPLS-AA parameter set [16, 17]. The interactions between atoms of different type (cross parameters) have been parameterized using geometric-mean Lorentz-Berthelot mixing rules. The parameters for the halogen and nitrate anions have been obtained by fitting their Born–Mayer potentials developed for crystalline and molten salts [2, 3] to the Lennard-Jones function. It is important to stress that the resulting Lennard-Jones parameters of these inorganic anions are quite different from those usually employed for chlorine or nitrate groups bonded to organic molecules or from those obtained for chloride or nitrate anions in aqueous solutions. The differences are apparent if we take for instance the interaction diameter of the chloride ion/chlorine atom, σ_{Cl}, in the three situations: 3.77 Å in molten salts/ionic crystals [2, 3, 26], around 3.5 Å in organic chlorine atoms [4, 15–17] and around 4.4 Å in aqueous solution [15].

The coulomb parameters describing the electrostatic forces acting on each ion are also determined using calculations of the electron density at the MP2 theoretical level and using sufficiently large basis sets (cc-pVTZ(-f)). The point charges placed at the center of each atom of the ion are then calculated from the electron density using an electrostatic surface potential methodology (CHelpG), in which the values of the atomic charges are optimized to reproduce the electrostatic field generated by the molecule. The most stable conformation obtained by geometry optimization of the isolated ion is generally taken as the reference but, in ions with multiple stable conformers, several ab initio calculations were performed, and an averaging process was implemented to define the charges in each atom. Since one of the objectives of the present force field is the possibility of parameter transfer within families of ionic liquids, the point charges attributed to the various molecular residues were subjected to different degrees of

approximation to find general trends that could then be applied along an entire series of analogous ionic liquids, without the need to perform quantum calculations and parameter fitting procedures for each ion within a family. These approximations include: (1) the charges on equivalent atoms are the same (for instance, the three hydrogen atoms of a methyl group will have averaged-out point-charges obtained from the ab initio data; (2) symmetrical or slightly asymmetrical rings exhibit similar charges on the "symmetrical" atoms (for instance, cf. Fig. 3, the C4 and C5 atoms of imidazolium ions or the ortho atoms of pyridinium ions); (3) the length of alkyl side chains does not affect the charge distribution on the high charge-density parts of the ions; (4) the charge distribution on carbon and hydrogen atoms situated in alkyl side chains three bonds removed from the high charge-density parts of the ion are given the corresponding OPLS-AA values for alkanes [16]; v) the charge in the atoms that establish the connection between the high charge-density parts of the ion and its alkyl side chains is adjusted empirically in order to respect the total charge of the ion (for instance, cf. Fig. 3, C_E in $[C_2 \, \mathrm{mim}]^+$ and C_2 in $[C_n \, \mathrm{mim}]^+$ with $n > 2$). Interestingly, these "junction" atoms exhibit the same charge in different families of ionic liquids [9, 10] (for instance both in 1-alkyl-3-methylimidazolium and in N-alkylpyridinium cations).

3 Validation

The CL&P model has permitted the access to the molecular properties of ionic liquids through computer simulation "virtual experiments". Before these results could be fully explored, it was necessary to validate the force field parameters by comparing experimental results to the in silico measurements. Ionic liquids have a non-measurable vapor pressure at room temperature conditions or indeed for the larger part of their liquid temperature range. This means that the traditional way to validate a molecular force field, to be used within the framework of statistical mechanics calculations in condensed phases, was somehow curtailed at the time the model was introduced. The validation procedure had to rely exclusively on volumetric data available for some selected ionic liquids [2, 3, 8–11]. This is not a comfortable situation, since it is better to have at least one property related to the length scale of the ions (volumetric data) and at least one related to the energy scale (such as the heat of vaporization). However, and since the force field had been built in a systematic way, it was possible to test it against series of homologous ionic liquids, by predicting their densities both in the liquid and in the solid states and comparing the results with experimental data (liquid density and X-ray diffraction data).

It must be stressed that in the case of solid state data, not only the volumetric data per se were validated but also the parameters defining the crystalline unit cells (in some cases simulations started with a spurious initial crystalline structure were allowed to evolve and equilibrate to the correct crystalline structure [2, 3]).

After the discovery that ionic liquids could be vaporized at moderately high temperature (around 550 K) and reduced pressure (around 500 Pa) [27], new experimental data related to their liquid–vapor equilibrium, and thus cohesive energy, became available, and it was possible to test the performance of the CL&P force field against different estimations of the enthalpy of vaporization of 1-alkyl-3-methylimidazolium bis(trifluoromethylsulfonyl) imide ionic liquids. Although the uncertainties both in the simulation and experimental results are still big at this stage due to the extremely difficult nature of the experiments and constraints imposed on the simulation methods (poor statistics of the simulation of an isolated ionic pair in the gas phase), it was possible to conclude that for this particular family of ionic liquids, the enthalpies of vaporization at room temperature are in the 120–160 kJ/mol range (with MD over-predicting the experimental results by 15–30%) [28]. It must be stressed that the use of the model is in all cases (density and calorimetric estimations) purely predictive, which means that deviations of a few percentage points in the case of density and a few tens of percentage points in the case of the enthalpies of vaporization are in fact quite reasonable.

Validation with another property related to the energy scale (surface tension data would be an obvious choice) proved unfeasible due to the lack of reliable experimental data and also to the large uncertainty associated with the results obtainable by MD simulation. However, very recent results have shown that it is possible to estimate correct surface tension values using the CL&P force field if sufficiently long simulation times are employed [29].

4 Past, present, and future developments

In order to improve transferability, the parameterization of ionic liquids in the CL&P force field was not meant be too specific because the objective was to deal with families of similar compounds that could be combined with different counter-ions. Homologous series may be expressed in the cation, such as in alkylmidazoliums, or in the anion, such as alkylsulfates. Force field parameterization was concentrated on parts of the molecules that were common to an entire family of ions, and strategies were adopted to add specific molecular residues or functional groups. Moreover, the anions and cations were modeled independently, meaning that an ionic liquid can be assembled from any available cation–anion combination (as illustrated in Fig. 1). This transferability implies different sorts of approximation. For example, it does not account for the possibility of charge-transfer effects between ions, an issue that was thoroughly discussed at different stages of force field development and that may not be very significant in condensed phases, where each ion has a large coordination number of neighbors of opposite charge. Also, polarization of electron clouds is not taken into account explicitly, although it was included a posteriori through a number of refinement schemes (see below). Inclusion of explicit polarization accelerates the microscopic dynamics obtained with the models but has little effect on the estimation of equilibrium or structural properties (cf. below).

Most force field parameterizations that appeared at approximately the same time of the CL&P model included either united-atom (UA) or all-atom (AA) descriptions based on extensions of well-established force fields like CHARMM [18], AMBER [15], or OPLS [16, 17] with extra parameters obtained from quantum calculations. As examples, we can cite the work based on the UA force field of Ed Maggin and co-workers [4] used to predict gas solubilities [30], the structure and dynamics studies by Mauro Ribeiro using his own UA force field [31], or the introduction by Maggin and co-workers of a new AA version of their own force field, used to predict the solubility of CO_2 in imidazolium-based ionic liquids [32]. These examples reflect a shift from united-atom models to all-atom models, attempts to encompass whole families of ionic liquids (breaking with the past tradition of modeling ionic liquids on a one-to-one basis), and some extra care in incorporating into the model some degree of flexibility (the description of the torsion movements of the ions are attempted in some cases). These advances show in fact a general convergence between the different approaches to force field development, in agreement with the basic postulates behind the development of CL&P (Fig. 1). At that time, a model based on force matching to Car-Parinello molecular dynamics simulation of ionic liquids was also introduced [33], but poor rendering of dispersion forces and the extremely small size and time scales of these simulations are major difficulties when treating ionic liquids and their slow dynamics.

More recent developments witnessed the emergence of the modeling of ionic liquid plus molecular solvent systems. In terms of force field development, this means that the premises of transferability/compatibility claimed by most models have to be valid not only between the ions of the ionic liquid but also between those ions and molecular (neutral) species. All-atom models based on traditional force fields (CHARMM [18], AMBER [15], or OPLS [16, 17]) have a clear advantage in addressing this problem since most of the neutral species are already parameterized within those force fields. Examples of ionic liquid mixtures and solutions studied by MD include: fluid-phase equilibria based on MD simulations using the CL&P force field [34, 35]; the discussion of solvation effects in ionic liquids [36] using MD results obtained using a CHARMM-based rigid force field [37]; the analysis of ionic liquid–solute interactions [38] and the analysis of the interplay between solvation and nanostructure [33], both based on the CL&P force field [2, 3, 8–11]; interpretations of large-angle X-ray scattering experimental data [39] using derived versions of the CL&P force field [2, 3]; Monte Carlo computations [40] of pure and binary gas isotherms in ionic liquids using parameters form the Maggin force field [41] to model the cations and the CL&P force field to model the bistriflamide anion [8]; or studies of the solvation of toluene in different ionic liquids in order to interpret their specific catalytic capabilities [42] using the CL&P force field [2, 3, 8].

In the latest years, ionic liquid modeling witnessed a further shift from bulk conditions to non-equilibrium or interfacial settings, and many of the newer models include polarization explicitly. Examples include the discussion of the interfacial structure of ionic liquids using x-ray reflectometry studies aided by a self-consistent mean field theory (SCMFT) model and MD simulations using a polarizable force field based on AMBER parameterizations [43]; the characterization of ionic liquids that can be used as electrolytes for batteries and super-capacitors (including the correct description of the transport of lithium ions in an ionic liquid media) using a polarizable, transferable, and quantum-chemistry-based force field denominated as Atomistic Polarizable Potential for Liquids, Electrolytes, & Polymers (APPLE&P) [44, 45]; the study of different properties of ionic liquids such as interfacial structure, self-diffusion, and viscosity using coarse-grain models, namely the effective force coarse-graining (EF-CG) method [46]; the analysis of ionic liquids under non-equilibrium conditions such as high electric fields [47] using an AMBER-based all-atom force field [5] and the EF-CG coarse-grained [46] force field; studies related to the structure of the electrochemical interface between a graphite surface and an ionic liquid using an all-atom AMBER-based force field [48]; the determination of self-diffusion coefficients in a series of ionic liquids using an adapted and extended version of the CL&P force field [49]; or, the estimation of different non-equilibrium properties of amino-functionalized ionic liquids using an all-atom AMBER-based force field [50].

Due to the complexity of ionic liquid phases, both in structural and dynamic terms, and the wide range of space and time scales that have to be considered in order to capture the behavior of these fluids, multi-scale approaches are a promising route. Here, information obtained at a more detailed level is used to impose consistency when building

a coarser model. The levels of description that can be connected in this way are electronic structure calculations, atomistic force fields, and coarse-grained models [51].

These developments have been complemented by other new contributions in terms of force field refinement given by Ludwig and co-workers [52] and Siehl [53]. The latter author introduced the possibility of modeling charge transfer between ions as a way to account for polarization effects in ionic liquids and their solutions; the former authors produced a refined version of the CL&P force field [2, 3, 8], capable of taking into account the specific interactions between the acidic hydrogen atoms of dialkylimidazolium cations and different anions. All these contributions represent valid routes to the development of force fields that are fine-tuned to certain classes of ionic liquids, where problems such as polarizability, charge transfer, hydrogen bonding, or the occurrence of specific interactions may hinder the use of more general models.

The development and use of different types of force field for ionic liquids can be summarized as follows: Different force fields with different degrees of resolution from coarse-grain to atomistic models have emerged in the past decade. All-atom models have been the most popular due to their general character and the straightforward way in which they can be integrated with vast force field parameter sets for common molecules. United-atom models (halfway between all-atom and coarse-grain models) are loosing popularity, with most simulations being run at the atomistic or coarse-grain levels. The latter type of models are specially suitable for non-equilibrium conditions but sometimes lack the detail necessary to the correct description of specific interactions, a situation that is particularly relevant in the case of mixtures of ionic liquids with molecular solvents. Quantum mechanical simulation methods represent the ultimate models in terms of "resolution" but are costly in terms of computing time and were only applied to a few systems.

On the other hand, in order to improve transferability (see Fig. 1), the parameterization of ionic liquids in systematic atomistic force fields like CL&P was not meant be too specific because the objective was to model families of homologous ions that could be combined with different counter-ions. This meant that, on one hand, the parameterization was concentrated on parts of the ions that defined entire homologous series and, on the other hand, that the anions and cations were parameterized independently. This sort of transferability implies different types of approximation. For example, it does not account for the possibility of charge-transfer effects between different ions (the charges are usually obtained ab initio from isolated ions). In the condensed phase where each ion is surrounded by neighbors of opposite charge such effect does not have a profound impact in terms of structure but will certainly play a role in terms of the intensity of possible specific interactions between ions and the formation of (transient) ion pairs. [52] Also, polarization of electron clouds is not taken into account explicitly, although was included a posteriori through a number of refinement schemes [43–45]. Inclusion of explicit polarization accelerates the microscopic dynamics obtained with the models but has little effect on the estimation of equilibrium or overall structural properties [54]. On the other hand, local structural effects or specific interactions can be affected by the modeling or not of polarizability effects. [55, 56].

The introduction of these two types of correction/approximation (charge-tranfer and polarization methods) represent two major routes to the refinement of systematic (either atomistic or coarse grained) models. In fact the two types of approximation/correction have been somewhat merged into a single procedure [54] that tries to account for the lack of polarizability effects in the CL&P by using an ad-hoc correction factor applicable to all parameterized point charges that mimics a charge-transfer process between ions. The resulting anions and cations—with overall charges around (-0.8 to -0.9) and ($+0.8$ to $+0.9$), respectively—interact via subdued electrostatic interactions which in turn lead to a better description of the available enthalpy of vaporization and volumetric data, faster dynamics and (possibly) a less stringent separation between polar and non-polar domains (nano-segregation) of the ionic liquids [57, 58]. The task of using this kind of procedure in a systematic and coherent way (it is for instance questionable the application of the correction factor to the charges of the alkyl side chains that are not subjected to noticeable polarizability effects) is one of the challenges to be addressed in future developments of the CL&P force field.

Acknowledgments JNCL acknowledges the FCT projects PTDC/QUI–QUI/101794/2008 and PTDC/ECM/103490/2008.

References

1. Canongia Lopes JN, Pádua AAH (2006) J Phys Chem B 110(7):3330
2. Canongia Lopes JN, Deschamps J, Pádua AAH (2004) J Phys Chem B 108(6):2038
3. Canongia Lopes JN, Deschamps J, Pádua AAH (2004) J Phys Chem B 110(30):11250
4. Hanke CG, Price SL, Linden-Bell R (2001) Mol Phys 99(10):801
5. de Andrade J, Boes ES, Stassen HJ (2002) J Phys Chem B 106(14):3546
6. Shah JK, Brennecke JF, Maginn EJ (2002) Green Chem 4:112
7. Margulis CJ, Stern HA, Berne BJ (2002) J Phys Chem B 106(46):12017
8. Canongia Lopes JN, Padua AAH (2004) J Phys Chem B 108(43):16893
9. Canongia Lopes JN, Padua AAH (2006) J Phys Chem B 110(39):19586
10. Canongia Lopes JN, Padua AAH, Shimizu K (2008) J Phys Chem B 112(16):5039
11. Shimizu K, Almantariotis D, Costa Gomes MF, Padua AAH, Canongia Lopes JN (2010) J Phys Chem B 114(10):3592
12. Bates ED, Mayton RD, Ntai I, Davis JH (2002) J Am Chem Soc 124(6):926
13. Pensado AS, Costa Gomes MF, Padua AAH (2011) J Phys Chem B 115(14):3942
14. Pensado AS, Costa Gomes MF, Canongia Lopes JN, Malfreyt P, Padua AAH (2011) Phys Chem Chem Phys 13:13518
15. Cornell WD, Cieplak P, Bayly CI, Gould IR, Merz KM, Ferguson DM, Spellmeyer DC, Fox T, Caldwell JW, Kollman PA (1995) J Am Chem Soc 117(19):5179
16. Jorgensen WL, Maxwell DS, Tirado-Rives JJ (1996) J Am Chem Soc 118(45):11225
17. Kaminski G, Jorgensen WL (1996) J Phys Chem 100(46):18010
18. MacKerell AD, Bashford D, Bellott M, Dunbrack RL, Evanseck JD, Field MJ, Fischer S, Gao J, Guo H, Ha S, Joseph-McCarthy D, Kuchnir L, Kuczera K, Lau FTK, Mattos C, Michnick S, Ngo T, Nguyen DT, Prodhom B, Reiher WE, Roux B, Schlenkrich M, Smith JC, Stote R, Straub J, Watanabe M, Wiorkiewicz-Kuczera J, Yin D, Karplus M (1998) J Phys Chem B 102(18):3586
19. Hitchcock PB, Lewis RJ, Welton T (1993) Polyhedron 12(16):2039
20. Gordon CM, Holbrey JD, Kennedy AR, Seddon KR (1998) J Mater Chem 8:2627
21. Halgren TA (1990) J Am Chem Soc 112(12):4710
22. Pádua AAH (2002) J Phys Chem A 106(43):10116
23. Bonifacio RP, Filipe EJM, McCabe C, Costa Gomes MF, Padua AAH (2002) Mol Phys 100(15):2547
24. Liu ZP, Huang SP, Wang WC (2004) J Phys Chem B 108(34):12978
25. Lopes JNC, Shimnizu K, Padua AAH, Umebayashi Y, Fukuda S, Fujii K, Ishiguro S (2008) J Phys Chem B 112(5):1465
26. Tosi MP, Fumi G (1964) J Phys Chem Solids 25(1):45
27. Earle MJ, Esperanca J, Gilea MA, Lopes JNC, Rebelo LPN, Magee JW, Seddon KR, Widegren JA (2006) Nature 439:831
28. Esperanca JMSS, Canongia Lopes JN, Tariq M, Santos LMNBF, Magee JW, Rebelo LPN (2010) J Chem Eng Data 55(1):3
29. Shimizu K, Pensado A, Malfreyt P, Padua AAH, Canongia Lopes JN (2012) 154:155
30. Kumelan J, Kamps APS, Urukova I, Tuma D, Maurer G (2007) J Chem Thermodyn 37:595
31. Urahata SM, Ribeiro MCC (2004) J Chem Phys 120:1855
32. Morrow TI, Maginn EJ (2002) J Phys Chem B 106:12807
33. Del Popolo MG, Lynden-Bell RM, Kohanoff JJ (2005) Phys Chem B 109:5895
34. Rebelo LPN, Canongia Lopes JNC, Esperança JMSS, Guedes HJR, Lachwa J, Visak VN, Visak ZP (2007) Acc Chem Res 40:1114
35. Canongia Lopes JN, Costa Gomes MF, Padua AAH (2006) J Phys Chem B 110:16816
36. Chowdhury PK, Halder M, Sanders L, Calhoun T, Anderson JL, Armstrong DW, Song X, Petrich JW (2004) J Phys Chem B 108:10245
37. Shim Y, Duan J, Choi MY, Kim HJJ (2003) Chem Phys 119:6411
38. Youngs TGA, Holbrey JD, Deetlefs M, Nieuwenhuyzen M, Gomes MFC, Hardacre C (2006) ChemPhysChem 7:2279
39. Fujii K, Mitsugi T, Takamuku T, Yamaguchi T, Umebayashi Y, Ishiguro S (2009) Chem Lett 38(4):340
40. Shi W, Maginn EJ (2008) J Phys Chem B 112(7):2045
41. Cadena C, Zhao Q, Snurr RQ, Maginn EJ (2006) J Phys Chem B 110(6):2821
42. Gutel T, Santini CC, Padua AAH, Fenet B, Chauvin Y, Canongia JN, Bayard F, Gomes MFC, Pensado AS (2009) J Phys Chem B 113(1):170

43. Lauw Y, Horne MD, Rodopoulos T, Webster NAS, Minofar B, Nelson A (2009) Phys Chem Chem Phys 11(48):11507
44. Borodin O, Smith GD (2006) J Phys Chem B 110:11481
45. Borodin O, Smith GD, Henderson W (2006) J Phys Chem B 110:16879
46. Wang Y, Feng S, Voth GA (2009) J Chem Theory Comput 5(4):1091
47. Daily JW, Micci MM (2009) J Chem Phys 131(9):094501
48. Kislenko SA, Samoylov IS, Amirov RH (2009) Phys Chem Chem Phys 11(27):5584
49. Tsuzuki S, Shinoda W, Saito H, Mikami M, Tokuda H, Watanabe M (2009) J Phys Chem B 113(31):10641
50. Liu X, Zhou G, Zhang S, Yao X (2009) Fluid Phase Equilib 284(1):44
51. Wang Y, Jiang W, Yan T, Voth GA (2007) Acc Chem Res 40(11):1193
52. Koddermann T, Paschek D, Ludwig R (2007) Chem Phys Chem 8(17):2464
53. Tanaka M, Siehl H-U (2008) Chem Phys Lett 457(1–3):263
54. Bhargava BL, Balasubramanian S, Klein M (2008) Chem Commum 29:3339
55. Bedrov D, Borodin O, Li Z, Smith GD (2010) J Phys Chem B 114:4984
56. Borodin O (2009) J Phys Chem B 113:11463
57. Smith GD, Borodin O, Li L, Kim H, Liu Q, Bara JE, Gin DL, Nobel RD (2008) Phys Chem Chem Phys 10:6301
58. Smith GD, Borodin O, Magda JJ, Boyd RH, Wang Y, Bara JE, Millerm S, Gin DL, Noble RD (2010) Phys Chem Chem Phys 12:7064

2 Springer

Theor Chem Acc (2012) 131:1143
DOI 10.1007/s00214-012-1143-9

REGULAR ARTICLE

Including many-body effects in models for ionic liquids

Mathieu Salanne · Benjamin Rotenberg ·
Sandro Jahn · Rodolphe Vuilleumier ·
Christian Simon · Paul A. Madden

Received: 28 April 2011 / Accepted: 11 September 2011 / Published online: 17 February 2012
© Springer-Verlag 2012

Abstract Realistic modeling of ionic systems necessitates taking explicitly account of many-body effects. In molecular dynamics simulations, it is possible to introduce explicitly these effects through the use of additional degrees of freedom. Here, we present two models: The first one only includes dipole polarization effect, while the second also accounts for quadrupole polarization as well as the effects of compression and deformation of an ion by its immediate coordination environment. All the parameters involved in these models are extracted from first-principles density functional theory calculations. This step is routinely done through an extended force-matching procedure, which has proven to be very successful for molten oxides and molten fluorides. Recent developments based on the use of localized orbitals can be used to complement the force-matching procedure by allowing for the direct calculations of several parameters such as the individual polarizabilities.

Published as part of the special collection of articles: From quantum mechanics to force fields: new methodologies for the classical simulation of complex systems.

M. Salanne (✉) · B. Rotenberg · C. Simon
UPMC Université Paris 06, CNRS,
ESPCI, UMR 7195, PECSA, 75005 Paris, France
e-mail: mathieu.salanne@upmc.fr

S. Jahn
GFZ German Research Centre for Geosciences,
Telegrafenberg, 14473 Potsdam, Germany

R. Vuilleumier
ENS, CNRS, UMR 8640, PASTEUR, 75005 Paris, France

P. A. Madden
Department of Materials, University of Oxford, Parks Road,
Oxford OX1 3PH, United Kingdom

Keywords Ionic liquids · Polarizable force field · Force-fitting

1 Introduction

Ionic liquids are a class of Coulombic liquids [1], which encompasses several types of compounds. They are often separated into just two categories, inorganic molten salts and room-temperature ionic liquids, although this may be too restrictive. For example, inorganic molten salts contain various families based on monoatomic anions, such as molten halides or molten oxides, but also other ones in which molecular anions are involved such as carbonates and nitrates. In the case of room-temperature ionic liquids, organic ions are involved, which opens the way for an almost infinite number of potential electrolytes exhibiting a wide range of physical and chemical properties [2].

Ionic liquids play an important role in many fields of chemistry and physics. Apart from the intrinsic interest for such systems, they have been studied by computer simulations in many contexts. For example, molten oxides are the main components of magmatic melts [3–6], molten fluorides and chlorides are investigated for their importance in many metallurgical processes [7–9] and for their potential use in the nuclear industry [10, 11], and room-temperature ionic liquids for their use in electrochemical storage applications [12–14] or their solvation properties [15–17].

In order to obtain the thermodynamic and transport properties of interest for all these applications, it is necessary to simulate large enough samples for a sufficiently long time, which is out of reach of ab initio molecular dynamics methods. This is done instead via classical simulations, in which interactions are described by an analytical force field derived from a model of the interactions between closed-shell species. Such a model [18] must account not only for the classical electrostatic interaction, but also for three interactions arising from the quantum nature of electrons. The exchange-repulsion or van der Waals (VdW) repulsion is a consequence of the Pauli principle, while the dispersion (VdW attraction) arises from correlated fluctuations of the electrons. Lastly, the induction term reflects the distortion of the electron density in response to electric fields, which are dominated by polarization effects but may also partially represent *incipient* charge transfer associated with bond formation between closed-shell species [19].

In this paper, we review some methodological advances which have been proposed in order to include all these effects, and particularly the many-body polarization ones, in accurate models of ionic liquids. The critical physical factor to appreciate is that the electronic densities of anions, in particular, are strongly affected by their interactions with their environment. For example, the oxide ion, which as an isolated ion is unstable by 8 eV with respect to autoionization [20], only exists in the condensed phase because of the confining potential exerted on its electrons by its neighbors [21, 22]. The confining potential compresses and stabilizes the oxide ion and allows us to treat its interactions in the condensed phase as those of a closed-shell species, like an inert-gas atom. However, a model that is *transferable* from one material to another, or even between different phases of the same material, must recognize that the confining potential, and hence the intrinsic properties of the oxide ion will change between them. Furthermore, in a dynamical context, fluctuations in the environmental potential due to the thermal motion of the neighbors mean that the ion's electron density is also fluctuating and this effect must be included in the interaction model. The effect is less severe for halide ions, where the isolated ion is stable, but where the importance of the environmental effect is indicated by the observation that the polarizability of the fluoride ion in crystalline LiF is 6.2 a.u. compared to 16 a.u. for the gas-phase ion [21, 22]. One would expect that for the large molecular anions involved in room temperature systems, the effect would be weaker still. Our considerations will be directed at the inorganic molten salt systems for which the environmental potential effects are pronounced, but with the compensating simplification that the complexities of molecular shape are avoided. However, we note that similar methods have recently been applied to develop polarizable models for room temperature systems [23].

In the first part of the present paper, we provide the expression for the interaction potential in the framework of two models with different complexity: the aspherical ion model, necessary for a transferable description of oxides,

and the simpler polarizable ion model. We then show how all the parameters involved in these models can be determined from first-principles density functional theory calculations through an extended force-matching procedure. In a third part, we show recent developments based on the use of localized orbitals, which can complement the force-matching procedure by allowing for the direct calculations of several parameters such as the individual polarizabilities.

2 Interaction potentials

The functional forms introduced below have been proposed previously on the basis of separate examinations of each contribution to the ionic interaction energy in a series of well-directed electronic structure calculations on the condensed phase. [24–26] The potential is best described as the sum of four different components: charge–charge, dispersion, overlap repulsion and polarization.

$$V^{\text{total}} = V^{\text{charge}} + V^{\text{dispersion}} + V^{\text{repulsion}} + V^{\text{polarization}} \tag{1}$$

Two of these terms will attract more of our attention, the repulsion and polarization terms. These contributions to the overall energy depend on ionic properties like the ionic radius, for the repulsion, or the ionic multipoles, for the polarization. As the electronic density of the ions adapt to the environment that acts on it through a confining potential or local electric fields, these ionic quantities change in a way so as to minimize the overall energy of the system. The response to the environment of these additional degrees of freedom such as ionic radius and multipoles gives rise to many-body effects in the potential energy of the system.

The first term corresponds to the electrostatic interaction between two formal charges,

$$V^{\text{charge}} = \sum_{i<j} \frac{q^i q^j}{r^{ij}} \tag{2}$$

where q^i is the charge of ion i. The use of formal charges underpins the description of the other interactions as characteristic of a closed-shell system [27] and the expectation that the interaction parameters should be transferable. The dispersion component includes dipole–dipole and dipole–quadrupole terms

$$V^{\text{dispersion}} = -\sum_{i<j} \left(f_6^{ij}(r^{ij}) \frac{C_6^{ij}}{(r^{ij})^6} + f_8^{ij}(r^{ij}) \frac{C_8^{ij}}{(r^{ij})^8} \right) \tag{3}$$

where C_6^{ij} (C_8^{ij}) is the dipole–dipole (dipole–quadrupole) dispersion coefficient, and f_n^{ij} are Tang-Toennies dispersion damping functions, [28] describing the short-range penetration correction to the asymptotic multipole expansion of dispersion [27], which take the following form:

$$f_n^{ij}(r^{ij}) = 1 - e^{-b_n^{ij} r^{ij}} \sum_{k=0}^{n} \frac{\left(b_n^{ij} r^{ij}\right)^k}{k!} \tag{4}$$

where the b_n^{ij} parameter sets the range of the damping effect. The two other terms that include many-body effects may differ depending on the systems of interest, as the magnitude of the response to the environment varies from one species to another. In the most complex functional form, the so-called aspherical ion model (AIM), which has been introduced to model oxide materials [3, 29, 30], the overlap repulsion component is given by

$$\begin{aligned} V_{\text{AIM}}^{\text{repulsion}} &= \sum_{i \in O, j \in M} \left(A^{ij} e^{-a^{ij} \rho^{ij}} + B^{ij} e^{-b^{ij} \rho^{ij}} + C^{ij} e^{-c^{ij} r^{ij}}\right) \\ &+ \sum_{i,j \in O, i<j} A^{ij} e^{-a^{ij} r^{ij}} + \sum_{i,j \in M, i<j} A^{ij} e^{-a^{ij} r^{ij}} \\ &+ \sum_{i \in O} \left[D^i \left(e^{\beta^i \delta \sigma^i} + e^{-\beta^i \delta \sigma^i}\right) \right. \\ &\quad \left. + \left(e^{\zeta^i |v^i|^2} - 1\right) + \left(e^{\eta^i |\kappa^i|^2} - 1\right)\right] \end{aligned} \tag{5}$$

where

$$\rho^{ij} = r^{ij} - \delta \sigma^i - S_\alpha^{(1)} v_\alpha^i - S_{\alpha\beta}^{(2)} \kappa_{\alpha\beta}^i, \tag{6}$$

and summation of repeated indices is implied. Sets O and M are the sets of oxide anions and metallic cations, respectively, so that the first three summations represent cation–anion, anion–anion and cation–cation short-range repulsions, respectively. In the cation–anion term, $\delta \sigma^i$ is a variable that characterizes the deviation of the radius of the oxide anion i from its default value, i.e. its "breathing". $\{v_\alpha^i\}$ is a set of three variables, for each anion, describing the Cartesian components of a dipolar distortion of the ion shape. Similarly, $\{\kappa_{\alpha\beta}^i\}$ is a set of five independent variables describing the corresponding quadrupolar shape distortions. $|\kappa|^2 = \kappa_{xx}^2 + \kappa_{yy}^2 + \kappa_{zz}^2 + 2(\kappa_{xy}^2 + \kappa_{xz}^2 + \kappa_{yz}^2)$, and the two interaction tensors $S_\alpha^{(1)} = r_\alpha^{ij}/r^{ij}$ and $S_{\alpha\beta}^{(2)} = 3 r_\alpha^{ij} r_\beta^{ij}/(r^{ij})^2 - \delta_{\alpha\beta}$ are also used in Eq. 6. The last summations include the self-energy terms, that is the energy cost of deforming the charge density of an ion. β, ζ and η are effective force constants determining how difficult it is for a particular ion to be deformed in a spherical, dipolar or quadrupolar way.

The magnitude and orientation of the spherical breathing, dipolar and quadrupolar deformations will be determined by minimization of this energy, with respect to these variables: they will therefore depend on the instantaneous positions of neighboring ions and consequently change at each timestep in a molecular dynamics run.

It was assumed throughout that these shape deformations do not affect significantly the anion–anion and cation–cation repulsions, which have been represented by simple exponentials as if the ions were effectively spherical in their interactions with other ions of the same type. We have also assumed that short-range cation distortion effects can be neglected in the case of oxide materials.

The polarization part of the AIM potential includes both dipolar and quadrupolar contributions

$$\begin{aligned} V_{\text{AIM}}^{\text{polarization}} &= \sum_{i,j} \left[\left(q^i \mu_\alpha^j g_D^{ij}(r^{ij}) - \mu_\alpha^i q^j g_D^{ji}(r^{ij})\right) T_\alpha^{(1)} \right. \\ &+ \left(\frac{q^i \theta_{\alpha\beta}^j}{3} g_Q^{ij}(r^{ij}) + \frac{\theta_{\alpha\beta}^i q^j}{3} g_Q^{ji}(r^{ij}) - \mu_\alpha^i \mu_\beta^j\right) T_{\alpha\beta}^{(2)} \\ &+ \left(\frac{\mu_\alpha^i \theta_{\beta\gamma}^j}{3} + \frac{\theta_{\alpha\beta}^i \mu_\gamma^j}{3}\right) T_{\alpha\beta\gamma}^{(3)} + \left. \frac{\theta_{\alpha\beta}^i \theta_{\gamma\delta}^j}{9} T_{\alpha\beta\gamma\delta}^{(4)} \right] \\ &+ \sum_i \left(k_1^i \mid \boldsymbol{\mu}^i \mid^2 + k_2^i \mu_\alpha^i \theta_{\alpha\beta}^i \mu_\beta^i + k_3^i \theta_{\alpha\beta}^i \theta_{\alpha\beta}^i + k_4^i \mid \boldsymbol{\mu}^i \cdot \boldsymbol{\mu}^i \mid^2\right) \end{aligned} \tag{7}$$

where $k_1^i = \frac{1}{2\alpha^i}$, $k_2^i = \frac{B^i}{4(\alpha^i)^2 C^i}$, $k_3^i = \frac{1}{6C^i}$ and $k_4^i = \frac{-(B^i)^2}{16(\alpha^i)^4 C^i}$. α^i, B^i and C^i are the dipole, dipole–dipole–quadrupole and quadrupole polarizabilities of ion i. $T_{\alpha\beta\gamma\ldots} = \nabla_\alpha \nabla_\beta \nabla_\gamma \cdots (r^{ij})^{-1}$ are the multipole interaction tensors [27], with the superindex indicating the order of the operator. They are computed using the Ewald summation technique [31, 32]. The instantaneous values of the dipole and quadrupole moments are, again, obtained by minimization of this expression. The charge–dipole and charge–quadrupole cation–anion asymptotic terms are also corrected for penetration effects [33] at short-range by using Tang-Toennies damping functions [28]

$$g_D^{ij}(r^{ij}) = 1 - c_D^{ij} e^{-b_D^{ij} r^{ij}} \sum_{k=0}^{n} \frac{\left(b_D^{ij} r^{ij}\right)^k}{k!} \tag{8}$$

$$g_Q^{ij}(r^{ij}) = 1 - c_Q^{ij} e^{-b_Q^{ij} r^{ij}} \sum_{k=0}^{n} \frac{\left(b_Q^{ij} r^{ij}\right)^k}{k!} \tag{9}$$

with D and Q standing for the dipolar and quadrupolar parts. It should be noted that the b_D (b_Q) parameters, which determine the distance at which the overlap of the charge densities begins to affect the induced multipoles, are the same for both anions and cations, while the c_D (c_Q) parameters, which are set to unity in the case of the dispersion interaction (Eq. 4), measure the strength of the ion response to this effect and therefore depend on the identity of the ion. The necessity of larger-than-unity values was indicated by ab initio calculations of the induced multipoles in distorted crystals. [26] The short-range induction corrections are usually neglected in both anion–anion and cation–cation interactions.

The AIM potential can be seen to contain several (seventeen per oxide anion) additional degrees of freedom (induced dipoles and quadrupoles, and ion shape deformations), which describe the state of the electron charge density of the ions. When calculating the forces on the ions in a molecular dynamics simulation, these electronic degrees of freedom should have their adiabatic "Born-Oppenheimer" values, which minimize the total potential energy, for every atomic configuration. The value taken by these extra degrees of freedom is then a complicated functional of the atomic configurations so that, even if each term in the total energy is written as a sum of individual or pair components, the total system energy includes many-body effects. We search for the ground-state configuration of the electronic degrees of freedom at each time step, using a conjugate gradients routine, i.e.

$$\left(\frac{\partial V_{AIM}^{repulsion}}{\partial \xi^j}\right)_{\{\xi^M\}} = 0, \quad \text{where} \quad \{\xi^M\} = \{\delta\sigma^N, v_\alpha^N, \kappa_{\alpha\beta}^N\}$$
(10)

$$\left(\frac{\partial V_{AIM}^{polarization}}{\partial \xi^j}\right)_{\{\xi^M\}} = 0, \quad \text{where} \quad \{\xi^M\} = \{\mu_\alpha^N, \theta_{\alpha\beta}^N\}$$
(11)

The convergence criteria typically require the energy at successive steps to differ only in the 8th significant figure for polarization and in the 10th for repulsion. The dynamics is thus similar to the so-called Born-Oppenheimer ab initio molecular dynamics. Since the ions remained fixed during the conjugate gradient process, many of the terms required to evaluate the energy do not need to be recalculated at each minimization step.

In the case of halide systems, we observed that a simpler functional form was sufficient to describe accurately the interactions: In that case, we have used the "polarizable ion model" (PIM), where the repulsion term does not include any ion shape deformation term anymore, which corresponds to the simple exponential form:

$$V_{PIM}^{repulsion} = \sum_{i<j} A^{ij} e^{-a^{ij}r^{ij}}$$
(12)

In addition, good representations of the halides are found with the polarization term only including the terms up to the dipoles,

$$V_{PIM}^{polarization} = \sum_{i,j} \left[\left(q^i \mu_\alpha^j g_D^{ij}(r^{ij}) - \mu_\alpha^i q^j g_D^{ji}(r^{ij}) \right) T_\alpha^{(1)} - \mu_\alpha^i \mu_\beta^j T_{\alpha\beta}^{(2)} \right]$$
$$+ \sum_i \left(\frac{1}{2\alpha^i} |\boldsymbol{\mu}^i|^2 \right)$$
(13)

With these simplifications, the potential now includes only three additional degrees of freedom per ion, associated with

the induced dipoles, which are calculated in a single minimization procedure:

$$\left(\frac{\partial V_{PIM}^{polarization}}{\partial \mu_\alpha^i}\right)_{\{\mu_\alpha^N\}} = 0$$
(14)

Such interaction potentials have been introduced by our group in the case of molten chlorides [34, 35] and fluorides [36–38]. They were also used by Trullas and co-worker to study a series of silver and copper halides [39–43], and the PIM model now is implemented in the CP2K code. [44] Going back to the case of oxides, when attention is restricted to a single phase or where similar materials are being compared, it is often sufficient to neglect the full complexity of the AIM model, and we have also successfully used some PIM-type potentials in the case of amorphous GeO_2 [45, 46] and of doped zirconia crystals [47–50]. The potential developed for SiO_2 by Tangney and Scandolo [51] on the basis of similar force-matching methods to those we describe below is also worth noting. Their model, which performs very nicely in reproducing the properties of various phases of SiO_2 [52, 53], differs from the PIM one due to the use of partial charges for the ions, which hinders its transferability to other silicate compounds.

3 Obtaining the parameters from force-matching and multipole-matching

In their pioneering work, Tosi and Fumi developed a series of force field parameters for alkali halides in the framework of the Born model [54, 55]. These parameters were chosen in order to reproduce solid-state data. Their potential has then been used in many simulation works involving alkali halides in the molten state [56–58]. This type of empirical approach has long been the standard procedure for parameterizing interaction potentials. A new source of data for fitting these potentials was made possible by the development of ab initio techniques. For example, the most commonly used potential for SiO_2 is based on Hartree-Fock calculations of a single H_4SiO_4 cluster and on several crystal properties [59]. Oeffner and Elliott used a similar approach to develop an interaction potential for GeO_2 [60].

The introduction of the Car-Parrinello method [61] opened the way to a new class of simulations, namely the ab initio molecular dynamics (AIMD) [62]. In AIMD simulations, may it be of the Car-Parrinello or of the Born-Oppenheimer type, the introduction of an analytical interaction potential is avoided through the use of density functional theory (DFT) calculations. This has an appealing "hands-free" aspect, but unfortunately these methods cannot yet access the time scales necessary to determine many material properties, in particular in ionic liquids for which transport properties are of primary importance. Nevertheless, AIMD simulations are now considered to be very accurate, and they can be used as a reference for classical MD simulations. The idea has therefore emerged to develop interaction potentials yielding a trajectory that *mimics* the one which would be obtained by AIMD. Since molecular dynamics is based on an iterative integration of Newton's equation of motion, this can be done if the forces generated by the interaction potential are the same as the DFT-calculated ones. This approach [63], which is now termed force-matching or force-fitting, has been used (and extended) for a decade to develop accurate force fields for ionic liquids in the framework of the PIM and of the AIM. We will now focus on practical details of these parameterizations.

DFT calculations on periodic systems are based on the Kohn-Sham (KS) method to determine the ground-state electronic wavefunction $\{\phi^0\}$ of a given condensed phase configuration containing N ions. This wavefunction is obtained by minimization of the Kohn-Sham energy E^{KS}. The force acting on each atom i, with position $\mathbf{r}^i$, is then extracted from:

$$F^i_{\alpha,\text{DFT}} = \frac{\partial E^{KS}[\{\phi^0\}]}{\partial r^i_\alpha} \tag{15}$$

The corresponding classical forces for the same condensed phase configuration are easily obtained for a given set of parameters from:

$$F^i_{\alpha,\text{classical}} = -\frac{\partial V^{\text{total}}}{\partial r^i_\alpha} \tag{16}$$

where the interaction potential could either derive from the PIM or the AIM, and the additional degrees of freedom minimize the total energy for this configuration. The idea behind force-fitting consists of finding the set of parameters, which will minimize the error made in the classical calculation with respect to the DFT one. If this error is minimal, the interaction potential can be considered to be of ab initio accuracy. An important concern when dealing with ionic systems is that it is often required to perform studies with varying conditions of temperature, pressure and compositions. A prerequisite of the interaction potentials therefore is to be *transferable* from one thermodynamic point to another, which can be achieved by determining the interaction potential parameters on a series of representative condensed phase configurations instead of one. In practice, this is done by minimizing the following expression:

$$\chi^2_F = \frac{1}{N_c} \sum_{j=1}^{N_c} \frac{1}{N_j} \sum_{i=1}^{N_j} \frac{|\mathbf{F}^i_{\text{DFT}} - \mathbf{F}^i_{\text{classical}}|^2}{|\mathbf{F}^i_{\text{DFT}}|^2} \tag{17}$$

where N_c is the number of configurations and N_j is the number of ions in configuration j. Another quantity that can

also be determined rather straightforwardly from both the ab initio and the classical calculation is the stress tensor $\boldsymbol{\sigma}$.[1] It is therefore possible to extend the fitting dataset by minimizing

$$\chi^2_S = \frac{1}{N_c} \sum_{j=1}^{N_c} \frac{\sum_{\alpha\beta} |\sigma^i_{\alpha\beta,\text{DFT}} - \sigma^i_{\alpha\beta,\text{classical}}|^2}{\sum_{\alpha\beta} |\sigma^i_{\alpha\beta,\text{DFT}}|^2} \tag{18}$$

In that case, relative weights then have to be assigned to χ^2_F and χ^2_S in the minimization procedure. Of course, all the parameters of the interaction potential are involved in the calculation of the force and of the stress tensor, so that they can all be fitted at this stage. There is nevertheless a risk of interplay between the various terms of the potential, so that one term will do the work which should be attributed to the other. This would contribute to a loss of transferability of the potential.

Consequently, we generalize the force-fitting to require that the induced multipoles on our ions in the model reproduce the ab initio ones and determine the parameters in $V^{\text{polarization}}$ from this requirement. In DFT calculations, it is possible to calculate the multipole moments of the individual ions thanks to the maximally localized Wannier function (MLWF) formalism [64]. The MLWFs provide a picture of the electron distribution around atoms which is easily interpreted from a chemical point of view. They are determined by unitary transformations of the KS eigenvectors

$$|\phi^w_n\rangle = \sum_{m=1}^{N} U_{nm} |\phi^0_m\rangle \tag{19}$$

where the sum runs over all the occupied KS states $\{\phi_n\}_{n\in[1,N]}$, and the unitary matrix U is determined by iterative minimization of the Wannier function spread Ω, which is defined as

$$\Omega = -\frac{1}{(2\pi)^2} \sum_{n=1}^{N} \log |s_n|^2; \quad s_{n,\alpha} = \langle \phi^w_n | e^{-i\frac{2\pi}{L}r_\alpha} | \phi^w_n \rangle \tag{20}$$

when periodic boundary conditions are applied, where α refers to the coordinates axes, x, y, and z.

A complete theory of electric polarization in crystalline dielectrics has been developed in recent years [65–67], which validates the calculation of the dipole moments of single ions or molecules from the center of charge of the subset of MLWF which are localized in their vicinity [68–70]. The MLWF centers are computed according to [68]

---

[1] In the case of a classical calculation involving additional degrees of freedom, the correct expression for the stress tensor (as well as for the heat current) is obtained by exploiting the Hellmann–Feynman theorem and ignoring any explicit derivatives of the additional degrees of freedom with respect to the ion positions.

$$r^w_{n,\alpha} = -\frac{L}{2\pi}\Im(\log s_{n,\alpha}) \tag{21}$$

and the induced dipole moment of a given ion i is defined, in atomic units, as

$$\boldsymbol{\mu}^i = -2\sum_{n\in i} \boldsymbol{r}'^w_n \tag{22}$$

where $\boldsymbol{r}'^w_n$ is measured with respect to the position of the ion. In ionic systems, the attribution of a Wannier center to a given ion can be done unambiguously. Several routes have also been proposed to determine the induced quadrupole moments. Aguado et al. have exploited the fact that the MLWFs are well localized inside the periodic cell and evaluate the components of the quadrupole on an ion i from the real-space integral of the charge densities of the MLWFs on that ion:

$$\theta^i_{\alpha\beta} = -2\sum_{n\in i}\int_{V^i_{\text{cut}}} d\mathbf{r}\, |\,\phi^w_n(\mathbf{r})\,|^2\,(3r'^i_\alpha r'^i_\beta - (r'^i)^2\delta_{\alpha\beta})/2 \tag{23}$$

Here, the integral runs over the space within a sphere of radius r_{cut} around i, and r'^i is the distance from $\mathbf{r}$ to the nucleus position $\mathbf{r}^i$ calculated with a minimum image convention [29]. This method is implemented in the CASTEP code. Another method consists in determining the second moment of each Wannier function:

$$\langle r_\alpha r_\beta\rangle_n = -\frac{L^2}{8\pi^2}\left(\ln\frac{|\,\langle\phi^w_n|e^{-i\frac{2\pi}{L}r_\alpha}e^{-i\frac{2\pi}{L}r_\beta}|\phi^w_n\rangle\,|^2}{|\,\langle\phi^w_n|e^{-i\frac{2\pi}{L}r_\alpha}|\phi^w_n\rangle\,|^2\,|\,\langle\phi^w_n|e^{-i\frac{2\pi}{L}r_\beta}|\phi^w_n\rangle\,|^2}\right) \tag{24}$$

The quadrupole moment of an ion is then given by [71]

$$\theta^i_{\alpha\beta} = -2\sum_{n\in i}\langle 3r'_\alpha r'_\beta - r'^2\delta_{\alpha\beta}\rangle_n \tag{25}$$

where $\langle r'_\alpha r'_\beta\rangle_n$ is measured again with respect to the position of the ion. The advantage of this method, which is implemented in the CPMD code, is that there is no need to define a cut-off radius around i (apart from the attribution of the Wannier center to the ion). Sagui et al. have extended this approach for the calculation of higher multipole moments [72]. The generalization of the force-fitting procedure to multipoles requires the minimization of the functions:

$$\chi^2_D = \frac{1}{N_c}\sum_{j=1}^{N_c}\frac{1}{N_j}\sum_{i=1}^{N_j}\frac{|\,\boldsymbol{\mu}^i_{\text{DFT}} - \boldsymbol{\mu}^i_{\text{classical}}\,|^2}{|\,\boldsymbol{\mu}^i_{\text{DFT}}\,|^2} \tag{26}$$

$$\chi^2_Q = \frac{1}{N_c}\sum_{j=1}^{N_c}\frac{1}{N_j}\sum_{i=1}^{N_j}\frac{\sum_{\alpha\beta}|\,\theta^i_{\alpha\beta,\text{DFT}} - \theta^i_{\alpha\beta,\text{classical}}\,|^2}{\sum_{\alpha\beta}|\,\theta^i_{\alpha\beta,\text{DFT}}\,|^2} \tag{27}$$

This fitting procedure has now been applied to an important variety of ionic systems. Here, we will describe the parameterization of the oxide and fluoride series, for which the AIM and PIM frameworks were chosen respectively.

An important issue that remains is how to include the dispersion terms (C^{ij}_6 and C^{ij}_8 coefficients) in the fitting procedure. It is well known that the dispersion interactions cannot be represented accurately by a local or a semilocal exchange-correlation functional, such as that provided by the commonly used LDA and GGA functionals [73]. We have therefore decided not to include the dispersion parameters in the fitting procedure, but rather to fix them to some given values. The choice of these values is discussed in the next section. Now the second important question is whether it is appropriate or not to include these dispersion effects when calculating the classical forces during the fitting procedure. Here, our choice depends on the nature of the functional used for the DFT calculations. It is well known that the LDA functional leads to overbinding effects (although those are not due to dispersion), while GGA functionals like PBE [74] or BLYP [75, 76] lead to underbinding effects. We have therefore chosen to include the dispersion effects in the calculation of the classical forces when using the LDA functional only (the effect is included, but the corresponding parameters are kept fixed so that they are not fitted during the procedure).

In the case of oxide systems, our objective was to develop a potential suitable for simulation of Earth materials under crustal to lower mantle conditions of temperature and pressure [3]. The most abundant species are those of the so-called CMAS system, i.e. the cations Ca^{2+}, Mg^{2+}, Al^{3+} and Si^{4+}. In this work, the LDA functional has been used for the DFT calculations: A detailed comparison between GGA- and LDA-based potentials was undertaken for Al_2O_3; this study showed that the description of the melt was much better for the latter [77]. The AIM functional form was used for the classical interaction potential. We can therefore summarize the fitting procedure this way:

1. Generation of a series of typical condensed phase configurations
2. DFT calculations on each of these configurations:

   (a) Determination of the ground-state wavefunctions, which give access to the ab initio forces and stress tensor components

   (b) Wannier localization, from which the ab initio induced dipoles and quadrupoles components are calculated

3. Determination of the dispersion coefficients (see next section)

4. Minimization of χ_D and χ_Q with respect to the parameters of the polarization term ($V^{\text{polarization}}$)

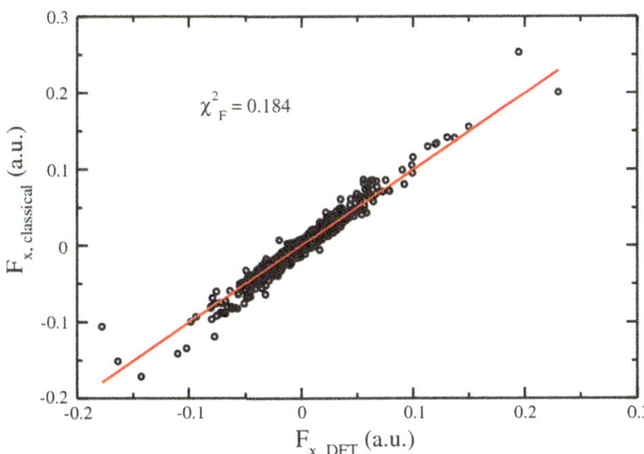

Fig. 1 The quality of the fits of the x-component of the forces on the ions to those obtained from the ab initio calculations is shown for a series of 19 configurations (including high- and low-pressure polymorphs of Al_2O_3, MgO, SiO_2, $MgAl_2O_4$, $MgSiO_3$, CaO, $CaAl_2Si_2O_8$ and two Al_2O_3 melts)

5. Minimization of χ_F and χ_S with respect to the parameters of the repulsion term $(V^{repulsion})^2$

The quality of the fits of the x-component of the forces on the ions to those obtained from the ab initio calculations is shown for a series of 19 configurations of approximately 100 atoms (including high- and low-pressure polymorphs of Al_2O_3, MgO, SiO_2, $MgAl_2O_4$, $MgSiO_3$, CaO, $CaAl_2Si_2O_8$ and two Al_2O_3 melts) on Fig. 1, and the parameters obtained for the CMAS system with this procedure are summarized in Table 1.

The potential reproduces well lattice parameters and elastic constants of various oxides and silicates of the CMAS system. Typically, the lattice constants from the simulations are up to a few percent lower than the experimental values, which is consistent with the use of LDA for the DFT reference calculations. The model also provides reasonable predictions for the thermal expansion and the volume compression of major phases of the Earth's mantle in a wide pressure range [3]. Liquid-state densities of oxide and silicate melts are usually within the error bars of the experimental data. The model performs somewhat less well for open network-forming structures close to the pure SiO_2 composition.

Applications of the potential include the investigation of structural and physical properties of solid and liquid oxides relevant in geological or technological context. A number of studies were concerned with the stability and mechanisms of phase transitions in magnesium silicates (see [80] for a summary). The atomic structure and

dynamics of magnesium and calcium aluminate liquids was studied in combination with neutron and X-ray scattering experiments using aerodynamic levitation techniques [81–84]. An example of comparison between simulated [78] and experimental [79] total structure factors of liquid $MgAl_2O_4$ is shown on Fig. 2. The almost quantitative agreement between measured and computed static and dynamic structure factors underlines the high accuracy and transferability of the potential, which was optimized using mostly solid-state configurations. The structural studies illustrate that molecular simulations are essential to interpret the diffraction data of multicomponent liquids since even two different experiments (neutron and X-ray diffraction) do not allow the unambiguous determination of some of the anion–cation nearest neighbor distances and cation coordination numbers [84]. Besides the structural model, the simulations provide insight into the relation between melt structure and physical properties, such as the viscosity or the self-diffusivities [77, 78]. The latter are important parameters, e.g., to understand the evolution of the early Earth, which may have been covered by a magma ocean composed of silicate liquid. Recent simulations of Mg_2SiO_4 liquids using the CMAS potential give new constraints on the thermodynamic and rheological properties of the melt in the relevant range of pressure and temperature [85, 86]. A number of ongoing projects make use of the CMAS potential to study solid–solid and solid–liquid interfaces, the structure of complex silicate melts in conjunction with X-ray absorption spectroscopy or with the interpretation of trace element partitioning data between solid and liquid silicates.

In the case of fluoride systems, we are mainly interested in determining physico-chemical properties of melts including many different components. We have therefore built a transferable PIM interaction potential, which includes Li^+, Na^+, K^+, Rb^+, Cs^+, Be^{2+}, Ca^{2+}, Sr^{2+}, Y^{3+}, La^{3+} and Zr^{4+} ions. In this work, the PBE functional has been used for the DFT calculations (unlike the case of oxide materials, no test has been made using the LDA). We can therefore summarize the fitting procedure this way:

1. Generation of a series of typical condensed phase configurations
2. DFT calculations on each of these configurations:
   (a) Determination of the ground-state wavefunctions, which give access to the ab initio forces components
   (b) Wannier localization, from which the ab initio induced dipoles components are calculated
3. Minimization of χ_D with respect to the parameters of the polarization term $(V^{polarization})$

---

² Note that all the χ_i values could be minimized together but the use of a two-step procedure permits to avoid some cancelation of errors in the parameterization.

Table 1 Parameters in the repulsive and polarization parts of the potential for the CMAS system

Ion pair	$O^{2-}-O^{2-}$	$Ca^{2+}-O^{2-}$	$Mg^{2+}-O^{2-}$	$Al^{3+}-O^{2-}$	$Si^{4+}-O^{2-}$
A^{ij}	1,068.0	40.168	41.439	18.149	43.277
a^{ij}	2.6658	1.5029	1.6588	1.4101	1.5418
B^{ij}		5,0532	59,375	51,319	43,962
b^{ij}		3.5070	3.9114	3.8406	3.9812
C^{ij}		6,283.5	6,283.5	6,283.5	6,283.5
c^{ij}		4.2435	4.2435	4.2435	4.2435
$b_D^{ij} = b_D^{ji}$		2.0261	2.2148	2.2886	2.1250
c_D^{ij}		3.9994	2.8280	2.3836	1.5933
b_Q^{ij}		1.5297	1.9300	2.1318	1.9566
c_Q^{ij}		1.6301	1.3317	1.2508	1.0592
C_6^{ij}	44.372	2.1793	2.1793	2.1793	2.1793
C_8^{ij}	853.29	25.305	25.305	25.305	25.305
$b_6^{ij} = b_8^{ij}$	1.4385	2.2057	2.2057	2.2057	2.2057
D	0.49566		β	1.2325	
ζ	0.89219		η	4.3646	
α	8.7671		C	11.5124	

All values are in atomic units. The dipole–dipole–quadrupole polarizability was set to 0

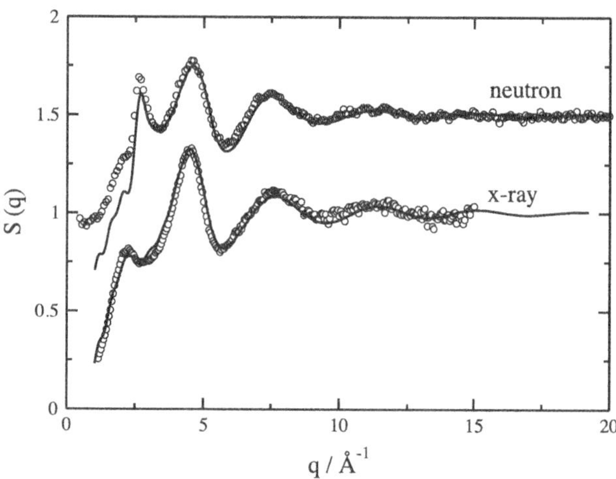

Fig. 2 Total static structure factors of $MgAl_2O_4$ at $T = 2,500$ K from MD simulation [78] (*lines*) compared to experimental neutron and x-ray diffraction data (*circles*) at $T = 2,423$ K. [79] The neutron structure factors are offset by a value of 0.5

4. Minimization of χ_F with respect to the parameters of the repulsion term ($V^{\text{repulsion}}$)
5. Determination of the dispersion coefficients (see next section)

Small differences are observed with the oxide materials situation: Due to the smaller number of parameters involved in the fitting procedure and the absence of quadrupole polarization effects in the PIM, only χ_D and χ_F had to be minimized in this case. In addition, the dispersion coefficients can be determined after the fitting step because the dispersion term is set to 0 when fitting the repulsion (due to the use of a GGA functional in the DFT calculations). Since we are mainly interested in systems which are in the liquid state, this procedure was in fact performed twice: Firstly, with a series of crystalline configurations, which led to a first set of parameters. And secondly with a series of liquid-state configurations of approximately 100 atoms generated from molecular dynamics simulations with the parameters of the first step. The quality of the fits of the x-component of the forces on the ions to those obtained from the ab initio calculations is shown for a series of 25 configurations (pure LiF, NaF, KF, CaF_2, SrF_2, YF_3, LaF_3 and ZrF_4, LiF-YF_3 and LiF-LaF_3 mixtures) on Fig. 3, and the parameters obtained for the fluoride system with this procedure are summarized in Tables 2 and 3.

A first test of the potential is provided by a simple comparison of the parameters within a series of chemical species. For example, in Fig. 4, we compare the repulsive term of the potential for various $M^{m+}-F^-$ ion pairs. Firstly, in the case of the alkaline cations, we observe that the repulsion wall is pushed toward longer distances in a way which is coherent with the increase in the ionic radius of the cation in this series. More interesting is the second test, which compares the period five elements Rb^+, Sr^{2+}, Y^{3+} and Zr^{4+}. All these ions have the same number of electrons and their ionic radius therefore do not differ a lot. The repulsion walls are very close one from each other and they become shorter-ranged when the cationic charge is increased, which is in agreement with chemical intuition.

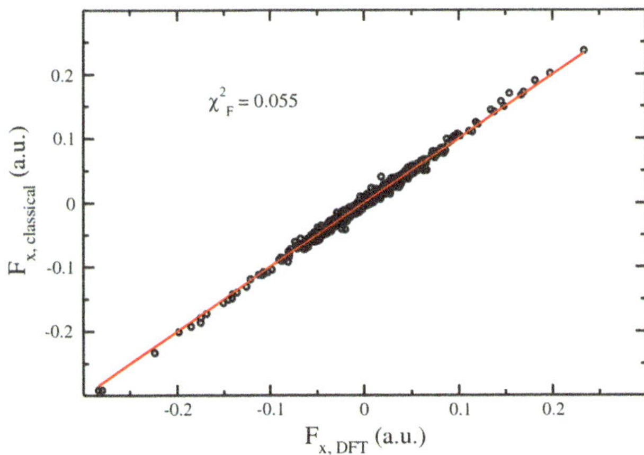

Fig. 3 The quality of the fits of the x-component of the forces on the ions to those obtained from the ab initio calculations is shown for a series of 25 configurations (pure LiF, NaF, KF, CaF$_2$, SrF$_2$, YF$_3$, LaF$_3$ and ZrF$_4$, LiF-YF$_3$ and LiF-LaF$_3$ mixtures)

Table 2 Fitted polarizabilities for the various ions in molten fluorides

Ion type	Polarizability (α^i)
F$^-$	7.9
Li$^+$	0
Na$^+$	1.0
K$^+$	5.0
Rb$^+$	8.4
Cs$^+$	14.8
Ca^{2+}	3.1
Sr^{2+}	5.1
Y^{3+}	3.8
La^{3+}	7.5
Zr^{4+}	2.9

All values are in atomic units

The interest in molten fluorides is mainly due to their involvement in numerous current (aluminum production) and proposed (molten salt reactors) technologies. They are often dangerous and corrosive materials, which are liquid at high temperatures, so that performing systematic experimental studies is difficult. The development of accurate models that are able to predict the physical properties over a wide range of conditions is therefore very valuable. Among all the molten fluorides, an enormous amount of experimental work has been done on LiF-BeF$_2$ mixtures and it is by far the best-characterized fluoride melt [88, 89]. For this reason, it was the ideal testing ground for evaluating our interaction potential development approach and we focused first on this system [37]. We could show that our model performed very well in reproducing the polymeric fluoroberyllate speciation and the vibrational properties in these melts [87]. Calculated dynamic properties (electrical conductivity, viscosity) also agree quantitatively with the experimental data: [90] To illustrate this, we show a comparison of the calculated [87] and experimental [88] viscosities of LiF-BeF$_2$ mixtures with varying compositions at $T = 873$ K on Fig. 5. The interfacial properties were also tested by calculating the surface tension from a simulation of the liquid–vapor interface. [91] In a second step, a series of simulation studies was carried out for other molten fluoride systems. This allowed us to predict heat-transport properties of LiF-NaF-KF and NaF-ZrF$_4$ mixtures [38] in order to evaluate their suitability to serve as primary coolant in the advanced high-temperature reactor concept [92]. Our simulations were also used to interpret EXAFS and NMR experimental data on fluorozirconate melts [93] and to calculate electrical conductivities [94], diffusion coefficients [95] and internal mobilities [96].

4 Beyond force-matching: direct calculation of parameters

Although the fitting procedure presented in the previous section has proven to be very successful, in particular in the context of the development of transferable potentials, it can sometimes be enhanced through the direct *calculation* of some of the parameters involved in the models. Among all the parameters, the dipolar polarizability α has an evident physical meaning since it measures the relative tendency of the electron cloud of the ion to be distorted from its normal shape by an electric field. Although one could think that the polarizability of a molecule or an ion is an intrinsic property, which remains unchanged from the gas phase to any condensed phase, this is not the case due to the effect of the confining environmental potential in the latter [21, 22].

Again, in order to determine the condensed phase polarizabilities from a direct DFT calculation, the MLWF formalism provides a very convenient route [97]. When a small external electric field $\mathcal{E}$ is applied to the system, the linear response may be characterized by an additional field-induced dipole moment $\delta\mu^i$ on each individual ion. It is convenient to think of the applied field as an optical field, in order to distinguish its effect from that of the static fields, which are caused by the permanent charge distributions of the molecules, and to think of $\delta\mu^i$ as the net-induced dipole which is oscillating at the optical frequency. For an electronically insulating material, the induced dipole can be written in terms of the total (optical frequency) electric field which acts on it,

$$\delta\boldsymbol{\mu}^i = \boldsymbol{\alpha}^i \cdot \left[\mathcal{E} + \sum_{j \neq i} \boldsymbol{T}^{(2)} \cdot \delta\boldsymbol{\mu}^j \right] \tag{28}$$

Table 3 Fitted parameters for the repulsion and polarization terms for molten fluorides

Ion pair	A^{ij}	a^{ij}	$b_D^{ij} = b_D^{ji}$	c_D^{ij}	c_D^{ji}
$F^- - F^-$	282.3	2.444	–	0.0	0.0
$F^- - Li^+$	18.8	1.974	1.834	1.335	
$F^- - Na^+$	44.9	1.927	1.831	2.5	0.022
$F^- - K^+$	138.8	2.043	1.745	2.5	−0.31
$F^- - Rb^+$	151.0	1.961	1.822	3.5	−0.44
$F^- - Cs^+$	151.1	1.874	1.930	3.4	0.49
$F^- - Ca^{2+}$	73.0	1.859	1.732	1.5	−0.31
$F^- - Sr^{2+}$	82.5	1.778	1.554	1.331	−0.33
$F^- - Y^{3+}$	87.4	1.832	1.847	1.966	−0.89
$F^- - La^{3+}$	161.6	1.867	1.614	1.348	−0.47
$F^- - Zr^{4+}$	62.6	1.74	1.882	1.886	−1.0
$M^{m+} - M^{m+}$	1.0	5.0	–	0.0	0.0
$N^+ - N^+$	5000.0	3.0	–	0.0	0.0

$b_6^{ij} = b_8^{ij} = 1.9$ for all the pairs. M^{m+} stands for any cation between Li^+, Na^+, K^+, Ca^{2+}, Sr^{2+}, Y^{3+}, La^{3+} or Zr^{4+}, while N^+ is Rb^+ or Cs^+. All values are in atomic units

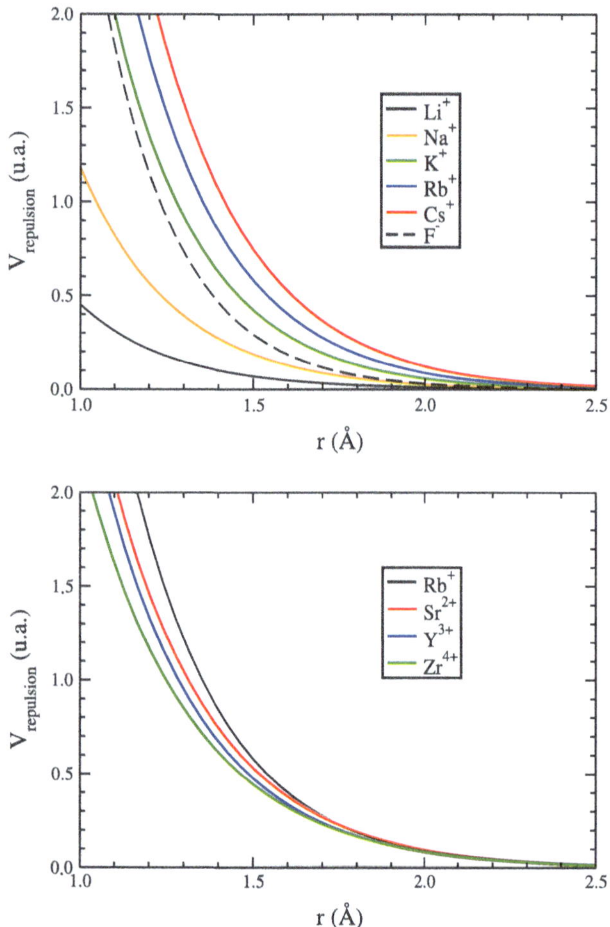

Fig. 4 Comparison of the repulsion terms obtained for various cation–fluoride pairs. *Top* Alkaline cations Li^+, Na^+, K^+, Rb^+ and Cs^+; *bottom* period 5 elements Rb^+, Sr^{2+}, Y^{3+} and Zr^{4+}

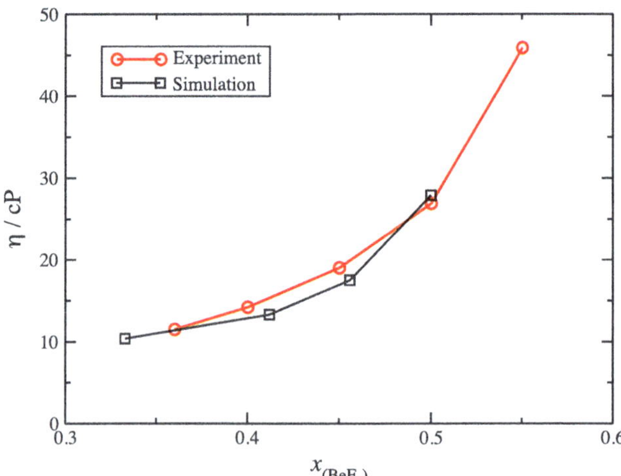

Fig. 5 Comparison of the calculated [87] and experimental [88] viscosities of LiF-BeF$_2$ mixtures with varying compositions at $T = 873$ K

where the sum runs over all polarizable ions $j \neq i$ in the system. In this equation, we have introduced the dipole polarizability tensor $\boldsymbol{\alpha}^i$ of ion i, for a particular condensed phase configuration. Also included is the dipole–dipole interaction tensor, $\boldsymbol{T}^{(2)}$, whose components are defined by $T_{\alpha\beta}^{(2)} = \nabla_\alpha \nabla_\beta \frac{1}{r^{ij}}$; in practice, for a periodic system, it will be computed using the Ewald summation technique [31, 32]. The first term on the right-hand side of the equation represents the direct contribution of the external electric field to the induced dipole; the second-term is the contribution of the reradiated electric fields due to the dipoles, which are induced in all the other ions ($j \neq i$) in the sample. In principle, higher-order induced multipoles also contribute to this expansion, but we will ignore them. In a uniform external field, the directly induced higher-order multipoles on spherical atoms and ions vanish, and even for molecules, their effect is expected to be much smaller than that of the dipoles.

In DFT calculations on periodic systems, the coupling between the external electric field and the electronic system is expressed through the macroscopic polarization of the periodically replicated cell [98, 99] and is defined using the Berry phase approach of Resta [100]. It is then possible to determine the new partial dipole moment for each species in the presence of a field, *via* another localization step. The field-induced dipoles are calculated from the difference between the total molecular dipoles in the presence and absence of the field.

Equation 28 can be inverted to determine the individual electronic polarizabilities for that particular condensed phase configuration. Consider the application of fields $\mathcal{E}^{(\alpha)}$, along each Cartesian direction $\alpha = x, y, z$, and denote by $\{\delta\boldsymbol{\mu}^{i,(\alpha)}\}$ the corresponding values of the induced dipole moments. The total field $\boldsymbol{f}^{i,(\alpha)}$ at each position $\boldsymbol{r}^i$ can be obtained from

$$\boldsymbol{f}^{i,(\alpha)} = \mathcal{E}^{(\alpha)} + \sum_{j \neq i} \boldsymbol{T}^{(2)} \cdot \delta\boldsymbol{\mu}^{j,(\alpha)} \tag{29}$$

which is conveniently evaluated from the electric field given by a dipolar Ewald sum. Finally, the polarizability tensor of ion i is given by

$$\boldsymbol{\alpha}^i = (\boldsymbol{F}^i)^{-1} \cdot \boldsymbol{\Pi}^i \tag{30}$$

where $\boldsymbol{F}^i$ and $\boldsymbol{\Pi}^i$ are second-rank three-dimensional tensors defined as

Table 4 Calculated ionic polarizabilities in the liquid phase

System	Ion type	Polarizability (α^i)
LiF	F^-	7.8
	Li^+	0.3
NaF	F^-	9.6
	Na^+	1.1
KF	F^-	10.7
	K^+	5.5
CsF	F^-	11.8
	Cs^+	16.3
NaCl	Cl^-	25.1
	Na^+	1.2
KCl	Cl^-	27.0
	K^+	5.5

All values are in atomic units

$$F^i_{\alpha\beta} = f^{i,(\beta)}_\alpha \tag{31}$$

$$\Pi^i_{\alpha\beta} = \delta\mu^{i,(\beta)}_\alpha \tag{32}$$

The results obtained for a series of molten fluorides and chlorides are summarized in Table 4. Important environmental effects occur when the nature of the liquid is changed. In the fluoride melts, for example, the polarizability of the fluoride anion shifts a lot, passing from 7.8 a.u. in molten LiF (which is very close to the value obtained from the force-fitting procedure, this is also the case for the cation polarizabilities) to 11.8 a.u. in CsF. This is due to differences in the confining potential, which affects the electron density around a given anion, and originates from both Coulombic interactions and the exclusion of electrons from the region occupied by the electron density of the first-neighbor shell of cations [21, 22, 101]. When passing from one cation to another (for example in the series $Li^+ \rightarrow Na^+ \rightarrow K^+ \rightarrow Cs^+$), two effects are then competing: On the one hand, the anion–cation distance increases, which results in a diminution of the confining potential, but on the other hand, the volume occupied by the cation electron density also increases, with an opposite effect on the confining potential. Here, the observed increase in polarizability with the size of the cation tends to show that the first effect is the most important. Indeed, the value obtained for CsF is approaching the free F^- anion polarizability, which is 16 a.u. These effects have also been observed in similar calculations performed on solid oxides [97] and protic solvents [102, 103]. They have indirectly been used for building quantitative Lewis acidity scales in inorganic materials [104, 105]. Such a result means that in order to build a completely transferable interaction potential, the model should allow the variation in the polarizability in response to the fluctuations of the environments. First steps have been made in that direction in the case of MgO only [30, 106].

Since the centers of the MLWFs allow one to derive atomic dipoles in the condensed phase and hence atomic polarizabilities, it is tempting to exploit further these localized orbitals, which are often interpreted in terms of non-bonding orbitals or lone pairs, to derive the remaining terms of the force field. A systematic procedure has been introduced recently for this purpose, in which a series of approximations leads to the attractive and repulsive van der Waals interactions from the knowledge of the geometric properties of the MLWF [107].

The central idea of this approach consists of assigning a localized electronic density considered as frozen around each nucleus, reconstructed from the associated MLWFs. The exponential decay of MLWFs in electronic insulators [108] suggests modeling them as Slater orbitals determined

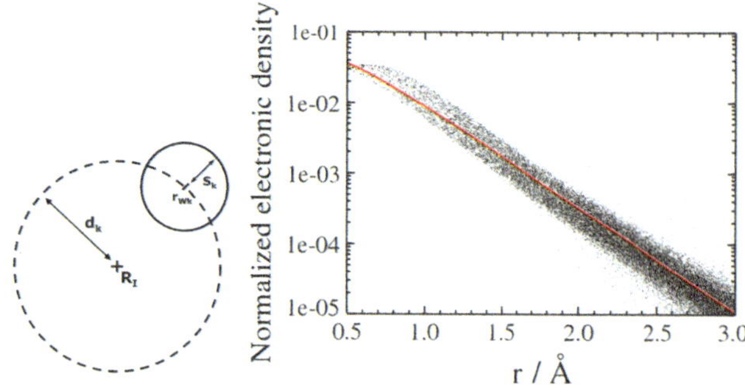

Fig. 6 Electronic density around Cl⁻ ions in molten KCl from DFT (scattered plot), compared to the analytic ansatz (Eq. 33), which involves only the spread and distance to nucleus of the MLWFs

Table 5 Properties of the Wannier orbitals: average spread S^i and distance d^i to the nucleus

System	Ion type	$\langle S^i \rangle$	$\langle d^i \rangle$
KF	F⁻	1.35	0.555
	K⁺	1.36	0.685
NaF	F⁻	1.32	0.560
	Na⁺	0.85	0.410
KCl	Cl⁻	1.92	0.869
	K⁺	1.36	0.680
NaCl	Cl⁻	1.90	0.877
	Na⁺	0.86	0.414

All values are in atomic units

solely by their spread S_k and center located at $\mathbf{r}_k^w$ determined from the localization procedure [109]. These two properties can be obtained for each type of MLWF from averages over liquid configurations. Each MLWF contributes to an electronic density $\rho_{Wk} = n_k \frac{\kappa_k^3}{8\pi} e^{-\kappa_k |\mathbf{r} - \mathbf{r}_k^w|}$ with $n_k = 2$ the number of electrons in the orbital and $\kappa_k = \frac{2\sqrt{3}}{S_k}$. In order to obtain orientationally averaged interaction potentials, one further proceeds to an orientational average of the density around each nucleus i (see Fig. 6), assuming an isotropic distribution of the N_W MLWFs centers with the result:

$$\rho^i(\mathbf{r}) = \sum_{k=1}^{N_W} \frac{n_k \kappa_k}{8\pi r_> r_<} e^{-\kappa_k r_>}$$
$$[(1 + \kappa_k r_>) \sinh(\kappa_k r_<) - \kappa_k r_< \cosh(\kappa_k r_<)] \quad (33)$$

where $r_> = \max(|\mathbf{r} - \mathbf{r}_k^w|, d_k)$ and $r_< = \min(|\mathbf{r} - \mathbf{r}_k^w|, d_k)$, with $d_k = |\mathbf{r}_k^w - \mathbf{r}^i|$ the distance of the kth MLWF center to the nucleus. The spread and distance to nucleus of Wannier orbitals for molten NaF, NaCl, KF and KCl are summarized in Table 5.

Figure 6 also compares the electronic density around Cl nuclei with the spherically averaged result given by Eq. 33. The agreement with the mean density is very good even at short distances, the scattering of the measured electronic densities being mainly due to thermal fluctuations not reproduced by the rigid density approximation. Exponential decay of the molecular density is also apparent over more than two decades and valid in the important region where the electronic densities of neighboring ions overlap. For one of the considered species (K⁺ in molten KF and KCl), the electronic density around the nucleus, derived from the MLWFs, is better described at large distances by a nucleus-centered Slater orbital (spread 1.30 a.u. in KF, 1.34 a.u. in KCl) than by Eq. 33. Equation 33 provides a more accurate description near the nucleus, but not further away.

Once the localized density around each nucleus in condensed phase is known, the repulsive van der Waals interaction, which contains a kinetic and an exchange-correlation part, can be derived using two approximations: the above-mentioned frozen density ansatz and the use of a kinetic energy functional. The repulsion energy $V^{\text{repulsion}}$ is computed at the DFT level from the superposition of the rigid electronic clouds around each atom [110, 111] and the resulting energy surface parametrized as a function of atomic coordinates. For each interatomic separation r, we compute

$$V^{\text{repulsion}}(r) = T[\rho_s, \nabla \rho_s] + \int \epsilon_{xc}[\rho_s, \nabla \rho_s]\rho_s \, \mathrm{d}^3\mathbf{r}, \quad (34)$$

where $\rho_s = \rho^i + \rho^j$ is the superposition of frozen atomic densities which have their centers separated by a distance r, T and ϵ_{xc} are the kinetic energy and exchange correlation functionals, respectively. Any form can be used for the kinetic energy functional T and the exchange-correlation energy density ϵ_{xc}, provided that the latter is the one used in the starting DFT calculations. However, the resulting $V^{\text{repulsion}}$ depends on their choice. The LLP kinetic energy functional [112] is generally the most accurate [111, 113]. Note that improvements of the functionals would automatically transfer to the resulting force fields. The resulting $V^{\text{repulsion}}$ generally decays as Ae^{-ar} in the region corresponding to the first-neighbor shell, thus justifying the analytical form of the repulsion term in the PIM and the AIM and providing the associated parameters.

As a first example, we applied this strategy to the case of NaF and NaCl melts, starting from DFT calculations with the PBE functional and using the LLP kinetic functional to compute the kinetic energy in Eq. 34. The relevance of this approach can be appreciated on Fig. 7, which reports the mean-square relative error on the forces χ_F^2 defined in previous section as a function of the Na⁺-Cl⁻/F⁻ repulsion parameters Ae^{-ar}, all other parameters of the force field remaining fixed. The sets obtained by this method stand

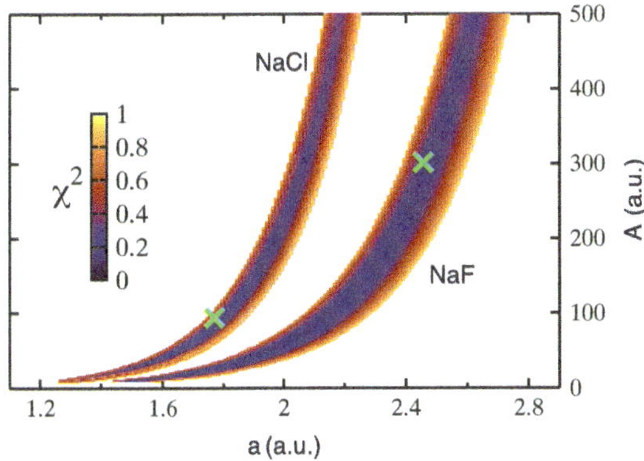

Fig. 7 Mean-square relative error χ_F^2 on the atomic forces compared to DFT (PBE functional) in the NaCl and NaF melts, as a function of the Na$^+$-Cl$^-$/F$^-$ repulsion parameters Ae^{-ar}. The set obtained by our method ($\times$) stands well within the $\chi_F^2 < 1.0$ isosurface in both cases

well within the narrow windows $\chi_F^2 < 1.0$, thus justifying a posteriori the frozen density approximation for these systems. The corresponding χ_F^2 are 0.241 and 0.363 for NaF and NaCl, respectively. In the latter, less favorable case, the error is not much larger than the value obtained with a force field of the same form (PIM) but with all parameters numerically fitted by force-matching ($\chi_F^2 = 0.118$), and significantly smaller than the one obtained with the popular Tosi-Fumi potential ($\chi_F^2 = 1.250$) [114].

Finally, we now turn to the problem of the dispersion interaction. As mentioned previously dispersion forces, are not easily captured at the DFT level. While the use of non-local functionals seems to be a promising approach [115], their computational cost has so far limited their use in AIMD simulations, for which empirical corrections such as the one developed by Grimme [116] are preferred. A promising alternative based on a simpler non-local van der Waals density functional has recently been proposed [117]. In the work concerning the oxide and fluoride series, we have used C_6^{ij} and C_8^{ij} values obtained from a limited set of ab initio (coupled-cluster or Møller-Plesset) values on in-crystal ions [118] and mixing rules to scale values from one material to another. It is also possible to derive systematically approximate parameters for the dispersion term defined in Eq. 3. Silvestrelli showed that MLWF provide again a convenient framework to tackle the dispersion effects and introduced the computation a posteriori of the $-C_6^{kl}/r_{kl}^6$ interaction between Wannier centers [109], where C_6^{kl} depends only on the spread of the MLWFs:

$$C_6^{kl} = \frac{3}{32\pi^{3/2}} \int\limits_{|\mathbf{r}| \le r_c} d\mathbf{r} \int\limits_{|\mathbf{r}'| \le r_c'} d\mathbf{r}' \frac{\sqrt{\rho_{Wk}(r)\rho_{Wl}(r')}}{\sqrt{\rho_{Wk}(r)} + \sqrt{\rho_{Wl}(r')}}$$

(35)

where the cut-off radius for each orbital depends on its spread as

$$r_c = [1.475 - 0.866 \ln S]S. \tag{36}$$

These coefficients are obtained from the MLWFs using the expression proposed by Andersson et al. [119] for the long-range interaction between two separated fragments. One can then obtain the dispersion interaction between a pair of atoms as the averaged sum over pairs of MLWFs (for k, l from different sites) [107]. Assuming an isotropic distribution of centers around the nuclei i, j at fixed distance $d_{k,l}$, we obtain to 2nd leading order: $V^{\text{dispersion}} = - \sum_{n=6,8} C_n^{ij}/(r^{ij})^n$ where the dispersion coefficients are:

$$C_6^{ij} = \sum_{k \in i, l \in j} C_6^{kl} \tag{37}$$

$$C_8^{ij} = \sum_{k \in i, l \in j} 5(d_k^2 + d_l^2) C_6^{kl} \tag{38}$$

We recall that the distances $d_{k,l}$ result from the orbital localization procedure and are not adjustable parameters. This approach thus provides two of the parameters entering in Eq. 3, leaving only the ones in the damping functions to be determined independently.

All the developments introduced in this section can be used in a procedure which can be summarized this way:

1. Generation of a series of typical condensed phase configurations
2. DFT calculations on each of these configurations:

   (a) Determination of the ground-state wavefunctions, which give access to the ab initio forces components
   (b) Wannier localization, from which the MLWFs spreads and positions are extracted and the ab initio induced dipoles components are calculated

3. Reconstruction of the electronic density around each ion (following Eq. 33)
4. Calculation of the $V^{\text{repulsion}}$ term parameters for each ion pair (following Eq.34)
5. DFT calculations on each of these configurations under an applied electric field $\mathcal{E}$ (one calculation for each direction), from which new sets of ab initio induced dipoles components are calculated
6. Calculation of the individual polarizabilities for each ion (following Eq. 30)
7. Minimization of χ_F and χ_D with respect to the damping parameters of the polarization term ($V^{\text{polarization}}$) only
8. Determination of the dispersion coefficients for each ion pair (following Eqs. 35 and 38)

This procedure, which leaves fewer parameters to be fitted than a straightforward application of force-fitting, has up to now been tested and validated in the simple case of

Table 6 Parameters of the PIM interaction potentials obtained for NaCl, NaF, KCl and KF from the alternative procedure

System	Ion pair	A^{ij}	a^{ij}	C_6^{ij}	C_8^{ij}	b_D^{ij}	c_D^{ij}	c_D^{ji}
KF	$F^- - F^-$	144.0	1.954	32.7	100.6	–	–	–
	$F^- - K^+$	163.3	2.052	34.3	133.1	2.052	6.4	−1.2
	$K^+ - K^+$	109.5	2.004	36.0	168.7	–	–	–
NaF	$F^- - F^-$	162.9	2.012	29.4	92.1	–	–	–
	$F^- - Na^+$	301.5	2.456	5.6	13.4	2.456	9.5	−0.4
	$Na^+ - Na^+$	599.0	3.176	1.3	2.2	–	–	–
KCl	$Cl^- - Cl^-$	187.9	1.601	333.0	2516.2	–	–	–
	$Cl^- - K^+$	90.2	1.636	102.4	623.6	1.636	3.8	0.4
	$K^+ - K^+$	244.8	2.183	35.9	166.3	–	–	–
NaCl	$Cl^- - Cl^-$	205.9	1.626	298.4	2294.1	–	–	–
	$Cl^- - Na^+$	94.6	1.770	14.2	66.5	1.770	3.5	0.8
	$Na^+ - Na^+$	546.4	3.1	1.3	2.3	–	–	–

For all the ion pairs, we have taken $b_6^{ij} = b_8^{ij} = a^{ij}$. All values are in atomic units

NaCl, NaF, KCl and KF [107]. The parameters are summarized in Table 6 (the polarizabilities are given in Table 4). The dynamic (diffusion coefficients, viscosity, thermal and electrical conductivity) and static (density) properties of these molten salts were tested against experimental data: An excellent agreement was obtained, showing again the predictive character of such first-principles-based approaches.

5 Conclusion

In conclusion, we have described in this paper two models of different complexity, which can be used to describe the interactions in ionic liquids. The first one, the polarizable ion model, includes only one many-body effect, namely the dipole polarization of the ions. This model was recently implemented in the CP2K code [44]. The second one, the aspherical ion model, not only extends the inclusion of polarization effects up to the quadrupole level, but also includes deformations of the anion in a spherical, dipolar or quadrupolar way for the calculation of the short-range repulsion. Depending on the system of interest and the quantities that have to be determined, one has to choose the most appropriate model.

These interaction potentials include several parameters that have to be determined for each ion type / ion pair. Since the final objective is to predict physico-chemical properties of the materials of interest, no experimental information should be included in the parameterization stage. Here, we have summarized the procedure, which was used for developing PIM potentials for molten fluorides and AIM potentials for molten oxides and consists in performing an extended force-fitting using ab initio density

functional theory results as reference data. In the last part, we show how the use of MLWFs can provide alternative ways for calculating some of the parameters.

Here, we have focused on inorganic molten salts. It is worth noticing that an increasing number of studies on room-temperature ionic liquids have involved models similar in spirit to the PIM (they mainly differ due to the choice of Lennard-Jones form to represent the van der Waals interactions) [120–124]. For the moment, the force-matching procedure was not used in these works, but we have recently extended it to the case of the 1-ethyl-3-methylimidazolium tetrachloroaluminate ionic liquid [23].

Acknowledgments S.J. was supported by the German Science Foundation (DFG) under grant no. JA 1469/4-1. M.S., B.R., R.V. and C.S. acknowledge the support for the French Agence Nationale de la Recherche (ANR), under grant SYSCOMM (ANR-09-SYSC-012) "Multiscale Simulation of Ions at Solid-Liquid Interfaces", and for PACEN (Programme sur l'Aval du Cycle Nucléaire) through PCR ANSF and GNR PARIS programs. We would like to thank Sami Tazi, Toon Verstraelen and Teodoro Laino for their help in implementing the PIM in the CP2K code.

References

1. Hansen JP, McDonald I (1986) Theory of simple liquids, 2nd edn. Academic Press, London
2. Hallett JP, Welton T (2011) Chem Rev 111(5):3508
3. Jahn S, Madden PA (2007) Phys Earth Planet Inter 162:129
4. Guillot B, Sator N (2007) Geochim Cosmochim Acta 71(5):1249
5. Guillot B, Sator N (2007) Geochim Cosmochim Acta 71(18):4538
6. Vuilleumier R, Sator N, Guillot B (2009) Geochim Cosmochim Acta 73(20):6313
7. Chen GZ, Fray DJ, Farthing TW (2000) Nature 407(6802):361

8. Ghallali HE, Groult H, Barhoun A, Draoui K, Krulic D, Lantelme F (2009) Electrochim Acta 54:3152

9. Groult H, Barhoun A, Briot E, Lantelme F, Julien CM (2011) J Fluorine Chem 132(12):1122

10. Salanne M, Simon C, Turq P (2006) J Phys Chem B 110(8):3504

11. Salanne M, Simon C, Turq P, Madden PA (2008) J Phys Chem B 112(4):1177

12. Siqueira LJA, Ribeiro MCC (2007) J Phys Chem B 111:11776

13. Borodin O (2009) J Phys Chem B 113(33):11463

14. Vatamanu J, Borodin O, Smith GD (2011) J Phys Chem B 115(12):3073

15. Maginn EJ (2007) Acc Chem Res 40:1200

16. Zhao W, Leroy F, Heggen B, Zahn S, Kirchner B, Balasubramanian S, Muller-Plathe F (2009) J Am Chem Soc 131(43): 15825

17. Kohagen M, Brehm M, Thar J, Zhao W, Müller-Plathe F, Kirchner B (2011) J Phys Chem B 115(4):693

18. Stone AJ (2008) Science 321(5890):787

19. Ribeiro MCC (2000) Phys Rev B 61(5):3297

20. Harding JH, Pyper NC (1995) Phil Mag Lett 71:113

21. Fowler PW, Madden PA (1984) Phys Rev B 29(2):1035

22. Fowler PW, Madden PA (1985) Phys Rev B 31(8):5443

23. Salanne M, Siqueira LJA, Seitsonen AP, Madden PA, Kirchner B (2012) Faraday Discuss 154:171

24. Wilson M, Madden PA, Pyper NC, Harding JH (1996) J Chem Phys 104:8068

25. Rowley AJ, Jemmer P, Wilson M, Madden PA (1998) J Chem Phys 108:10209

26. Jemmer P, Wilson M, Madden PA, Fowler PW (1999) J Chem Phys 111(5):2038

27. Stone AJ (1996) Theory of intermolecular forces. Oxford University Press, Oxford

28. Tang K, Toennies J (1984) J Chem Phys 80(8):3726

29. Aguado A, Bernasconi L, Jahn S, Madden PA (2003) Faraday Discuss 124:171

30. Aguado A, Madden PA (2004) Phys Rev B 70:245103

31. Aguado A, Madden PA (2003) J Chem Phys 119(14):7471

32. Laino T, Hutter J (2008) J Chem Phys 129(7):074102

33. Wang B, Truhlar DG (2010) J Chem Theory Comput 6(11): 3330

34. Wilson M, Madden PA (1993) J Phys Condens Matter 5:6833

35. Wilson M, Madden PA (1994) Phys Rev Lett 72(19):3033

36. Foy L, Madden P (2006) J Phys Chem B 110(31):15302

37. Heaton RJ, Brookes R, Madden PA, Salanne M, Simon C, Turq P (2006) J Phys Chem B 110(23):11454

38. Salanne M, Simon C, Turq P, Madden PA (2009) J Fluorine Chem 130(1):38

39. Bitrian V, Trullas J (2006) J Phys Chem B 110(14):7490

40. Bitrian V, Trullas J, Silbert J (2007) J Chem Phys 126(2): 021105

41. Bitrian V, Trullas J (2008) J Phys Chem B 112(6):1718

42. Alcaraz O, Bitrian V, Trullas J (2011) J Chem Phys 134(1): 014505

43. Bitrian V, Alcaraz O, Trullas J (2011) J Chem Phys 134(4): 044501

44. CP2K developers group. http://cp2k.berlios.de

45. Marrocchelli D, Salanne M, Madden PA, Simon C, Turq P (2009) Mol Phys 107(4–6):443

46. Marrocchelli D, Salanne M, Madden PA (2010) J Phys Condens Matter 22(15):152102

47. Norberg ST, Ahmed I, Hull S, Marrocchelli D, Madden PA (2009) J Phys Condens Matter 21(21):215401

48. Marrocchelli D, Madden PA, Norberg ST, Hull S (2009) J Phys Condens Matter 21(40):405403

49. Norberg ST, Hull S, Ahmed I, Eriksson SG, Marrocchelli D, Madden PA, Li P, Irvine JTS (2011) Chem Mater 23(6):1356

50. Marrocchelli D, Madden PA, Norberg ST, Hull S (2011) Chem Mater 23(6):1365

51. Tangney P, Scandolo S (2002) J Chem Phys 117(19):8898

52. Liang Y, Miranda C, Scandolo S (2006) J Chem Phys 125:194524

53. Giacomazzi L, Carrez P, Scandolo S, Cordier P (2011) Phys Rev B 83:014110

54. Fumi FG, Tosi MP (1964) J Phys Chem Solids 25:31

55. Tosi MP, Fumi FG (1964) J Phys Chem Solids 25:45

56. Lantelme F, Turq P, Quentrec B, Lewis JWE (1974) Mol Phys 28(6):1537

57. Lantelme F, Turq P (1982) J Chem Phys 77(6):3177

58. Anwar J, Frenkel D, Noro MG (2003) J Chem Phys 118(2):728

59. van Beest BWH, Kramer GJ, van Santen RA (1990) Phys Rev Lett 64(16):1955

60. Oeffner R, Elliott S (1998) Phys Rev B 58(22):14791

61. Car R, Parrinello M (1985) Phys Rev Lett 55(22):2471

62. Editorial (2010) Nat Mater 9:687

63. Laio A, Bernard S, Chiarotti GL, Scandolo S, Tosatti E (2000) Science 287:1027

64. Marzari N, Vanderbilt D (1997) Phys Rev B 56(20):12847

65. King-Smith R, Vanderbilt D (1993) Phys Rev B 47(3):1651

66. Vanderbilt D, King-Smith R (1993) Phys Rev B 48(7):4442

67. Souza I, Wilkens T, Martin RM (2000) Phys Rev B 62(3):1666

68. Silvestrelli PL, Parrinello M (1999) Phys Rev Lett 82(16):3308

69. Bernasconi L, Wilson M, Madden PA (2001) Comput Mater Sci 22(1–2):94

70. Bernasconi L, Madden PA, Wilson M (2002) Phys Chem Commun 5(1):1

71. Silvestrelli PL, Parrinello M (1999) J Chem Phys 111(8):3572

72. Sagui C, Pomorski P, Darden T, Roland C (2004) J Chem Phys 120(9):4530

73. Hult E, Rydberg H, Lundqvist B (1999) Phys Rev B 59(7):4708

74. Perdew JP, Burke K, Ernzerhof M (1996) Phys Rev Lett 77:3865

75. Becke A (1988) Phys Rev A 38(6):3098

76. Lee C, Yang W, Parr RG (1988) Phys Rev B 37(2):785

77. Jahn S, Madden PA, Wilson M (2006) Phys Rev B 74:024112

78. Jahn S (2008) Am Mineral 93:1486

79. Hennet L, Pozdnyakova I, Cristiglio V, Cuello GJ, Jahn S, Krishnan S, Saboungi ML, Price DL (2007) J Phys Condens Matter 19:455210

80. Jahn S (2010) Acta Cryst A 66:535

81. Krishnan S, Hennet L, Jahn S, Key TA, Saboungi ML, Madden PA, Price DL (2005) Chem Mater 17:2662

82. Jahn S, Madden PA (2007) J Non Cryst Solids 353(32–40):3500

83. Pozdnyakova I, Hennet L, Brun JF, Zanghi D, Brassamin S, Cristiglio V, Price DL, Albergamo F, Bytchkov A, Jahn S, Saboungi ML (2007) J Chem Phys 126:114505

84. Drewitt JWE, Jahn S, Cristiglio V, Bytchkov A, Leydier M, Brassamin S, Fischer HE, Hennet L (2011) J Phys Condens Matter 23:155101

85. Adjaoud O, Steinle-Neumann G, Jahn S (2008) Chem Geol 256(3–4):184

86. Adjaoud O, Steinle-Neumann G, Jahn S (2011) Earth Planet Sci Lett 312(3–4):463

87. Salanne M, Simon C, Turq P, Heaton RJ, Madden PA (2006) J Phys Chem B 110(23):11461

88. Cantor S, Ward WT, Moynihan CT (1969) J Chem Phys 50(7):2874

89. Vaslow F, Narten AH (1973) J Chem Phys 59(9):4949

90. Salanne M, Simon C, Turq P, Madden PA (2007) J Phys Chem B 111(18):4678

91. Salanne M, Simon C, Turq P, Madden PA (2007) C R Chim 10:1131

92. Forsberg C (2005) Prog Nucl Energy 47(1–4):32

93. Pauvert O, Zanghi D, Salanne M, Simon C, Rakhmatullin A, Matsuura H, Okamoto Y, Vivet F, Bessada C (2010) J Phys Chem B 114(19):6472
94. Salanne M, Simon C, Groult H, Lantelme F, Goto T, Barhoun A (2009) J Fluorine Chem 130(1):61
95. Sarou-Kanian V, Rollet A-L, Salanne M, Simon C, Bessada C, Madden PA (2009) Phys Chem Chem Phys 11:11501
96. Merlet C, Madden PA, Salanne M (2010) Phys Chem Chem Phys 12:14109
97. Heaton RJ, Madden PA, Clark SJ, Jahn S (2006) J Chem Phys 125:144104
98. Umari P, Pasquarello A (2002) Phys Rev Lett 89(15):157602
99. Umari P, Pasquarello A (2005) Int J Quant Chem 101:666
100. Resta R (1998) Phys Rev Lett 80:1800
101. Jemmer P, Fowler PW, Wilson M, Madden PA (1998) J Phys Chem A 102(43):8377
102. Buin A, Iftimie R (2009) J Chem Phys 131(23):234507
103. Salanne M, Simon C, Madden PA (2011) Phys Chem Chem Phys 13:6305
104. Duffy JA, Ingram MD (1971) J Am Chem Soc 93(24):6448
105. Angell CA (2009) J Solid State Electrochem 13:981
106. Aguado A, Madden PA (2005) Phys Rev Lett 94(6):068501
107. Rotenberg B, Salanne M, Simon C, Vuilleumier R (2010) Phys Rev Lett 104(13):138301
108. Brouder C, Panati G, Calandra M, Mourougane C, Marzari N (2007) Phys Rev Lett 98(046402)
109. Silvestrelli PL (2008) Phys Rev Lett 100(5):053002
110. Gordon R, Kim Y (1972) J Chem Phys 56(6):3122
111. Barker D, Sprik M (2003) Mol Phys 101(8):1183
112. Lee H, Lee C, Parr RG (1991) Phys Rev A 44(1):768
113. Tran F, Wesolowski T (2002) Int J Quant Chem 89:441
114. Ohtori N, Salanne M, Madden PA (2009) J Chem Phys 130(10):104507
115. Zhu W, Toulouse J, Savin A, Ángyán JG (2010) J Chem Phys 132(24):244108
116. Grimme S (2004) J Comput Chem 25(12):1463
117. Vydrov OA, Van Voorhis T (2010) Phys Rev A 81:062708
118. Fowler PW, Knowles PJ, Pyper NC (1985) Mol Phys 56(1):83
119. Andersson Y, Langreth DC, Lundqvist BI (1996) Phys Rev Lett 76(1):102
120. Yan T, Burnham CJ, Del Pópolo MG, Voth GA (2004) J Phys Chem B 108(32):11877
121. Chang TM, Dang LX (2009) J Phys Chem A 113(10):2127
122. Yan T, Wang Y, Knox C (2010) J Phys Chem B 114:6905
123. Bedrov D, Borodin O, Li Z, Smith G (2010) J Phys Chem B 114:4984
124. Schroder C, Steinhauser O (2010) J Chem Phys 133:154511

The manufacturer's authorised representative in the EU is Springer
Nature Customer Service Centre GmbH, Europaplatz 3, 69115 Heidelberg,
Germany. If you have any concerns regarding our products, please
contact ProductSafety@springernature.com

Printed and bound by CPI Group (UK) Ltd, Croydon, CR0 4YY
07/01/2026
02030937-0003